Why we must go to Mars. Part One: The King's Valley

GREG ORME

Dedication

This book is dedicated to a wonderful author and friend Sir Arthur C. Clarke, he is deeply missed. Though he never believed in Faces on Mars, he inspired many of my ideas. Also I dedicate this to Carl Sagan, who inspired me to search for extraordinary evidence to back up these extraordinary claims.

TABLE OF CONTENTS

Acknowledgements

The author wishes to thank members of the Society for Planetary SETI Research for discussions about the evidence in this book. SPSR however does not endorse books by its members. I would also like to thank Peter K. Ness, a good friend and sounding board on this subject for over 10 years, he also contributed several papers in the Appendix I urge you to read. There are too many to thank, but I have appreciated through 17 years the feedback from other researchers and also from fair minded skeptics who helped to expose weak points in this evidence.

Introduction

I started investigating this subject in 1993, since then the amount of evidence has increased and so I have moved toward a more detailed hypothesis. Hopefully this book will present a compelling case that some of these formations may be artificial. The title of the book "Why we must go to Mars" is provocative, but it is true for many reasons. We must go to Mars one day, to colonize it and as a springboard to go to the other planets and the stars. One day perhaps there will be a billion people living on Mars and the wealth in minerals there may rival that of Earth. This book is mainly concerned with another reason why we must go to Mars, that there may be ancient artifacts there from another sentient race. The exploration of these possible artifacts I believe can only be good for humanity; if they are real then they may impel us to do what we should have done already, form a colony on Mars and start turning it into a livable environment.

Nowadays the idea of alien civilizations is more plausible with the discovery of many planets around nearby stars[i] including one[ii] in a habitable zone. The main theory proposed here is that our solar system was visited by aliens perhaps over a billion years ago. In my opinion this is the most logical interpretation of the evidence though there is a small possibility sentient life may have evolved on Mars. This might have occurred during a short time frame when the climate was hospitable to life, perhaps with the aid of panspermia[iii] bringing advanced DNA from Earth to accelerate this process. In this book I will present some evidence based in the King's Valley with the best known anomaly being the Crowned Face, one of the Faces on the cover of this book. I compare this to two other well-known Martian faces, namely the Cydonia Face and the Meridiani Face discovered by Terry James. I also examine the idea of whether hypothetical aliens could travel from another star to visit our solar system, whether it was possible or desirable for aliens to make a trip to come here, or whether they might have sent an Artificially Intelligent[iv] probe capable of starting a colony. Other books will cover over 15 different areas with possible artifacts on Mars including the well-known Cydonia region. Because of the need to split up the large amount of evidence into several books it will seem fragmented at times, for example in this book I try and explain the most likely scenario according to the evidence but much of this evidence will be in future books.

In this book I also examine the pros and cons of researching these hypothetical ruins on Mars, for example if they do exist then what are the benefits to us to go and look at them. Much of this argument will be similar to that about Martian colonization generally, many feel that the expense outweighs the possible benefits and some will no doubt believe that we don't really need to go and look at alien artifacts on Mars whether they exist or not. I also look at a landing site near the King's Valley and how such an expedition and eventual colony of Mars for archeological purposes could be mounted. This expedition is initially set near the Crowned Face to explain some of the goals and pitfalls in such an endeavor but in later books I will examine landing sites near other anomalies. Currently I see 15 formations on Mars with a good potential for being artificial in that natural explanations seem to be unlikely or impossible.

Methodology is also discussed, how the images taken of these sites and others can be analyzed accurately including some of the most common mistakes in how photo compression can distort features and make them seem more artificial. Also an explanation is given of how this data can be falsified, so that the objective can be not to prove artificiality which is nearly impossible from photos but to disprove natural ways these anomalies could have been formed. This makes it much easier because we can use our knowledge of geology to explain many odd features as natural and to exclude others from being naturally formed which would make them artificial. It also helps to reduce our reliance on the visual impact of these formations, merely arguing that because some formations look artificial then they must be. Many things in nature appear to be artificial by accident, such as when we sometimes see faces in clouds[v]. On the other hand some objects on Mars do seem extremely unlikely to be formed randomly like faces in clouds; I then try to exclude this possibility by examining recurring patterns in these formations. It is more unlikely for the same kinds of artificial patterns to repeat over and over such as Faces there having similarities to each other. We might see faces in clouds but rarely the same face over and over yet on Mars there are similarities between the Faces there which imply a common template in their formation.

Another problem that often arises with this research is the evidence doesn't fit well with people's preconceptions of what alien ruins should look like. This may have been influenced by science fiction[vi] notably 2001: A Space Odyssey where a technologically superior artifact is found on the Moon. However just because the first artifacts we have found are very simple such as Faces doesn't mean there aren't ones there with higher technology. We simply have found so far what is easiest to see in photos from an orbiting camera, the great age of these possible artifacts might also ensure that anything made of metal hundreds of millions of years ago has long since disappeared. If a billion years after humanity became extinct someone took photos of the Earth from orbit whatever was left would give a highly misleading idea of our civilization as well.

I also try and construct a more precise model of what may have happened to the builders of these Faces, the evidence tends to point to an alien visitation in the remote past but could also possibly include the evolution of an indigenous species. The latter seems highly unlikely when we look at Mars today but this life may have become extinct a billion years ago, perhaps from the Hellas impact as a planet killer. If so then evidence of previous life there might not be able to be found except with a manned expedition looking for fossils. More evidence for this will be in a later book but basically the most likely theory is that a billion or more years ago aliens or an artificially intelligent probe arrived in our solar system. They terraformed[vii] Mars with meteor impacts and created some of the well-known geological features there as a by-product of these impacts heating up the planet such as Valles Marineris[viii], Tharsis[ix], and ice deposits from former poles where the Martian Rovers[x] are now. They may also have helped to terraform our own planet Earth and seeded it with their own DNA. This might explain why the Faces look somewhat like humans; we may have evolved from the same DNA.

This possible terraforming on Mars would have been started by aiming a meteor at the South Pole in the Argyre impact which created floods, rivers and melted a large sea in the Northern Lowlands[xi]. It also melted the frozen atmosphere at the Poles increasing the air pressure so the water would stay liquid and form clouds and rain. Over a long period of time this caused the Poles of Mars to move in a wide arc to their present day position, this movement mostly across the current equator of Mars. This spread water over many areas, kept the atmosphere from refreezing and presumably allowed much of it to be temporarily converted to Oxygen which would not refreeze while volcanoes create by the impacts added CO_2 and other gases to the atmosphere making it thicker. The outpouring of magma[xii] heated up Mars and because the volcanoes were near the Poles by design they melted ice and the atmosphere to maintain the shallow oceans and atmospheric pressure. Over time this terraforming wore off and the Hellas[xiii] impact may have been an attempt to restart it, or it could have been a random planet killer that ended the lives of the colonialists or the indigenous Martians if they evolved there. Then the planet would have cooled over many millions of years, the atmosphere returned to mostly CO_2 to freeze at the Poles again and Mars became the apparently lifeless planet we see today. Any life introduced there or having evolved would have been covered up by volcanic flows or organic compounds destroyed by radiation[xiv] removing all traces of it. The geological evidence for this polar wander is in the Appendix in a chapter called A History of Mars. It's likely this scenario happened but it is impossible to prove that it was initiated by visiting aliens. It was very convenient though and the positions of many of these anomalies make it more likely they were being built not long after this warming of Mars began.

The possible colonisation of Mars is hard to estimate in size but was likely quite small, perhaps less than a hundred thousand humanoids at its height because of the small number of buildings and monuments. At some stage this colony probably failed perhaps because the environment was too toxic[xv], the radiation was too dangerous because of a lack of a Martian magnetic field[xvi], or impacts such as Hellas killed the inhabitants. The possible reasons aliens would build Faces are many, for example for religious reasons or to honour leaders like we did with Mount Rushmore[xvii]. I use the idea that these Faces would likely have common themes to exploit similarities between the Faces as evidence for artificiality; a symbolic representation of using Dual Faces seems to recur at three sites namely the King's Valley, Cydonia, and the Face at Meridiani.

The likeness between us and the Martian faces implies a similar DNA, so they may have seeded our own planet with their own life, our DNA may have come from the remnants of a second failed colony here on Earth, or perhaps panspermia of life coming here from Mars on meteorites eventually brought this alien DNA here and we evolved from it. It is possible then that we are descended from these aliens, that an early evolving race seeded itself to many other planets with their own life signature for eventual colonization. It may then be similar to our own plans of colonization[xviii], Bracewell probes may have been sent out to many planets by these hypothetical aliens to terraform them, some of these inevitably would have failed and perhaps Mars was one of these failures. NASA has plans to do something similar when the technology is available, namely to send out in to interstellar space probes with artificial intelligence and a payload of DNA in the form of plant and animal life. Such a probe would attempt to seed other planets in this way so that by the time we could get there the probes would have created a suitable ecosystem for us to live in.

The probe sent to us, assuming it existed, was probably directed by Artificial Intelligence because it would probably be too difficult to travel between stars with a manned ship. People would be too heavy to accelerate to a significant fraction of the speed of light to make the journey and also they would be too difficult to keep alive for possibly centuries of space travel. This would also explain the lack of signs of more recent visitation, a random probe might not have been able to signal its success to its makers or the aliens that sent the probes may have died out in the enormous amount of time between this event and the current day. It would be virtually impossible to move a civilization away from danger such as nearby supernovae or a changing sun, perhaps they knew their civilization was doomed and sent out probes to continue their civilization in some way. Because this is so long ago it's likely the original civilization is extinct, run out of resources and confined to its own planet, otherwise it would be so advanced we would see signs of it with SETI[xix]. The reasons for aliens visiting us or sending probes to our solar system are pure speculation of course, I am just pointing out that there are some plausible reasons why this might happen.

Such a hypothesis is quite startling, the evidence is still thin but the proof is arguably there. Because this proof is barely sufficient in my opinion it must be carefully laid out, the format of this book is more like a forensic attempt to prove a difficult crime like Sherlock Holmes. Small pieces of evidence are often highly significant and can often be impossible to form geologically, it is necessary then to look at small details to prove this case. It would be ideal if some artefact was found that was so compelling that all doubts were laid to rest, this hasn't happened and it may never happen. So it is important to look at the evidence more deeply because if we send a manned mission to these sites 50 years late then we may deeply regret looking at the evidence so frivolously now. Now is the time to decide how credible this evidence is, then additional images of these areas can be taken to get further proof and robotic rovers sent to these sites.

This book will also present a case as to why artefacts on Mars would be a good reason for us to establish a colony there and do archaeological research on these formations. For example there may be technological knowledge we can learn from their builders such as how to travel faster than the speed of light, they may show where they came from so we can try to contact them, and we may be able to terraform Mars the same way if we study the geology there or we find a historical record of how they did it. There are also possible dangers in this exploration as has been theorized in science fiction, the faces could be a honeypot[xx] which cause us to try and contact someone who then attacks us. At worst we would be colonizing Mars sooner which can only benefit the human race over time, eventually there might be a billion people living there.

It may also be none of this evidence turns out to be artificial, I believe this book proves artificiality but I may be wrong. This can only be finally determined by sceptical but fair analysis of this evidence. I simply believed after 17 years of research that a sufficient amount of evidence had been accumulated by me and others to present to the reader to make up their own mind. Extraordinary claims can require extraordinary evidence[xxi] but it should be remembered that Carl Sagan[xxii] himself, the originator of this saying, also theorized about aliens visiting our solar system in the past. Really no one knows how likely such a thing could be, we mainly tend to think of this as almost impossible because of the vast distances and time involved in such a journey. With the discovery of so many planets around other stars the motivation of aliens to visit other worlds is likely to match our own desires to colonize other stars. If we can imagine visiting other worlds then it is likely aliens somewhere would also conceive of this, and maybe they did.

There is little point for mainstream science to be hostile to the possibilities described here. If none of these formations turn out to be artificial they still create a lot of interest in going to Mars, which can only help science in its goal to expand from Earth to other planets and the stars. If these formations are artificial they would create unprecedented opportunities and funding for science, the urge to create a colony on Mars to investigate them might be irresistible. There is little evidence of public hostility to the idea of extra-terrestrial[xxiii] life according to the reactions from people in the search for extra-solar planets. The opportunities for scientists on Mars would be vast, such as in geology, archaeology, mining, agriculture, etc. If this alien terraforming theory is correct we could transform Mars like they did, perhaps making parts of Mars able to support life within a hundred years.

In a sense the ability of mainstream science to theorize about alien visitation has been corrupted by the nature of the evidence. If there had never been any such evidence uncovered and no evidence of UFOs it would be natural for scientists to consider the idea of artefacts in our solar system hypothetically, for example Arthur C. Clarke wrote about this in 2001: A Space Odyssey and this was well received by scientists. The fringe[xxiv] of science often has many theories; some can be vindicated over time and become part of the mainstream while others become discredited. Often the mainstream[xxv] is forced into the position of rebutting poorly tested and deliberately faked evidence and this tends to make more legitimate and honest evidence look bad. For example when someone fakes crop circles and UFO images then it tends to cast a shadow over the more sensible idea of aliens wanting to visit and colonize other planets.

However it is one thing to agree that aliens might want to visit other worlds but quite another to prove they visited our own.

There is still a distinct possibility that all the evidence in this book is just natural formations, that there are no alien ruins on Mars at all. Over 17 years I have tried to discard dubious or ambiguous evidence but it is still quite possible that chance could have created these formations though I consider this to be highly unlikely. There should not be a rush to judgement for or against this book, it is good to be sceptical not only of the evidence for artificiality but also of poorly thought out explanations that faces on Mars can form naturally just because people see faces in clouds or on pieces of toast. If there are no faces on Mars then this book might well end up as no more than a hopefully entertaining work of science fiction. I have tried to present the evidence here honestly because I believe it deserves to be discussed, many of the ideas can seem ridiculous but they are also things that could have happened as mainstream science would admit. I would also urge people to see this as an attempt to logically investigate the concept of xenoarcheology[xxvi], the study of alien ruins, in a scientific way.

The purpose of this series of books is to prove the artificiality of at least one of these formations, it is therefore not important if some or even most of them turn out to be natural formations. If one is artificial, for example one Face then the books will have achieved their aim.

Chapter One

A history of my Martian artefact research

I found the Crowned Face on 9th July 2000; it was first imaged by the Mars Orbiter Camera in June 1999. It is shown in Figure 1; you can look at it yourself by following this link:

http://www.msss.com/moc_gallery/ab1_m04/images/M0203051.html

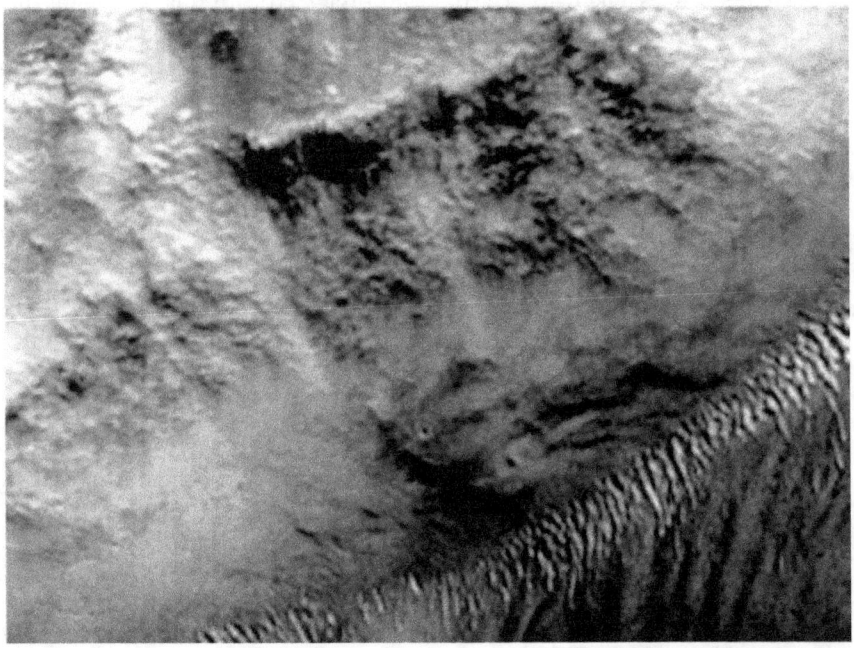

Figure 1

At the time images were usually embargoed for 6 months or more before being released to the public, this was to give some planetary scientists time to write peer reviewed papers on geological formations found. Usually the new images would be released in bulk at the Malin Space Systems website and then various researchers would go through them. I started doing this around 1994 after seeing photos of the Cydonia Face. I then looked at most of the photos from the Mars Global Surveyor when they were released to look for other possible artefacts, but also because it was such a wonderful experience to look at another planet like this. Over this time I found a lot of interesting possible artefacts which I published on my website ultor.org, the name Ultor being the Roman God of Mars. Other images that seemed to be natural geology I published on another website harmakhis.org. I also have a website on the enigmatic Martian Spider formations at martianspiders.com. All might be worth looking at for new images and possible artefacts I post there.

I still remember finding the Crowned Face in Figure 1 as stunning, the shock in seeing how face like it was after having spent thousands of hours of looking at mostly natural geology with a few interesting formations. At the time the Cydonia Face had been reimaged and to many people this made it seem less likely to be artificial. I disagreed with this, I thought that many features could be falsified as naturally formed, i.e. that natural geological processes could not make the Cydonia Face. However it was understandable for many people to become disheartened by it[xxvii]. Looking for possibly artificial structures on Mars has always been a frustrating experience where it was never certain if it was a total waste of time or not. My view was that Mars was so interesting geologically that the time looking at images was well spent even if nothing artificial looking was found. This attitude probably stopped me from becoming disheartened and giving up long before this. Looking at so many pictures also paid off to some degree when I was looking at the South Pole of Mars and saw photo M0804688, the first and most impressive image of the Martian Spiders. These spiders can be seen on my website martianspiders.com, in an upcoming book I will discuss all the different types, the controversies that arose from their similarities to vegetation, how Arthur C. Clarke became interested in them, how I co-wrote a peer reviewed paper[xxviii] on the spiders, and a proposed solution to how they are formed.

As more photos were taken by the Mars Orbiter Camera more artificial looking formations came to be discovered by myself and other researchers, so the case for artificiality on Mars seesawed as more of these other anomalies were found while Cydonia was being reimaged. Many formations appeared much more natural looking[xxix] than expected[xxx] such as the D&M Pyramid and the Fort. Many people expected them to look very artificial like constructed buildings we might see in one of our own cities and were very disappointed by the scarcity of evidence for artificiality. Discussions on the subject took place on several forums on the Internet, often these became very heated which is usually the case when there is little evidence but fixed opinions on both sides. The sceptics generally believed the reimaging of the Cydonia area closed the case on artificial formations on Mars[xxxi], according to them there weren't any. The illusion that there was a face on Mars was said to be caused by the low resolution of the original Viking images of the area, this allowed people to imagine features that weren't really there in the shadows. This is called pareidolia and people are said to be hard wired to do this, we evolved for example to see animals partially hidden in vegetation. The theory is this ability is useful in nature but when we are exposed to too many random patterns we begin to see things that aren't really there because the mind keeps trying to find something in this randomness. This may be why people become gamblers for example; they believe they can see patterns in cards, horse racing, roulette, etc. However sometimes there are real patterns in data that are hard to find and prove, this is how scientific research progresses.

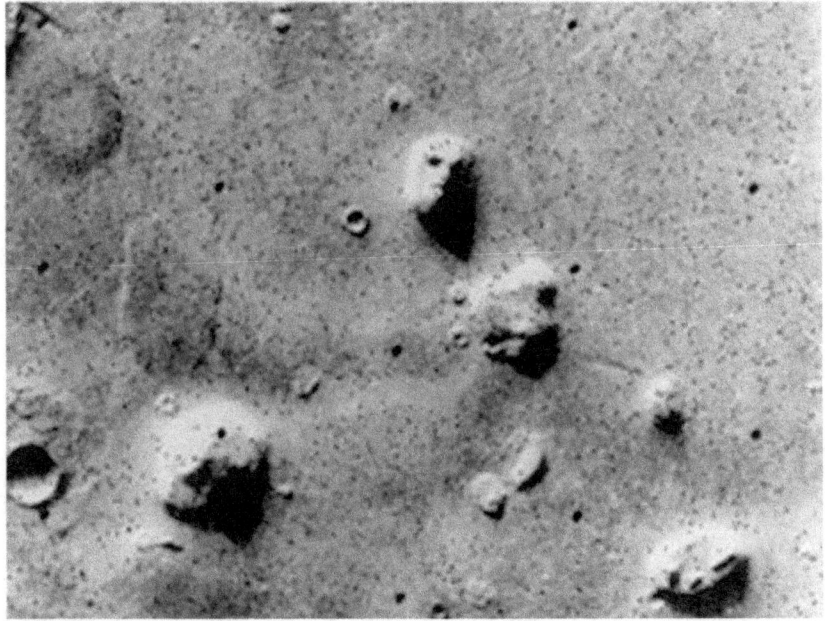

Figure 2

This was one of two images of Cydonia taken by Viking which showed the Face there. Even in retrospect it seems entirely reasonable for people to point out the obvious, that it looks like a face and that we didn't build it. The lack of additional knowledge caused speculation about this to rage until the area was reimaged by the Mars Orbital Camera, the result was Figure 3 below.

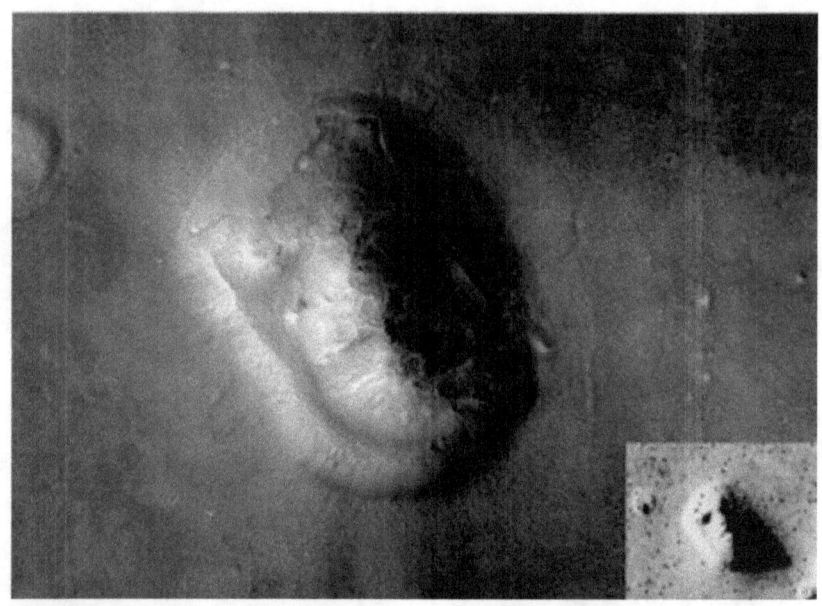

Figure 3

Comparing the two images of the face it becomes obvious that the mind anticipated features that were not there in real life. However this disappointment may have obscured many artificial looking aspects which will be analysed in an upcoming book.

One problem with this kind of research is that people develop an intuition after looking at tens of thousands of images of Mars and an artificial looking formation stands out in ways that is hard to explain or justify logically. This of course was not considered real evidence by the sceptics who considered researchers merely had an overactive imagination or were charlatans. With the trading of insults the two sides developed a rift that persists to this day, it is rare for the two sides to discuss the subject calmly and logically. In my view sceptics have generally done science a disservice with their attitude to this subject usually restricting their input to applying the Giggle Test, i.e. if it seems funny to them then it is not real. Usually however science evolves through discoveries that would seem funny to scientists in the past, Quantum Mechanics would have seemed absurd to physicists in the nineteenth century. The idea of this and future books is to try and present the evidence not as a polemic to try and convince people and hide all the weak points of the case, but to show that there exists quite compelling evidence and let the reader make up his own mind.

Soon after I discovered the Crowned Face I posted it to a forum on SPSR, which is an organization I joined devoted to looking for signs of xenoarcheology, or as it describes the subject, Planetary SETI. It has tried to focus on publishing peer reviewed scientific articles on various Martian anomalies and debating their merits in private. Often this enabled discussions to be more productive because to publically advocate artificial structures on Mars was considered to be detrimental to anyone's career in planetary science. SPSR does not endorse this book in any way; it does not endorse the publications of its members. The Crowned Face became well known to a lot of people around the world because of a press conference given by Tom van Flandern of SPSR who passed away a few years ago, the conference is seen here:

http://www.youtube.com/watch?v=5u-20g7Bwdw

This gave the face an international exposure; it was shown on Good Morning America, Regis Philby, London Daily Mail, the BBC, and CBS-TV. Fox included it in a half hour special on UFOs and related subjects where the reaction to it seemed to be the same mix of scepticism and enthusiasm that were found on forums on the subject. It was also in the New York Post.

Initially the name King Face was suggested by Tom van Flandern because of the obvious crown on the Face, but many researchers thought the face seemed feminine so Tom van Flandern then suggested the more androgynous Crowned Face name. I coined the name King's Valley because of their being several crowned faces there in a valley. The huge amount of media interest was interesting because it showed that enough people found it visually convincing to create a viral experience in the media and on the Internet, but also that it was so face like they would try to determine its gender. Of course if the face was real we have no way of knowing if the original model for the face was male or female, or even if the aliens or indigenous Martians had different sexes. Often with Martian anomalies there is a large difference of opinion from sceptics and cynics who claim not to see anything to more suggestible people who claim to see far too much in natural geology.

The area was first imaged here:

http://www.msss.com/moc_gallery/ab1_m04/images/M0203051.html

After this the King's Valley was reimaged by Malin Space Science System here:

http://www.msss.com/moc_gallery/ab1_m04/images/M0303483.html

This only shows part of the Crowned Face but shows other possibly artificial formations in the King's Valley, discussed later in this book. Then the valley was reimaged again here:

http://www.msss.com/moc_gallery/e07_e12/images/E11/E1100360.html

This shows more of the opposite side of the valley and some other possible Faces. Finally the main Crowned Face was reimaged again by HiRise in two photos almost identical, one shown here:

http://www.uahirise.org/ESP_018223_1830

These two photos show far more detail of the King's Valley and the Crowned Face, arguably increasing the credibility of most of the anomalies there, they will be analysed later in the book.

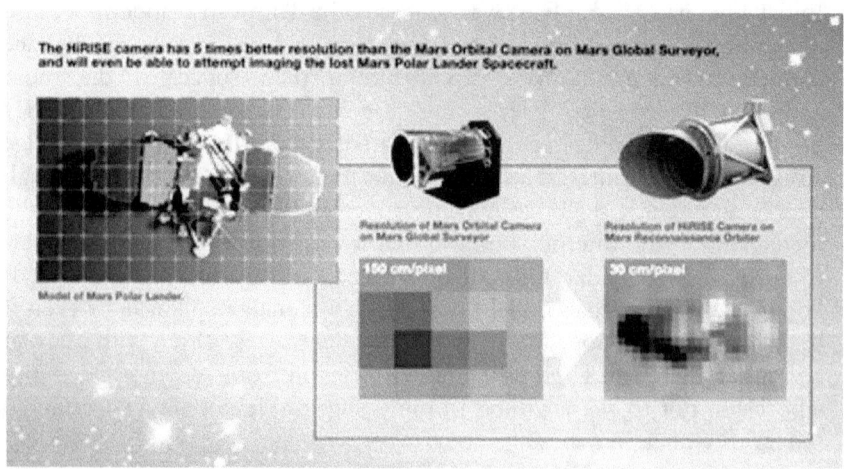

Figure 4

The HiRise camera is far more powerful than the Mars orbital Cameral, this enables much more details of these anomalies to be analysed. As will be shown, the results in the King's Valley are quite remarkable.

Chapter Two

A theory of artificiality

The data so far is pointing towards a consistent theory of these anomalies, this will be fully explained over a series of future books but the basics can be outlined here. The most likely time frame for this alien visitation or evolved Martians is over a billion years ago. The reason for this early date is the Argyre impact[xxxii], a large meteor which hit Mars. This impact may have been done deliberately by alien visitors to warm up the planet, otherwise a random meteor impact may have caused a chain of events that warmed Mars for long enough to help sentient life to evolve there. Some uncertainty exists as to the actual date of this impact because if natural then it should have occurred in the Late Heavy Bombardment in the formation of the solar system several billion years ago, but there is no reason why it could not have happened later. If the impact was caused by aliens to terraform Mars then it might have happened much later than the early bombardment but it is no doubt far more ancient than our own civilization and probably predates any complex life on our own planet. I use the time of a billion years ago not to try and pinpoint any particular time accurately but to emphasize it was in the very remote past. It is probably impossible to work out how old these artifacts are, assuming some are artificial, until we eventually colonize Mars.

The reason this impact is interesting from the point of view of terraforming is that it may have been a shallow impact that hit the South Pole at the time, i.e. it came in at a very low angle rather than from directly overhead. This theory is more fully explained in the Appendix with a chapter called a History of Mars. It outlines the geological evidence for this being a shallow impact and how the shock waves would have warmed the planet by causing extensive volcanism and the subsequent moving of the Poles. Later in the book I discuss how this impact could be worth trying to emulate to terraform Mars. This is worth investigating even if the impact was a coincidence because of the effects it may have produced.

It would be logical to terraform a planet with meteor impacts, the same idea has been studied by the Mars Society[xxxiii] and in science fiction books. The idea is the impact produces heat which warms up the planet, in a shallow impact the shock wave moves for longer on the surface and this heats up much more of the surface to radiate heat into the atmosphere rather than just below ground. It also causes more fractures in the surface to allow volcanic magma to come up and form volcanoes such as Tharsis and Elysium Mons. Heat from underground also comes up to the surface heating the atmosphere, along with outgassing thickening the air with CO_2 and sulfurous gasses. The longer this heat is concentrated on the Poles the more it prevents the air and water immediately refreezing back there, so aiming meteors at the poles would heat more of the frozen water and CO_2. When the water ran off from the Poles it created a sea in the Northern Lowlands, this however needed a sufficient air pressure to stay as liquid water, a situation nearly impossible[xxxiv] currently on Mars. The idea then is the additional CO_2 sublimated from the Poles thickens the atmosphere and stops the water in the seas from returning too quickly to the Poles. These seas would then create clouds, rain, snow, etc. which would be necessary for life to take root. Currently on Mars the air pressure is too low for liquid water to exist except fleetingly, this is a result of the triple point of water where with low enough air pressure it sublimates or boils directly into water vapor from ice without ever becoming water. So the additional air pressure from melting the CO_2 at the Poles can allow more liquid water to survive on Mars.

This life introduced for terraforming would be tailored for such a harsh environment, the Mars Society and NASA have already considered doing the same to terraform Mars[xxxv]. The most likely source for this life would be from the alien home world; it may be that life from there was seeded onto Mars and the Earth. When the colony failed over a long period of time this life evolved, perhaps also because it was engineered to evolve more quickly, and eventually it produced animals and humans that look something like the Faces of the aliens. Because they also evolved from the same kind of basic life it is only necessary for this to produce animals with two eyes, a nose and a mouth like Earth animals for there to be a resemblance between the Faces and ourselves. Life elsewhere in the galaxy may also look similar to us because ours is extensively based on a simple mathematical sequence called Fibonacci numbers. These form for example the tree like shape of blood vessels in animal bodies, internal organs can look like plants such as cauliflowers looking like brains and kidneys shaped like beans. The dimensions of a face have been shown to relate to the Golden Mean and Fibonacci numbers, Leonardo da Vince used this in his paintings. If alien plant life also evolves based on

Fibonacci numbers such as with roots and branches in plants then this may transfer to animals and lead to alien humanoids that look like us.

Of course there is no way to know if alien life was seeded on Mars and we evolved from it, but it shows there are plausible explanations for our resemblance to the Martian Faces. A good example of terraforming is the Mars trilogy science fiction books by Kim Stanley Robinson, one of the founders of the Mars Society where Mars is seeded with life. It is necessary for life to convert the CO_2 atmosphere to more Oxygen because while CO_2 will freeze in tiny amounts on our own Poles because of its relatively high freezing point, Oxygen freezes at a much lower temperature than would occur on Mars. If the air pressure is low on Mars however this reduces the amount of CO_2 that will freeze even below its own freezing temperature[xxxvi]. The race would then be on to stabilize the atmosphere to retain heat before the effects from the impact and volcanoes wore off and caused Mars to freeze up again, initially however the high amounts of CO_2 in the atmosphere should give a greenhouse effect. The theory here is that aliens may have done all this with meteors long ago to either try to colonize Mars or create enough atmosphere and heat for a short stay. Technically this would be much easier to do to Mars than the technology involved in travelling here from another star, the question is whether the impacts of Argyre and Hellas, and perhaps Isidis and Chryse could have successfully terraformed Mars. There may also have been more water and air available on Mars then to work with, much has been lost because of the solar wind and a lack of a magnetic field[xxxvii].

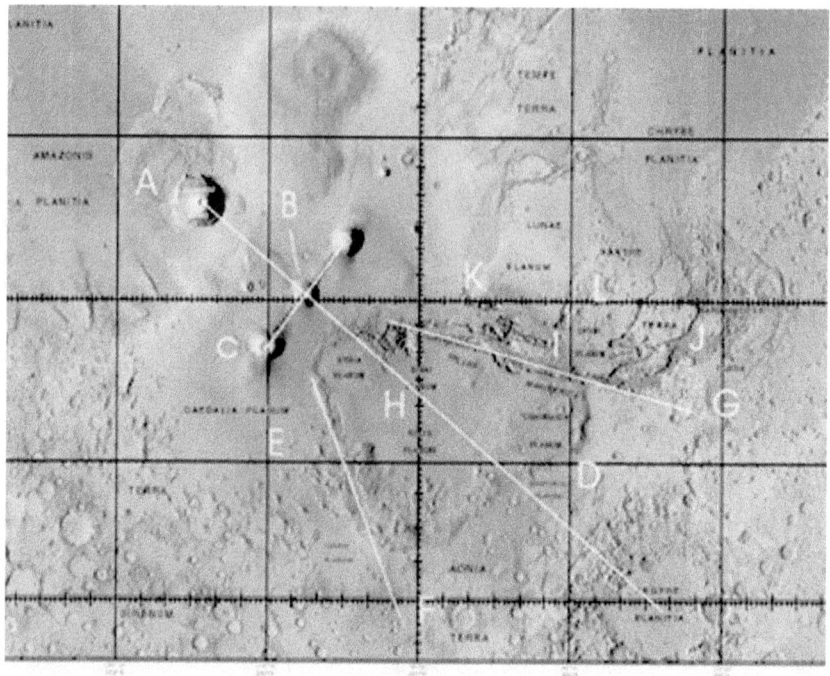

Figure 1

*This is a section of the Martian surface. Argyre Planitia is seen on the bottom right, this was the impact crater. If it was a glancing impact then the shock wave would travel along the surface towards **H**, the heat from this would have created fractures in the ground such as Valles Marineris at **G** and caused volcanism to raise this area. The darker areas in the center denote a higher elevation and as can be seen the 4 large mountains at **A**, **B**, and **C** namely Olympus Mons and the three Tharsis Montes volcanoes line up with this crater. At **I** there would have formed a channel for water melting on the pole to flow outwards as well as through **J**, **K**, and **L** into Chryse Planitia.*

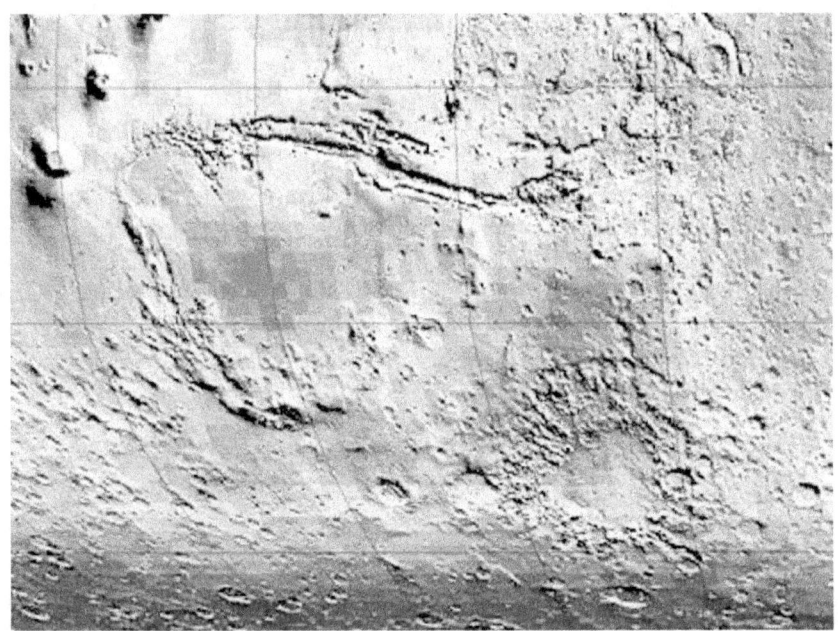

Figure 2

Argyre Planitia, the impact crater is shown again towards the bottom right. The darker areas are drier[xxxviii], they contain less water ice than the lighter areas while the dark area at the bottom is ice associated with the current South Pole. The image was originally color but needed to be converted to grey scale for this book, the link above takes you to the color image of this. The dark areas give an impression of radiating out from the crater towards the Tharsis volcanoes which can be seen in the top left. This would be caused by the shock wave heating the surface and creating the rifts and mountains shown. It can also be seen how these dry areas are on one side of the crater implying this shallow impact. By calculating the angle of this dark area radiating out from Argyre the approximate angle of impact could be estimated. We could do the same thing by aiming a meteor like this at the current North Pole of Mars letting the ice there melt and fill the Northern Lowlands, where the former Martian sea existed. At the same time this would sublimate much of the CO_2 frozen at the Poles and give Mars a temporary increase in air pressure.

To get all these benefits from a single impact would be a good result for these aliens, they would just have to nudge a large rock from the Asteroid Belt into a collision course with the Martian South Pole to give a glancing impact then wait for the planet wide effects like storms and Mars quakes to settle down enough to not be dangerous. This Pole was just to the East of Tharsis Montes at the time though of course these mountains did not exist as they would have been created by the impact. The heat from the impact melts the ice and CO2 at the South Pole and antipodal volcanism created another three volcanoes Elysium Mons, Hecates Tholus, and Albor Tholus next to the North Pole to heat up the ice and CO2 there. The Argyre impact would then produce polar wander, where the Pole moves to a different position because Mars would become unbalanced as large amounts of mass form on the Poles as these volcanoes grew. The Poles then tend to move so these volcanoes end up on the equator because of the centrifugal force, this causes water to be spread over large areas of Mars as the Pole melts when it moves over a warmer surface. The ice left behind is moved to lower latitudes because the Pole moves away from it and so it melts forming water runoffs, floods, rivers, rain, etc. A shallow meteor impact on one Pole combined with antipodal volcanism creating three other volcanoes on the opposite pole is a nearly ideal way to heat up Mars, the polar wander then is a good way to spread the water widely across it. Much of this water would go underground into an artesian system or freeze rather than return to the poles as rain or snow so if the temperature stays high long enough then higher oxygen levels could allow plants to survive and continue to terraform Mars. While there is no way of knowing for sure whether visiting aliens created this sequence of events it holds promise for our own efforts to terraform the planet. If we could duplicate these events with well-aimed meteor impacts then it may dramatically shorten the time it takes to make Mars habitable. More is explained on this later.

One problem with this idea is we don't know how long this terraforming would take, and hence whether the time scales involved would be practical for aliens. If they were colonizing Mars then presumably they having come all they to our solar system would be able to wait for however long it required. It is unlikely someone could come from another star with people being awake constantly so the likely alternatives to this are suspended animation or the colonists, flora and fauna would be frozen as seeds as the equivalent of our eggs and sperm to germinate in artificial incubators. It may have taken Mars hundreds of years or even thousands of years to settle down enough to colonize directly, we have no way of knowing this. Presumably the polar wander would take place over a long period of time, likely longer by millions of years than a probe or prospective colonists would be prepared to wait around but this polar wander could be occurring while they were actually on the ground with little effect on them. The other possibilities are that indigenous Martians evolved and there were no visiting aliens, or there were both, i.e. aliens visited indigenous Martians. In the case of Martians evolving this may have been accelerated by a chance shallow impact at Argyre and the resulting warm period on Mars allowed life there to rise to the level of sentience before becoming extinct when the planet cooled as the heat from the volcanoes ran out.

This also implies that the colonization lasted a long time, but this seems less likely because a colony that survived a million years for example should have left vast traces of its existence. For example our Earth civilization is arguably less than ten thousand years old and we have radically altered the planet in that time. It may be then that this polar wander occurred quickly or these aliens observed Mars from their home world as appearing more habitable after the Argyre impact and decided to come here after seeing this, much as we are now trying to do with extra solar planets by seeing if they are in a habitable zone and what their atmospheres are like. Aliens may have seen Mars on the edge of a habitable zone with a CO_2 atmosphere and decided to come here to visit or colonize Mars or other planets[xxxix] of our solar system.

One problem with the polar wander theory is that the various anomalies found on Mars tend to be clustered around a great circle which may have been their Martian equator. If so then the Poles would have wandered a great distance according to the polar wander scenario, this would make it more likely the aliens came after the polar wander and Argyre impact had happened, that a first visit by a robot probe or other kind of ship started the terraforming and then a colony ship came much later, or a colony only built these formations after being on Mars for a long time. These 15 formations I believe may be artificial tend to cluster in a nonrandom pattern in a mainly temperate like zone near an old Martian equator. This was first noticed by the deceased Tom van Flandern who saw that the Cydonia Face was on the equator of a former Martian Pole position near the Hellas impact crater, but this Pole in the wander path[xl] would only occur as a last stage before the current Pole position and far from its original position prior to the Argyre impact. He also noticed that the face was aligned at right angles approximately to this equator.

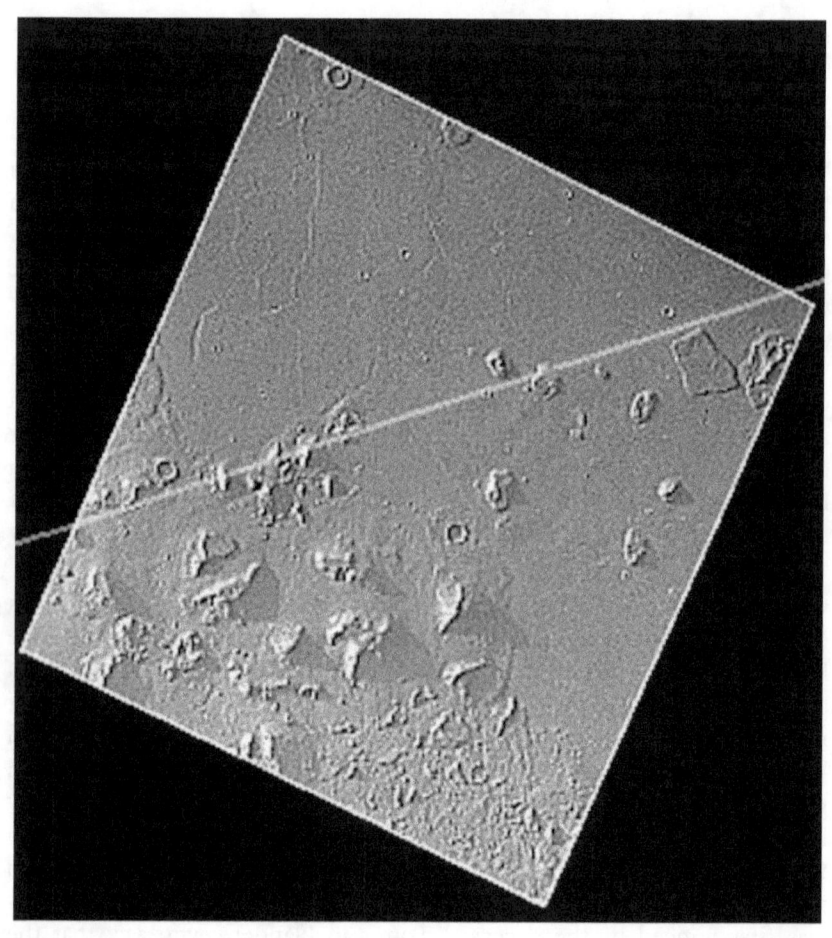

Figure 3

The white line represents an approximate equator with a South Pole east of the impact crater Hellas.

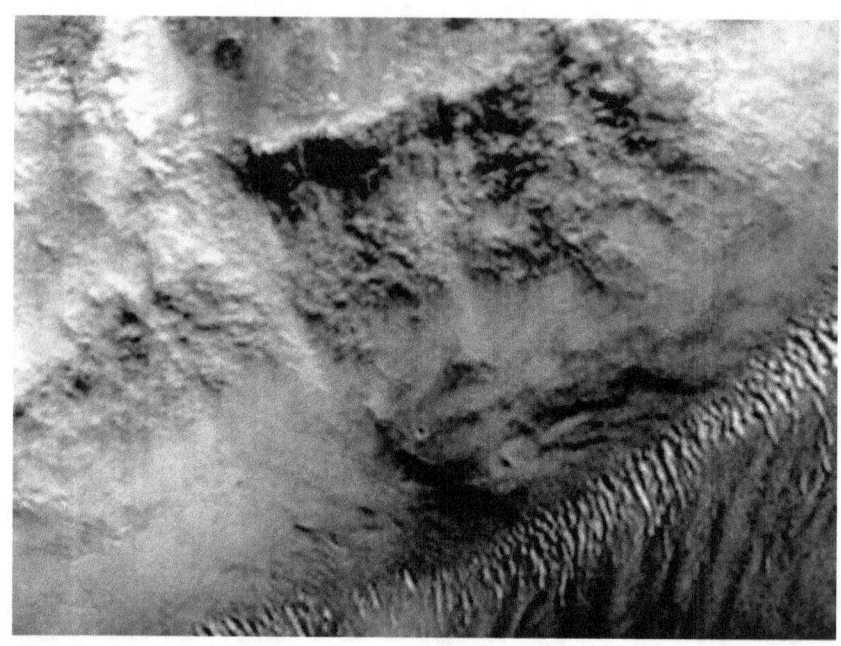

Figure 5

This is the Crowned Face. The area around this formation is the main focus of this book; the other 14 anomalous sites will be discussed in depth in future books.

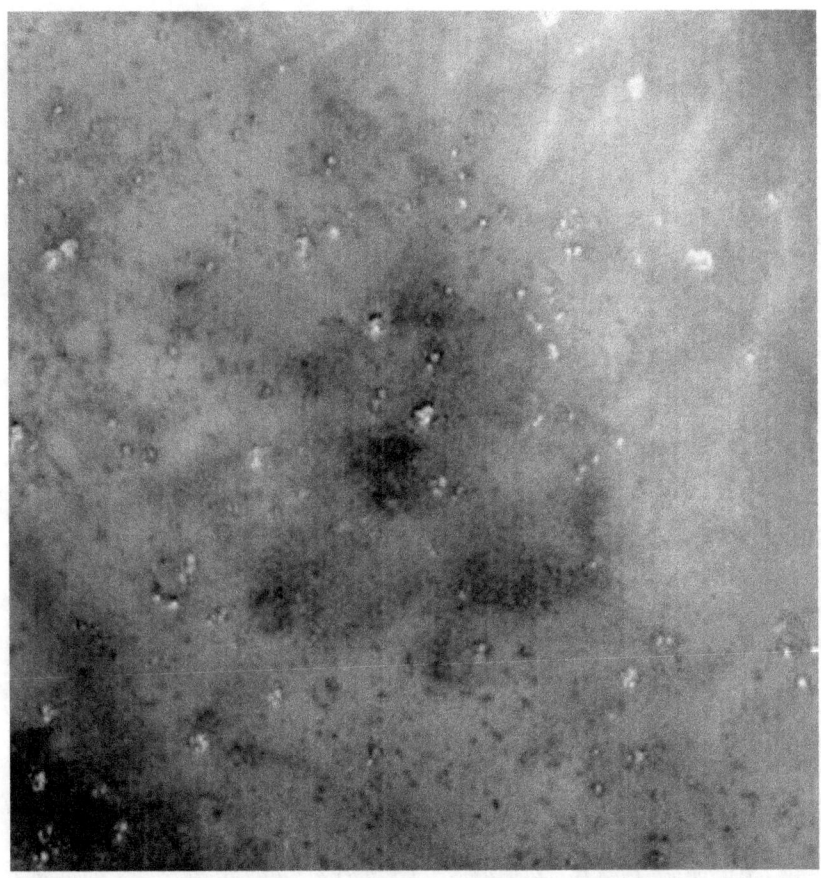

Figure 6

Nefertiti (found by J. P. Levasseur of SPSR) looks like a woman's head looking to the right. She appears to be wearing a hat like the famous Nefertiti sculpture hence the name. It is likely this is far older than ancient Egypt however; probably it was constructed well before there was any substantial life on Earth. The shape of the hat is not necessarily a coincidence; nearly every kind of hat shape has been used on Earth at one time so any hat covering on these Faces will likely resemble one of them.

Most of the other anomalies are near this equator giving rise to the theory that the Martian climate was still cold and near the equator was the most habitable area. They are also at close to the same elevation which is not unlikely considered Mars is so flat but most are either on the edge of a former sea or on islands in it. This can be seen for example in Figure 7 below where the dark area represents a low elevation that may have been filled by a mixture of water and mainly ice.

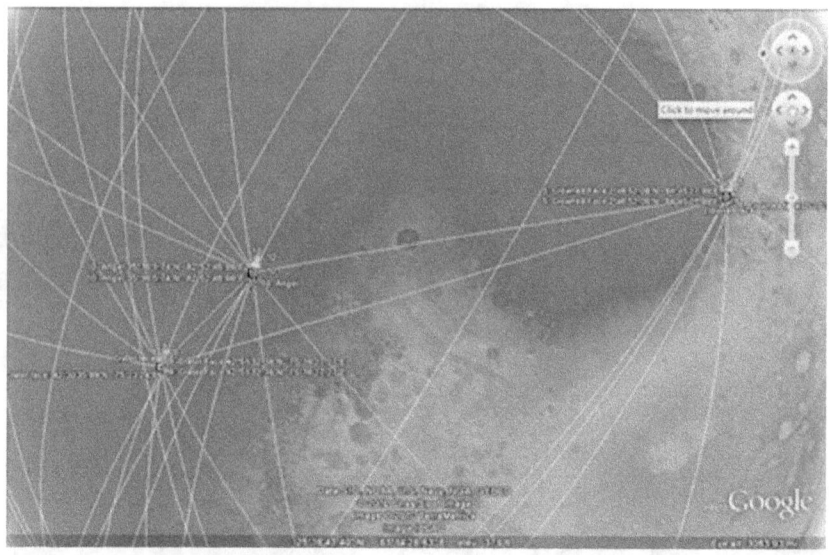

Figure 7

This shows three of these formations called the Angel, the Crater Face and the Crowned Face. As can be seen they appear to be on the edge of a dark depression which would have contained water in the past[xli] though it is not possible to be sure they were on the edge of this sea or just near it. If they did turn out to be very close to the shoreline this would be a testable prediction as the sea would be a likely place for cities to be built near. The Angel formation and the Crater Face will be analyzed in an upcoming book while the Crowned Face is analyzed here. The lines are drawn between 15 anomalous formations to determine if there is any mathematical significance to their positions. There need not be, hypothetical builders might have located them near natural resources for example but there are some mathematically improbable aspects to their positions. For example all of them appear on degenerate triangles, i.e. where three formations are in a line so that a straight line can be drawn through them. Others appear to be at right angles to each other. Because of the complexity of the mathematics this will be examined in a later book.

From our own experience we might expect that hypothetical alien colonists or evolved Martians preferred to live near a sea, most of the Earth's inhabitants live near an ocean or river. This water may have been used for irrigation if rain was uncommon or perhaps because life would survive more easily in water because of radiation so fishing was a major source of food. They might also have been amphibious, having evolved underwater and in the case of indigenous Martians only recently having moved onto the land similar to how we evolved. Forms of seaweed would also be more protected from radiation so it might have been a source of food. There is also some evidence of someone having created dams on the edges of craters to collect water; we still see water leaking out of the sides of craters and ravines on Mars today[xlii]. An artesian basin[xliii] would have formed as liquid water seeped into the soil and this would tend to leak out of the sides of craters and other depressions[xliv]. Other indications are that some craters could have been modified to hold water from a dam presumably to catch rain, or they could be constructed dams that look to us like craters but without the normal rounded shape and inclined crater walls.

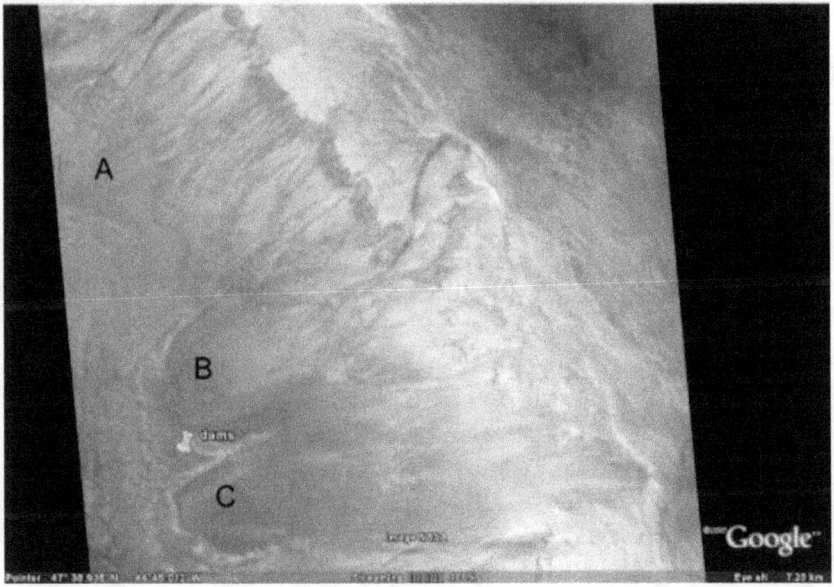

Figure 8

This shows three possible dams in a crater near Cydonia perhaps made to catch water. It is unlikely these are natural because they don't generally form where water ravines occur in craters elsewhere on Mars. More of these possible dams will be analyzed in an upcoming book.

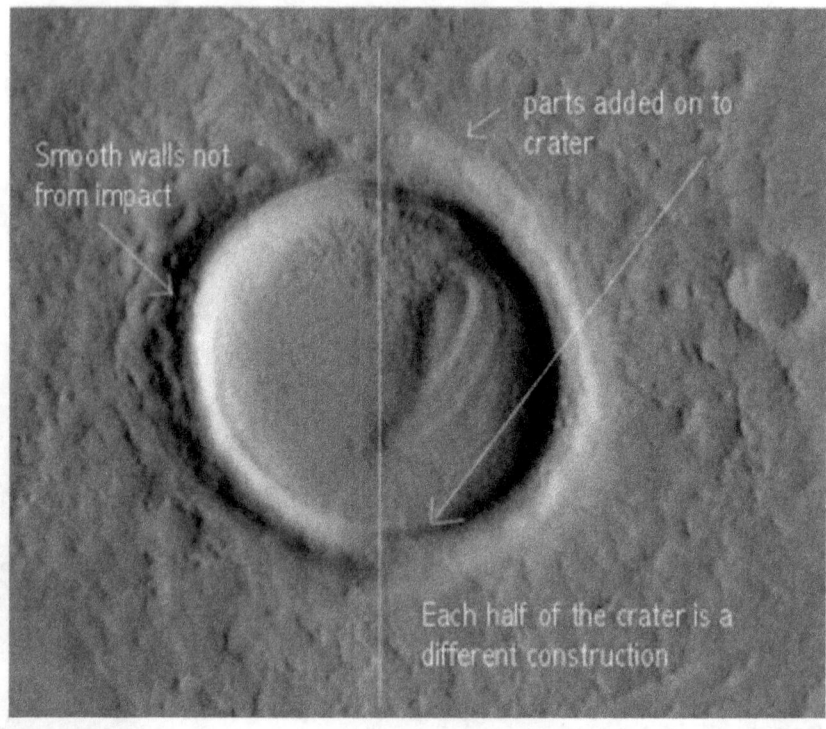

Smooth walls not from impact

parts added on to crater

Each half of the crater is a different construction

Figure 9

This shows a crater near the Angel formation that appears to have been altered away from its natural shape. This should be impossible to occur from the impact of a meteor[xlv] because the shock wave doesn't smooth out the crater walls or make an asymmetric shape. Half the crater is smooth with thinner walls and the other half much rougher with thicker walls, also the floor of the crater is flat rather than concave. It may have been altered to make a reservoir assuming rain was falling on Mars then. This will be more fully explained in an upcoming book.

Mars has a disadvantage for terraforming in that it has a weak magnetic field[xlvi] to protect it from the solar wind[xlvii] so even with this terraforming creating a thicker atmosphere the radiation would tend to be hazardous for most life[xlviii]. Living things such as small animals like tardigrades[xlix] and plants such as algae[l] in the ocean formed after the Argyre impact might survive but general life would tend to stay underground, live in water or under ice and snow coming to the surface for short periods during the day to avoid this radiation. It is hard to imagine indigenous life evolving sentience on Mars but if it did then it probably lived underground avoiding the radiation. The atmosphere may have been thicker then and Oxygen would tend to form Ozone which also protects life against ultra violet radiation. If so then we might find more evidence of habitation in caves and some formations which appear natural may be hollow, the signs of amphibious or underground life might be next to impossible to find without a manned expedition. The same might be for visiting aliens, it is logical and efficient to live underground or put large amounts of rock and earth[li] between living humanoids and the sun's radiation. This is our own plan for colonizing Mars.

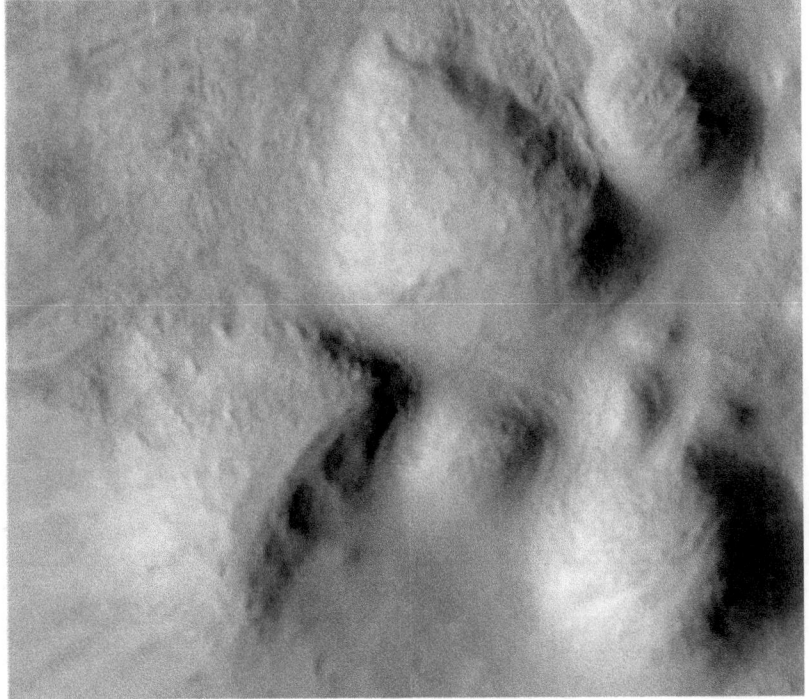

Figure 10

This shows possibly hollow structures in image SP240503. The hill on the bottom left has three openings in it which indicates they are caves or the whole hill is hollow. Since the other hills are odd geometric shapes they may be an example of ancient constructions, there are many of these left on Mars in various states of collapse. These will be more fully analyzed in an upcoming book.

The Hellas impact may have been another attempt at terraforming, either to reheat Mars for established colonies or aliens may have arrived with the

Pole in this area and caused this impact. The impact would have been at a shallow angle like Argyre, this can be seen from the elliptical shape[lii] of the crater below. By aiming the shock wave at the South Pole at the time this would have melted more ice and CO_2 to warm the planet, create rain and refill the Northern Lowlands with water. It may also be unrelated to terraforming though the impact being close to the Pole points to a possible connection. This Pole is consistent with the faces found on the Great Circle, perhaps a former Martian Equator. Assuming then the Pole was related to this equator then this Hellas impact would have melted the Pole water and CO_2 as described. The impact might also have killed off the colonists or indigenous Martians by the strength of its impact.

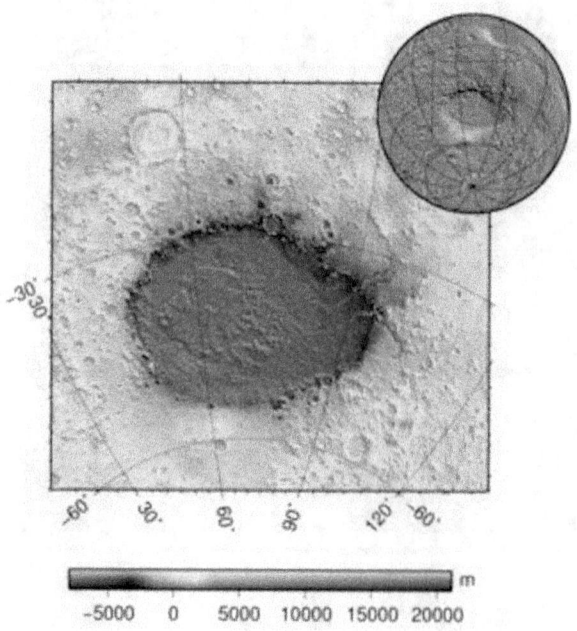

Figure 11

This is Hellas Planitia, a large meteor impact crater. The anomalous formations are clustered around an equator which has a Pole near here, just like Argyre crater is near a former Pole position. More on this can be read in the Appendix with the History of Mars.

There are many similarities in these formations which suggest a common style of building, for example some have mounds around them in angles that are consistent with surveying in a primitive way. The builders might have lined up formations by building mounds of rocks and taking measurements on them to create the needed angles much as a surveyor would in Ancient Egypt, in the Roman Empire or even today. To line up these mounds two would be built and then a man would move to a position so that he was directly in a line with the first two mounds and then a third mound built there, this is basically how the Romans built straight roads in Britain[liii]. To build a large monument like a Face these mounds could be formed in different angles so surveyors on the construction could tell if their work was anatomically correct compared to a scale model. By placing these mounds around the formation such as appear around the Cydonia Face then any position on it could be measured by the angle to these mounds, the angle of the sun using a plumb bob over a position and perhaps weak remnants of a magnetic field may have been enough to swing a compass needle.

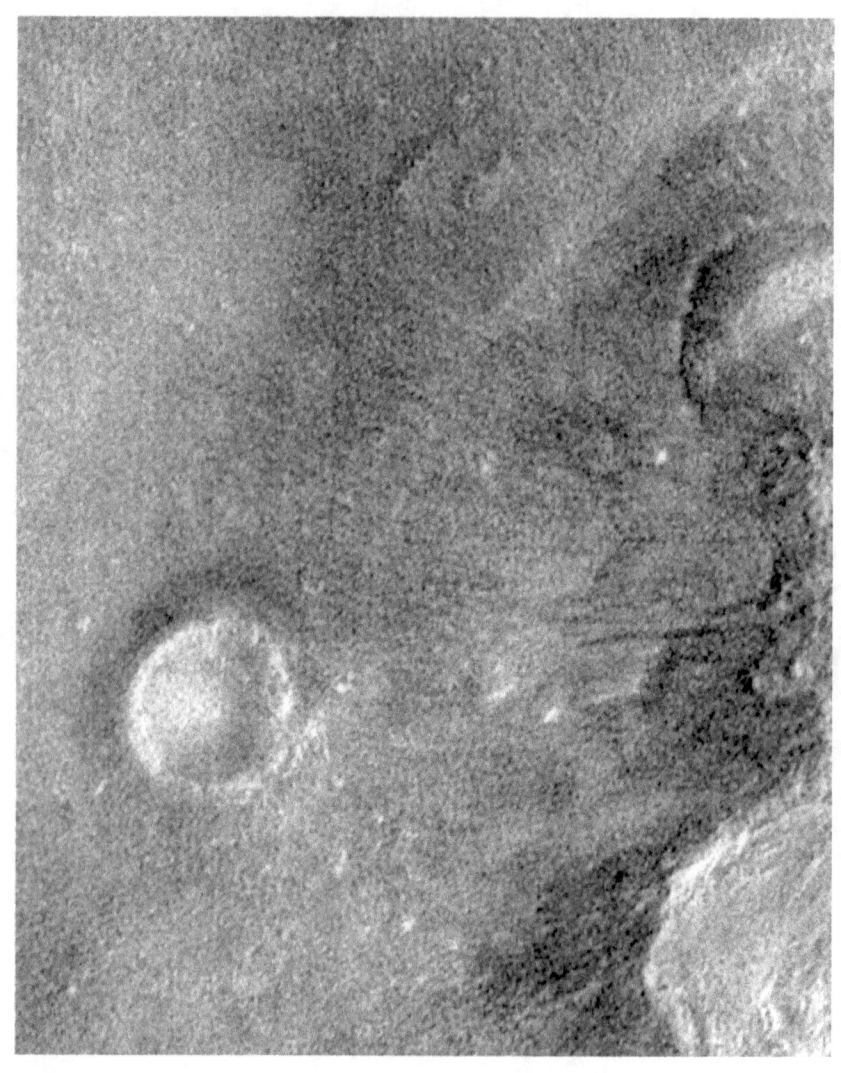

Figure 12

These mounds are close to the Cydonia Face and contain complex mathematical patterns, whether by chance or design.

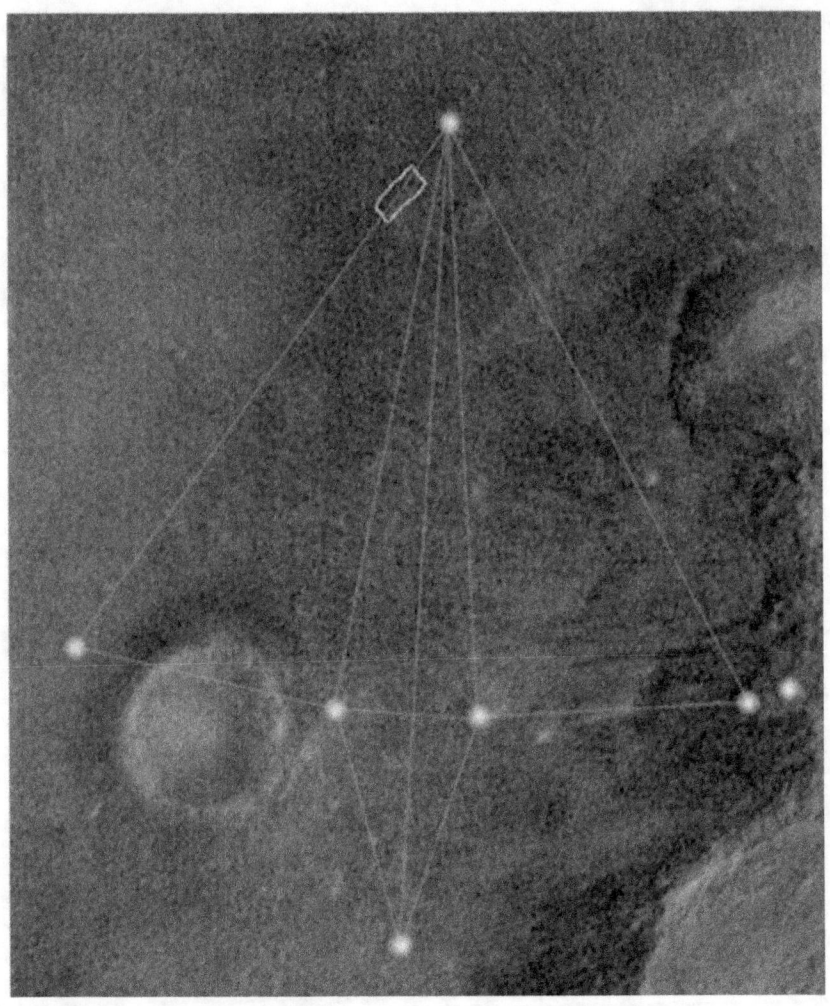

Figure 13

Here an almost symmetrical arrow shape containing right angles pointing to the Cydonia Face is shown by drawing lines between the mounds. Cydonia will be analyzed comprehensively in an upcoming book.

This level of technology seems very primitive compared to that available from aliens able to traverse the distances from another star, it may point to an indigenous race of Martians no matter how unlikely this appears. If so then we would expect to find primitive artifacts there similar to what we might excavate from Ancient Egypt such as pottery and primitive language records in stone or clay. It is unlikely any kind of cloth or paper

43

would survive but in a dry area underground without much air it is hard to estimate. One example of how dryness can enable preservation was in the City of the Sharp Nosed Fish but of course this is just a couple of thousand years not hundreds of millions of years or more. The lack of erosion on many parts of the Crowned Face argues for a significant amount of artifacts to be found, for example the eyes of this Face are still clearly visible despite abundant evidence of their age compared with faults in the area. Parts of the Faces that fell off would be valuable artifacts because they might be able to be put back on the Faces; they might also have designs or paint on them from the original preserved by being buried under dust. These geological signs of their antiquity will be explained in an upcoming book.

It may be the mounds such as in Figure 13 were left to increase the chances the formations would be recognized as artificial for as long as possible or they may also have some symbolic purpose. They also represent a problem in that these mounds have not eroded away around the Cydonia Face and yet the Face itself must be highly eroded to be plausibly artificial. It may be that the Face was made of more perishable materials such as being shaped by mud or clay in an outer layer while the mounds are piles of rocks which are more resistant to erosion. Such questions are difficult to answer with improved imagery, perhaps different materials on the Face might be detectable from space but the top layers of the Face might have been washed away. For example there may have been tsunamis in the area after a meteor impact which damaged the Face; it was probably just above the water in the Northern Sea. This might have washed away the added on materials which would prove artificiality or perhaps covered it with a layer of mud or dust that conceals them, in either case a manned expedition with all its expense might be needed to resolve the question. The mounds might also have been intended to represent constellations in the sky or have measured the seasons like our Stonehenge; this is something our ancestors did so it is plausible indigenous Martians might have done the same. The mound patterns would be unlikely to match star positions today because the positions of the stars would have shifted radically over hundreds of millions of years or more[liv]. We might never know for sure if they were intended to represent constellations.

Other formations are notable by their absence on Mars; we see little evidence of roads and quarries to get the rock and soil necessary to build these artifacts. This may imply underground tunnels were used to get around without exposing themselves to radiation but is nonetheless suspect for the whole concept of artificiality generally. That is, the idea that Faces and ruins on Mars might be built with little evidence of quarrying and transporting of materials implies they are in fact just natural structures, unless they were much tidier than construction on Earth. One possibility is that the ground was usually covered with ice and roads built on this ice naturally disappeared. Hypothetical aliens could move these materials using advanced technology but then it begs the question of why they were building primitive structures like Faces. A quarry by its nature would be of softer materials and might get washed away or covered in floods but the carving of large blocks of stone to build these Faces like the Sphinx or the Pyramids should leave clear evidence of quarries with the resolution of the photos we have. It may also simply be that the builders were tidy and didn't want a quarry or roads to spoil the visual artistry of the Faces.

One reason for this lack of cities and infrastructure could be the flag and footprints theory of visitation, that aliens visited Mars but did not colonize it for long if at all. They just marked it with records of their visit to show their home world and other people that evolved here later or also visited our solar system. Faces like this might indicate for example that this solar system was owned by a particular race or Empire; these Faces would last much longer than any electronically based beacons. It might also indicate the planet had been roughly terraformed perhaps along with our own, seeded with life, and then they moved on. An artificially intelligent probe might be capable of doing this construction, terraforming, and then sending information back to a nearby star of its progress. For example they might use laser light to give a message of its progress and prospects for colonization, we may have picked up signs of signals from another world using lasers[lv]. This would explain the lack of signs of colonization we would expect with the level of technology to build these formations, just enough to thoroughly mark the planet as having been claimed.

If so then we might expect to find signs of this claiming when we colonize Mars ourselves, an Empire bold enough to publically claim a solar system should leave enough clues as to who they were, where they came from, and their standing to make such a claim. Otherwise a claim would be impossible to verify, but a flag and footprints style claim might not give much information about who these aliens were. It is likely any hypothetical aliens would be afraid to leave signs of their location on a distant world to last for millions of years unless they were powerful enough to protect themselves against who was likely to find it. That begs the question though of why someone so powerful would not leave more signs of their visit, perhaps because terraforming and colonization is much harder than we can imagine in our position. Another possibility is that these formations are intended only for us, to be a sign for us to visit these formations and learn something, such as that our world was seeded with life by them, that there is or was a star Empire we might come in contact with, as bait to make us show our presence to someone, as a warning not to go outside our solar system, etc. In that case they need not be built to a high standard, just be good enough to arouse our curiosity enough to go and look. A hypothetical builder might have considered the possibility that we would initially discount them as being artificial but calculate that eventually we would colonize Mars and hardly be able to avoid investigating these areas. It may also be they could not conceive of our being so skeptical about their handiwork, and so did not see the point in making them more convincing from space.

Another paradox is our being able to recognize some of these formations as Faces and ruins; this implies a similarity to our own experiences what arguably should not be happening if aliens or indigenous Martians could evolve any kind of random appearance. For example there might be hills on Mars that appear natural to us but represent faces of aliens too unusual for us to recognize as possible life forms. Faces on Mars look similar to our own, this is highly suspicious and many authors have rightly pointed out this implies the human tendency to see faces in inanimate objects like clouds and rocks[lvi] and so these faces must be natural. However this coincidence might make a face on Mars be less likely to be artificial based on mere appearance, but it does not affect objects that can be falsified as not possible to form naturally. For example if a face or other formation cannot be formed by natural geological processes for some reason then it proves artificiality but trying to make a case based on a subjective impression of what they look like is probably doomed to failure. It does make it more important to look for falsifiability rather than just relying on a subjective impression of a humanoid Face though this kind of evidence is the most compelling. As said earlier, the humanoid Faces could have a natural explanation if aliens seeded Mars and Earth with their own DNA, this might evolve to have creatures with faces that have two eyes a nose and mouth. It is plausible that our DNA did evolve this way because nearly all animals, even fish have two eyes, a nose, and a mouth on a head in approximately the same positions relative to each other. To say that faces on Mars might be just as likely to have three eyes as two ignores the fact that no animals on Earth evolved with three eyes even though it would seem to confer an evolutionary advantage. There may be reasons why animals on nearly any world would evolve faces vaguely similar to our own; at least it is not evidence against artificiality that these Faces look somewhat like us. It is also important to realize that these faces appear humanoid to us but they are also just as likely to be animal like, there are thousands of Earth species with similar facial features to the Crowned Face such as squirrels or bats for example. We tend to discard these because the nose of most animals is different from a human nose, but such aliens might look vastly different to us but just have an animal face with a similar nose. Another aspect is the recurring motif of a split or pair of Faces sharing one eye and with two nose and two mouths, this might be the real face of these aliens and we change this with our preconceptions to two faces. We have no real evidence that primitive DNA from hundreds of millions of years ago would evolve into humanoid looking creatures all over again, except the somewhat circular argument that is explains the similarity between the faces on Mars and ourselves, but we also don't know that it wouldn't.

Another interesting part of this puzzle is the Spaceman image which will be discussed more fully in an upcoming book.

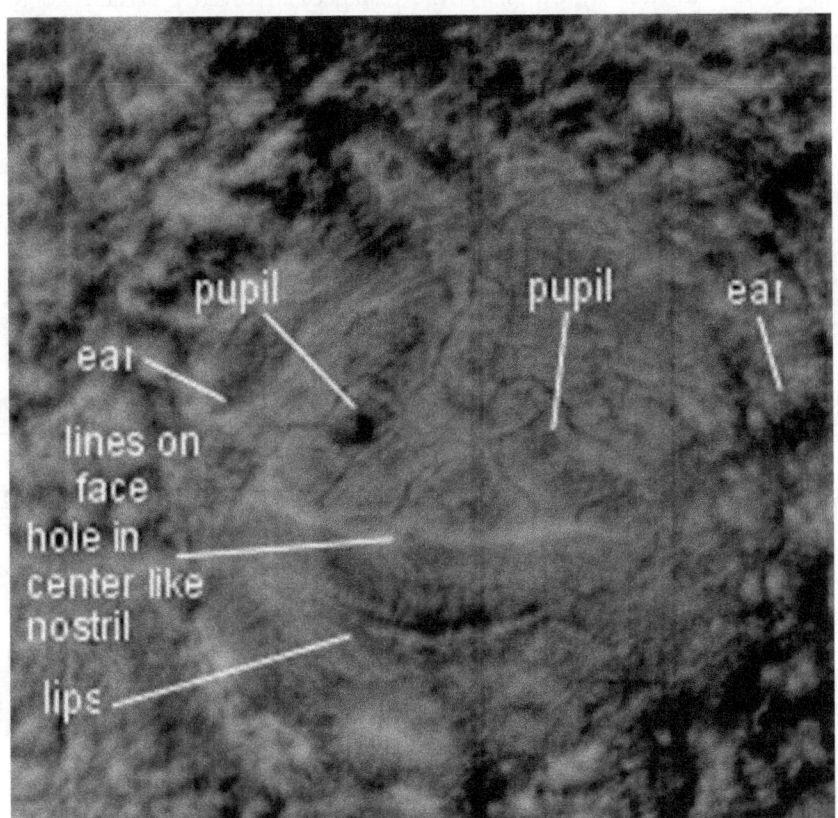

Figure 14

The spaceman formation has a more alien face but appears to interact with a humanoid face nearby. The two spaceman formations, very similar to each other, will be analyzed in an upcoming book. This face has also evoked skepticism presumably because it was not humanoid enough, while the Crowned Face is criticized because it is too humanoid looking.

This alien is more reassuring to some as it has only a single nostril and no protruding nose but it appears alien and artificial. It also appears on what would have been an island near the old equator so this ties it in with the possible terraforming of Mars, or perhaps aliens visited and met or even wiped out these indigenous Martians. The Spaceman creature reaching his hand out to a bowed humanoid implies a partnership of races which in turn implies one race visiting another inhabited planet, making an alliance or slaves, then travelling to a third solar system namely our own. One problem with this hypothesis is the distances involved for aliens to be doing this, it implies faster than light travel is possible in which case we should have vastly more evidence of other alien races dropping in on us. The long distances between stars should preclude one representative of a race meeting another for the sheer cost of such an expedition. If such events do occur then they conflict with the information we receive from SETI, that there are no communications we can detect from other stars. So there is a conflict between the apparent lack of technology in Martian anomalies compared with the ability to travel between the stars, this is even more problematic with this idea of aliens visiting each other and then coming here. It implies a star Empire of different races colonizing and terraforming worlds like Mars, but then disappearing long ago without a trace except for on Mars. It also implies that a hypothetical alien can get on a spaceship and stay alive long enough to travel to another star, provide themselves with food and air in the meantime, or if in suspended animation be able to propel a spaceship fast enough to do this. According to our best understanding this is all but impossible to do because of the huge amounts of power required, the possibility of hitting grains of dust at light speed would destroy any ship we could imagine, and the journey to another star would likely be longer than any lifespan of a creature we would compare to. If they are so different as to live long enough to travel between the stars, then why do they look like us at all? Another possibility, other than the simple one of the Spaceman being a natural formation, is that the humanoid face was a genetically engineered creature made for colonizing Mars that died out but somehow DNA from this came to Earth and evolved to form humanoids namely ourselves like this human carving on Mars. An additional problem is there are two

Spacemen formations in different locations quite similar to each other, one might have a coincidental likeness to a face but Mars should not be randomly making faces that look like each other. This would be like seeing the same face appear in a cloud more than once. The evidence on Mars is consistent with a few interpretations but they raise more questions than they answer.

The other problem that arises is the similarity to primitive Earth like societies to these hypothetical Martian ruins, which implies theories similar to Von Daniken and Zecariah Sitchin. However if these formations are as old as they appear then they could not possibly be related to ancient civilizations on Earth unless aliens were visiting our solar system for perhaps a billion years. They should then have left more traces than some formations on Mars unless there is a kind of Prime Directive about contacting us, but then why leave faces on Mars at all? We shall see in an upcoming book some possible artifacts on the Moon which seem to reveal too much for an alien race so keen to hide their visits. Having such formations on Mars and the Moon implies a spacefaring civilization which rules out the indigenous Martian theory because if Martians evolved highly enough to get off Mars then there should be much more evidence of this evolving civilization in their ruins. More likely any humanoid face with some kind of ornaments would have looked like a particular race on Earth because there is so much diversity in cultural art.

Figure 15

Akhenaten has a similar nose and eyes; there is also a crown but few Egyptian characteristics on the Crowned Face.

The resemblance of the Crowned Face to Akhenaten is notable, as analyzed in an upcoming book, even more so by the resemblance of the Nefertiti formation to the real life Nefertiti with the same kind of hat in

50

both cases. This implies a theory where aliens visited Akhenaten and Nefertiti, assisted them to make faces on Mars that only appear to be ancient and then the aliens departed or are still nearby. Such a message might be left for us to see eventually, go to Mars and then because we would be a spacefaring civilization be mature enough to know about our visitors and to receive an additional message or information about technology on Mars. They may have made these faces on Mars look like they were from this time in our history because this is what they thought we would recognize but this ignores the high likelihood these formations are in fact ancient, for example the Cydonia Face discussed extensively in an upcoming book must be extremely highly eroded because a builder would not have left it in that condition. So there is again the problem of two visits; one visit so long ago that an alien race should have either died out or be so dominant they are easily observed with SETI research. Either that or there may be a secretive nearby space Empire watching us under some kind of prime directive like from Star Trek, but why then construct messages on Mars like this instead of just communicating when it suits them? The simple answer may be that they were aliens and their habits incomprehensible to us, we have enough trouble understanding other races on Earth to find fault with the customs of aliens just because they may have looked somewhat like us[lvii].

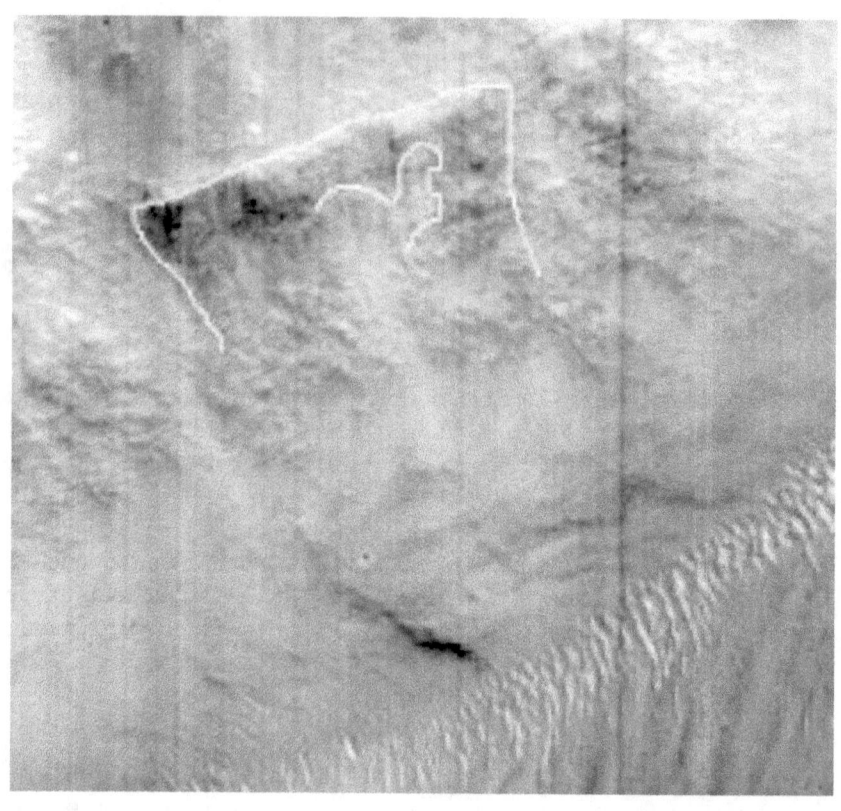

Figure 16

There may be an Asp like ornament on the crown of the Crowned Face like with Akhenaten, the crown is also similar. Akhenaten had a long face and nose like the Crowned Face. None of the Faces on Mars seem to look like the Sphinx nor do any of the possible buildings resemble the Egyptian Pyramids with four sides.

As can be seen the evidence is somewhat contradictory. This could be one of the indications of randomness in that some or all this evidence is false and there are no artifacts on Mars, however this is balanced by so many common features about these possible artifacts that this implies nonrandom common processes in their formation.

Chapter Three

Orientation of the Crowned Face to the old equator

In this section I plotted the Cydonia Face, the Crowned Face and Nefertiti on a Great Circle and calculated where the South Pole would have been if this was an equator. One reason as mentioned earlier is that Hellas might have been an attempt at terraforming by creating a shallow impact aimed at the South Pole at the time, heating up Mars while building the faces around the equator. In Figure 1 below Hellas Crater is the dark area in the upper left corner. This Pole would be close enough for Hellas to conceivably have been used for terraforming, but in Figure 2 below the Pole is further away from Hellas, too far to be melted by the impact directly. It may be after the impact the Pole moved to the West to the position where the faces were constructed on its equator, and then later as Mars cooled the Pole moved south to its current position.

In the photo below I first marked the former South Pole position from the polar wander path proposed by Sprenke and Baker[lviii], this is better explained in the History of Mars paper in the Appendix. This Pole position is only approximately known. The lines in Figure 1 point to various anomalous formations on Mars; these have some interesting mathematical relationships between them, analyzed in an upcoming book.

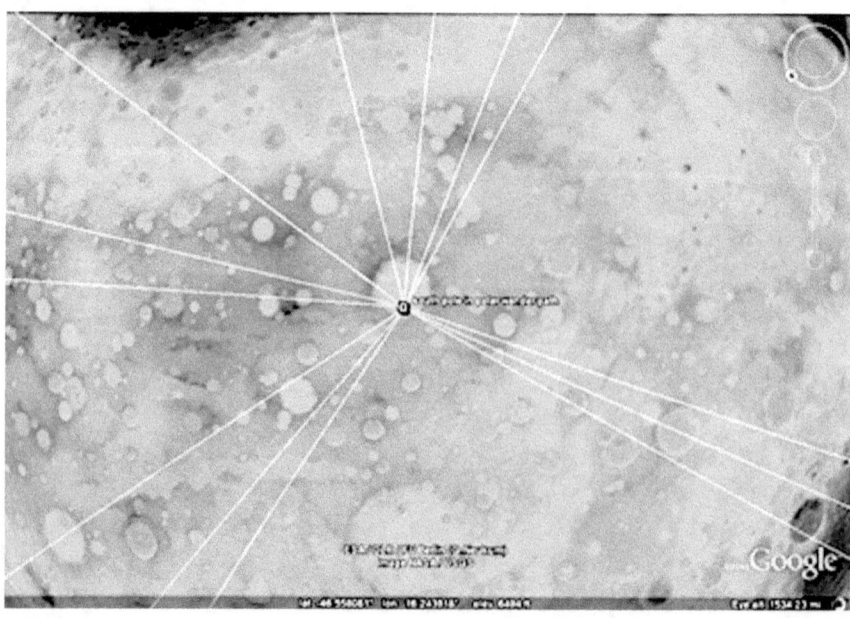

Figure 1

The large crater at this South Pole is Kaiser Crater.

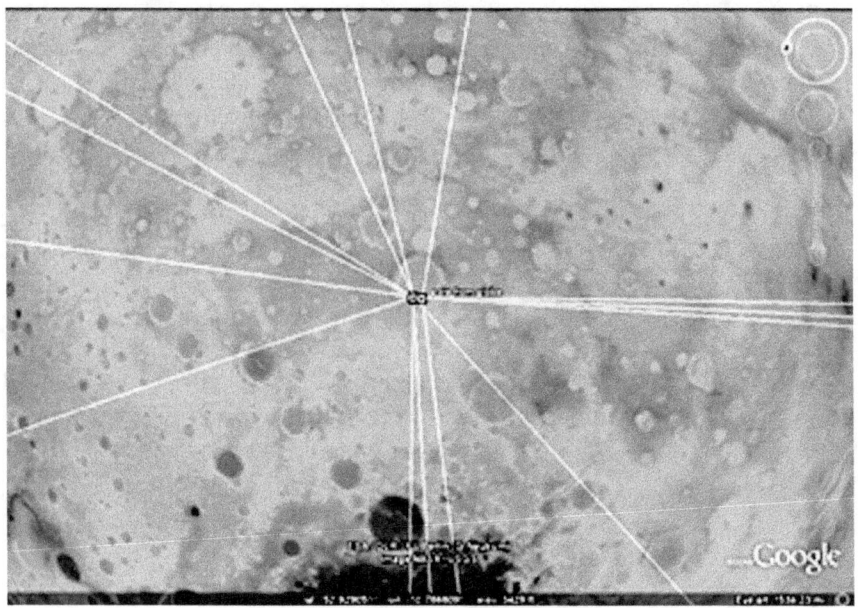

Figure 2

To work out the South Pole from the Great Circle that Nefertiti, the Crowned Face and the Cydonia Face fall on is difficult mathematically, the easiest way was to mark them on a globe and then move a tape with a length ¼ of the circumference of the globe in an arc from all three anomalies. Where they intersect is close to the South Pole position at 52S 10W. There are many variables which make this imprecise, Mars is not perfectly spherical and there are no formulae readily available to work this out on its exact shape. Also many areas of Mars have been deformed over time with meteor impacts and perhaps shifting areas similar to tectonic plates so we may not know the exact shape of Mars at the time until Mars is colonized.

The two South Pole positions are quite close to each other, the estimate by Sprenke and Baker is based on elliptical craters near the Pole and the terrain affected by ice, this however can be inexact. The elliptical craters form because asteroids tend to move in a disc around the sun called the Plane of the Ecliptic, when they hit near a Pole then this automatically gives them a shallow impact and therefore an elliptical crater. This may also mean that the Argyre and Hellas shallow impacts were not unusual, the Northern Lowlands may also have been carved out by grazing impacts from larger meteors[lix]. There are 6 possible artifact sites on or near this Great Circle so its position is imprecise to some degree, Mars not being perfectly circular might mean that these three sites falling on a Great Circle is partially a coincidence with this high level of precision. However it is plausible that this clustering around a great circle implies an equator since no longitudinal Great Circle would match a former Pole position unless it referred to the movement of the Pole from the area of Tharsis. This is explained in the Appendix with my polar wander paper there.

By keeping the same orientation and moving on Google Mars to the King's Valley it is possible to determine its orientation compared to this old Pole, this makes it possible to determine what shadows the sun made on these faces as it moved through the day and the seasons, this will be more fully analyzed in an upcoming book. There is a limitation in the position of these faces in that the valley probably could not be changed too much to accommodate more useful shadows on them but a good orientation can mean the changing sun angle would highlight certain parts of the faces.

Figure 3

This is worked out from the Pole position calculated by polar wander rather than by measuring the pole position from the Great Circle, however there is not enough difference between them to change this analysis.

In Figure 3 down is due south when the Crowned Face lay on the proposed Martian equator with the Cydonia Face and the Nefertiti formation. This would mean that as the sun rose in the east it would shine from the right side of the image then move over to the left at dusk. With the seasons the sun angle would change according to the axial tilt at the time. It's likely then that these faces were able to show the seasons, Face One here above the Crowned Face (analyzed later on page 171) might stand out more when the sun was higher in the Northern Hemisphere with their Summer Solstice, the Crowned Face or Face Two would show most at the Equinox and Face Three (analyzed on page 185) here below the Crowned Face would perhaps show more at the Northern Hemisphere's Winter Solstice. Note in Figure 3 Face One stands out very poorly but Face Two stands out with more shadows, the sun is coming from approximately the South East here. In the Mars Orbital Camera photo M0203051 Face Three has some deeper shadows and may have stood out more later in the year. It is only a coincidence that the shadows in Figure 3 approximate a Northern winter because the Pole is now in a different position, but it enables us to see what the faces would have looked like in that season. The idea of shadows changing with the seasons is similar to Stonehenge and many other ancient formations on Earth, the Sphinx is also said to be aligned to the Solstice[lx]. The shadows on its face may also be designed to change with the seasons, though this does not seem to have been investigated[lxi].

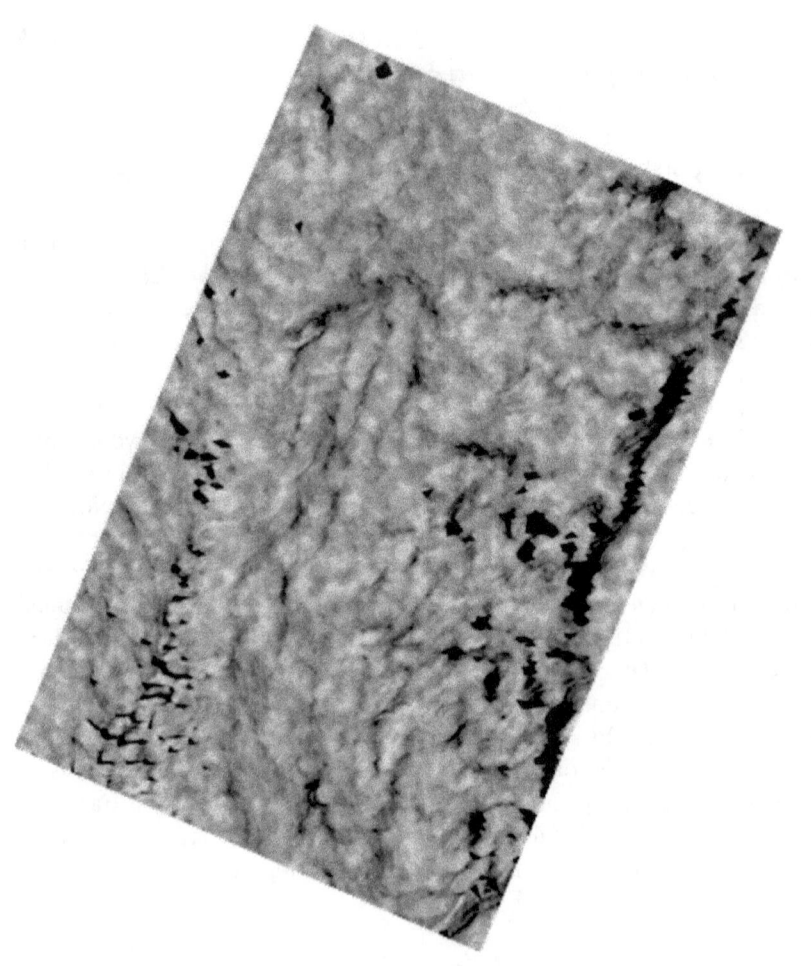

Figure 4

In this simulation of the terrain the sun is coming from approximately North East which would be mid-morning in summer in the Northern Hemisphere. The Crowned Face is on its side with the crown pointing to the right. More analysis of this will be done in an upcoming book but as can be seen the shadows are different compared to in Figure 3 which would be winter for the Northern Hemisphere.

Figure 4 is a shape from shading model of the Crowned Face with the sun set at the desired angle.

Cydonia, Nefertiti and the Crowned Face are on a Great Circle as if they were on the equator at the time when constructed. The Cydonia Face is approximately at right angles to this old Pole position; this will be analyzed in an upcoming book on Cydonia.

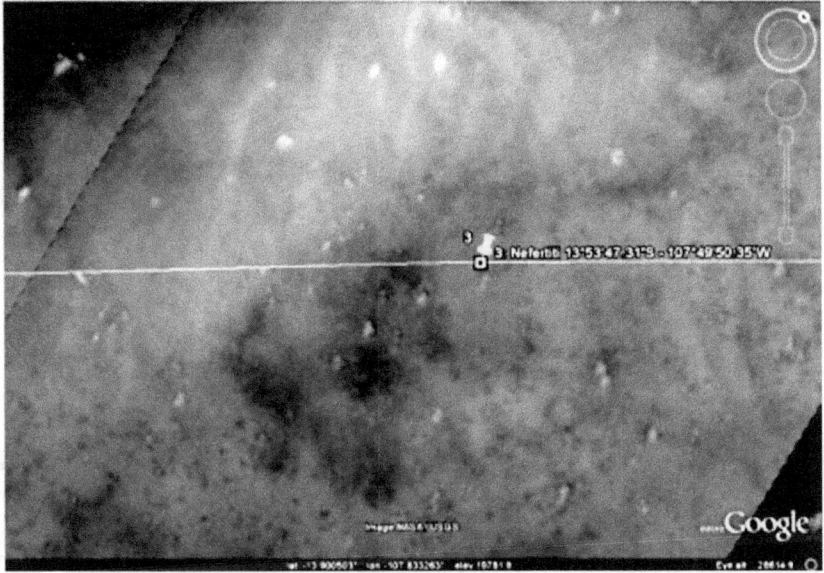

Figure 5

The Nefertiti Face is seen faintly here, this is from Google Mars rather than an enhanced image. The horizontal line is the Great Circle, probably a former equator so south is straight down. It does not seem to be exactly aligned north-south though that depends on exactly how the face is intended to be orientated. This face and three others in the area will be fully analyzed in an upcoming book. The evidence that these faces are intended to be aligned relative to the Poles is weak; the Cydonia Face seems to be but Nefertiti is not exactly North-South. The Crowned Face is not aligned to North-South but the shape of the valley constrains this. It may be that if more faces are found later that this theory of alignment to the Poles can be better tested.

Chapter Four

A Theory of Terraforming Mars

The main inspiration for this idea was the research done on polar wander. This led to a theory which can be seen in the appendix called A History of Mars. More will be written about this polar wander path and how it affects the location of possible artifacts in a future book. The theory of this paper as mentioned in the previous chapter is that the Argyre impact was a very shallow impact and its shock wave travelled partially along the Martian surface onto the South Pole of Mars at the time. This led to the Pole's ice and frozen CO_2 melting and the formation of Tharsis Montes, Valles Marineris and Olympus Mons.

It may also be that the same kind of meteor impact could be used today to terraform Mars today. The advantage of this method is (if the History of Mars theory is correct) we can see what the effects of a similar meteor impact had in the past. So by recreating this event at one of the current Martian poles we might expect a similar sequence of events as happened then. This might include melting both Poles, forming a liquid water ocean, and increasing the atmospheric pressure by outgassing from volcanoes as well as sublimated CO_2 from the Poles. The volcanism from this impact might well take centuries to build more giant Martian volcanoes, in the meantime this will provide heat on Mars near the Poles which would constantly melt any ice and CO_2 that redeposited there. It may be possible to build a base nearer these volcanoes initially to take advantage of their heat.

The basis of this terraforming would be to duplicate a similar event to the Argyre impact on Mars. Normally when a meteor hits Mars the shock wave goes into the ground in the direction the meteor was travelling, and the heat from the impact warms deep down where the energy is wasted for terraforming. Also if the objective is to form volcanoes to heat Mars then the shock wave underground is not useful for this as the heat energy from the impact normally stays underground.

A shockwave is strongest in front of a moving object. For example if a moving car uses its horn the sound waves have more energy as well as being higher in frequency in the direction of motion of the car than behind it, this is called the Doppler effect. The difference with a shallow impact is that the shock wave cone is strongest in the direction the meteor is moving and because the angle is so shallow on impact much of the shock wave cone doesn't go into the ground but part of the shockwave is above ground and part is below ground[1], therefore part of the shockwave travels on the surface. The outline of this shock wave might have been as shown in Figure 1. While the angle of impact is not known exactly it should be able to be worked out from geological data around the impact site, assuming this theory is correct. So an observer might have seen the Argyre meteor coming from low on the horizon heading towards the South Pole and strike a glancing blow on impact.

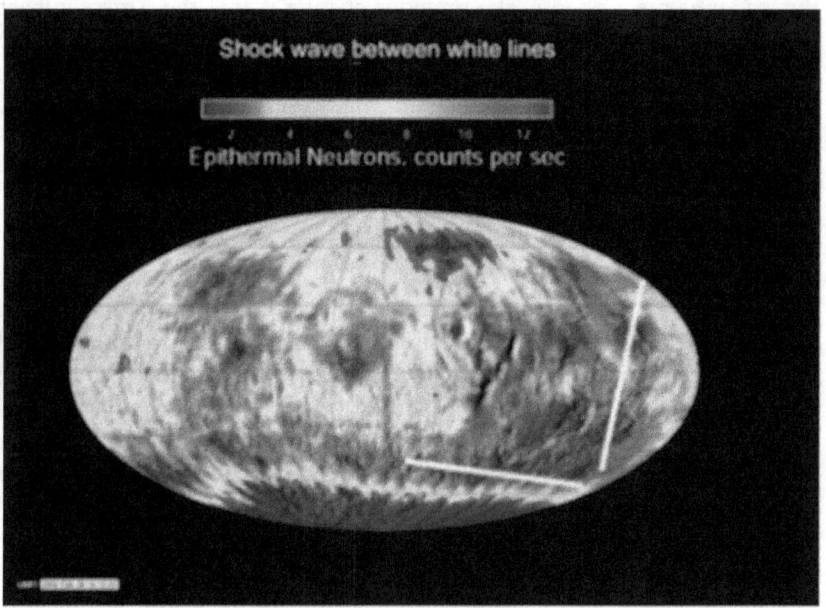

Figure 1[2]

The darker area between the lines indicates the ground contains less water ice; this may be because the impact dried the ground out with the heat from the shock wave. This radiates out from the Argyre impact crater approximately according to the white lines and the length of this dark boundary is longer than the diameter of Mars. The dark area at the South

---

[1] http://www.psi.edu/~betty/obliqmelt.html

[2] http://mars.jpl.nasa.gov/odyssey/gallery/latestimages/latest2002/march/Fig2_NS_Global.html

Pole by contrast contains large amounts of ice. It may be easier to see this in color with the link attached to the image. The shallow impact can also be seen in Figure 2[3], where a darker drier area again radiates out from Argyre crater.

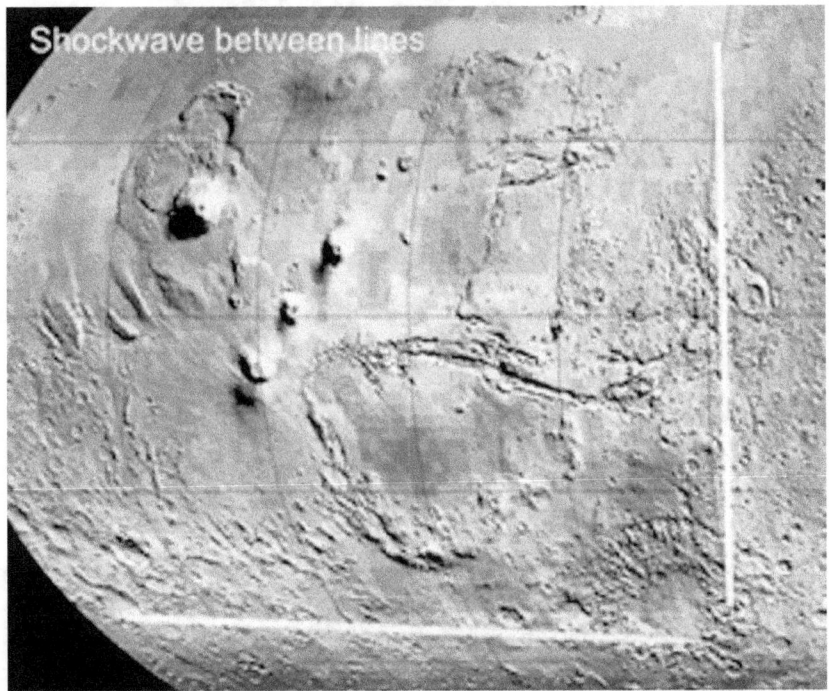

Figure 2

In Figure 3[4] below the drier area is represented by dark grey as radiating out from Argyre Crater.

---

[3] http://www.lanl.gov/orgs/pa/News/cover_epi.jpg

[4] http://www.lpi.usra.edu/meetings/sixthmars2003/pdf/3253.pdf

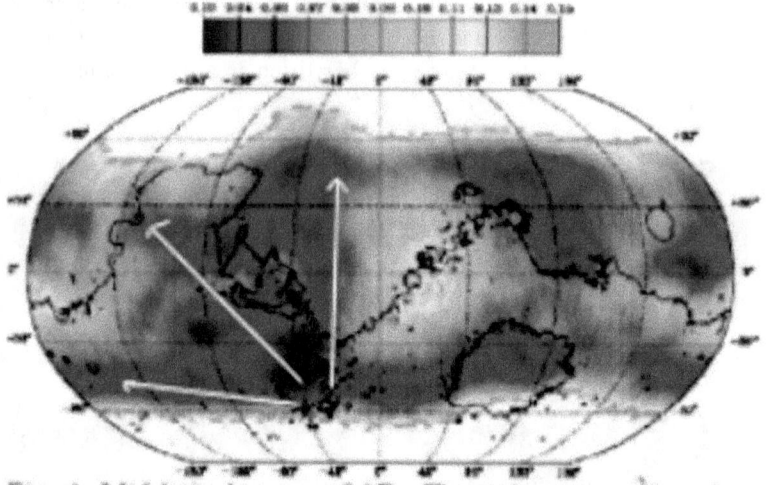

Fig. 4. Mid-latitude map of $\frac{\varsigma}{\varsigma}$. The units are cm^3/g. A contour of topography at 0 km elevation is superimposed on the map.

Figure 3

It is proposed that the strongest part of the shock wave from Argyre Crater was in the direction of its motion towards Olympus Mons. This warmed and dried the ground in a fan shaped area comprising Solis Planum, Tharsis Montes and Olympus Mons enough to drive some water out of it, so it was likely hotter than the sublimation temperature of water for the air pressure at that time. A similar smaller dried area also radiates out from Hellas Crater indicating it may also have been a shallow impact, though not as great as Argyre.

For this kind of terraforming of Mars to be done by us the size of the asteroid and its speed would have to be selected carefully. As can be seen here an enormous amount of energy was added to this area of ground, and this would be needed to heat the ground without blowing off large pieces of Mars rubble which might threaten to hit Earth later. So the angle of impact would have to send this debris outwards in the solar system, which may have been how some comets were formed in the past. For example icy rocks from the Argyre and Hellas impact as well as those impacts forming the Northern Lowlands may have reached escape velocity and have gone into an elliptical orbit out of the solar system as comets. Since Argyre and Hellas may have impacted near a Pole some of these rocks might contain large amounts of ice, as seen in some comets.

Heating such a large portion of the Martian surface hot enough to drive water out of it can translate to more heat in the atmosphere, which gives a window of opportunity for terraforming. Some of this heat comes from the kinetic energy of the meteor's momentum, as mass times velocity, but it also comes from opening up fractures and rifts allowing hot magma from underground up to the surface. So a shallow impact if it creates enough fractures can release far more energy than just from the impact itself.

In the History of Mars there was theorized to be a Pole[5] around the center of Figure 4. Argyre crater is seen in the lower right. The section of this crater pointing toward the center of the photo is shallower; this is consistent with a shallow impact with the shock wave travelling to the top left.

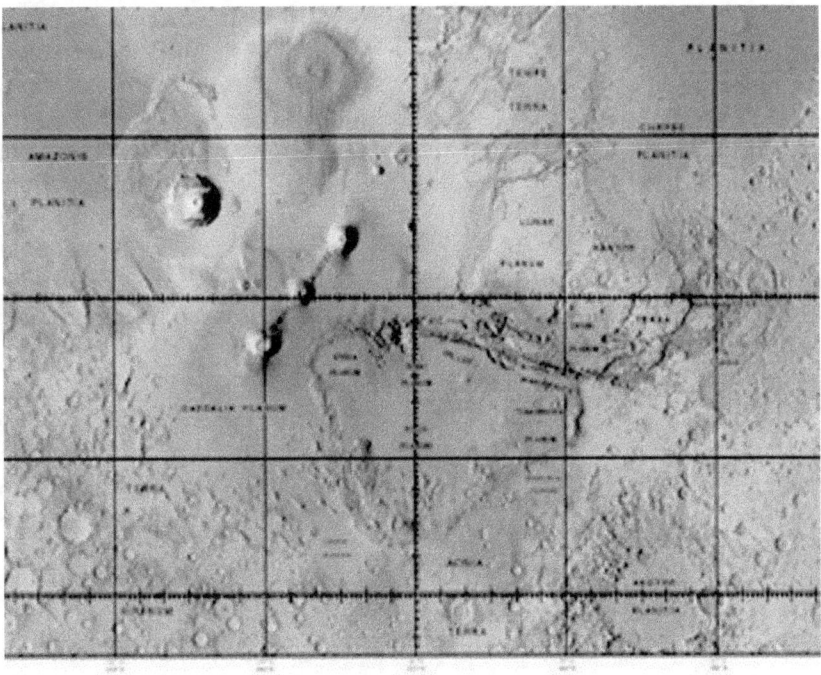

Figure 4

---

[5] http://www.aas.org/publications/baas/v31n4/dps99/40.htm

The shock wave was probably aimed at the South Pole of Mars at that time, so the darker higher areas shown in the center of Figure 4 would have been caused by volcanic activity from the impact. As the History of Mars paper proposes, the Pole then moved eastward to Meridiani Planum, then in a southerly direction to near Hellas Crater and eventually to its current position. Thus this single impact may have caused the Pole to move over much of Mars. This would also act to spread more water over the planet because the Pole changes the terrain as it moves and water from the Pole melts behind it. Using a similar impact to terraform Mars might also cause polar wander and spread more water around Mars. The heat from the Argyre impact directed at the Pole would have melted the water and CO_2 on it, and as proposed in the History of Mars paper much of this water ran into Chryse Planitia. We might then expect a similar amount of floodwater from a meteor impact aimed at the current Pole.

The North Pole might be preferable as a target for a meteor impact today because floodwater from the melting Pole would go into a depression called the Northern Lowlands which was a former sea. So if the ice at the current North Pole melts and starts to fill this depression then the hotter climate at lower latitudes from the impact and volcanoes forming will slow down its refreezing. The Pole is higher than the surrounding terrain because it is built of ice deposited there, so water that melts from the Pole would tend to sit around it or at lower latitudes. A Hadley Cell should start to form because of the temperature difference between the hotter equator and the still cold parts of the Poles; this would perhaps produce rains and trade winds across this expanse of water.

The shock wave cone of the terraforming meteor impact would be similar to those in Figures 2 and 3 (the angles of the fan shape would depend on the angle of impact). The Pole would tend to retain heat for a long time after the impact and remelt CO_2 and water ice that returned later to the Polar cold trap. For example the melted water would create large seas and the sublimated CO_2 would create a thicker atmosphere and this would tend to create weather transporting water vapor back to the Poles. So the heat from the shock wave and crater there should prevent the Pole reforming for a time, as it may have done after the Argyre impact.

So to terraform Mars the idea would be to do something similar, to aim a meteor to impact at a shallow angle near the current North Pole on Mars so the impact should likely occur at round 70-80 degrees north. Ideally it may be better to aim this shock wave at Chasma Boreale.

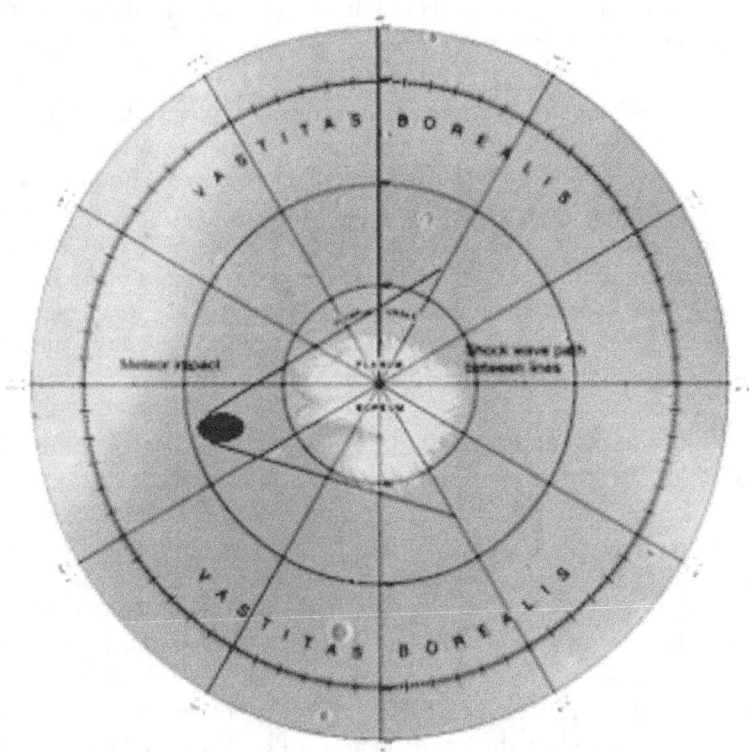

Figure 5

In Figure 5 the dark ellipse is a suggested location for the meteor impact site, and the dark lines show an estimated shock wave aimed at the North Pole around Chasma Boreale[6] [7]. Melting floodwaters from the Polar ice should flow back down this Chasma onto the hotter impact site, this floodwater would hit the impact crater and this heat would convert much more of it into steam. Ideally there would be a flood of water into the Northern Lowlands in this area and the impact crater would continue to boil this water into clouds and create rain elsewhere on Mars. Because the

---

[6] http://www.lpi.usra.edu/meetings/lpsc2003/pdf/1854.pdf

[7] http://www.lpi.usra.edu/meetings/lpsc2006/pdf/1363.pdf

Pole is higher than the surrounding terrain, from its ice sheet, then part of the shock wave above ground would impact the side of this ice and tend to fragment and melt it.

Elysium Mons is one of 3 mountains directly on the opposite side of Mars to the Argyre impact, likely formed by antipodal volcanism[8]. One possible event is that, according to the History of Mars paper mentioned earlier, the shock wave from the Argyre impact travelled through the Martian core and created a fracture in the ground, releasing magma and forming these mountains. One reason to believe this occurred is that the drier areas from the shockwave radiating out from Argyre are longer than the diameter of Mars. So if a shock wave could travel this far on the surface it should be able to travel the same distance through Mars to the other side.

So in terraforming Mars a volcano might form near the South Pole opposite the Northern impact crater, and this would provide some warming on the South Pole as well, melting water and draining into the surrounding lowlands and perhaps into Hellas Crater forming a Southern sea as it may have been in the past[lxii]. Hellas is so deep the air pressure is much higher at the bottom so water that ran into there should vaporize less quickly than other parts of Mars so a lake or sea is likely. This heat would also persist because magma from this volcano would continue to make it grow and add heat to the area; also a volcano like this should release gases to thicken the atmosphere further. Elysium Mons and her two sister volcanoes are quite large so if they were formed by antipodal volcanism then we might expect a similar result from our own terraforming. It also illustrates the large amount of additional heating of Mars created by the Argyre impact; enough to release enough magma build volcanoes this large. Using the same kind of impact antipodal heating might be substantial and melt much of the huge ice reserves of the South Pole, forming a sea in the Prometheus Impact Basin.

Meteors

The shallow impact directed at the Martian poles could also be done by several smaller meteors from different angles. One advantage of this is the effect of each meteor might be checked and subsequent meteors deflected away if not needed or there was too much effect from previous impacts. Also smaller meteors might be easier to deflect onto the desired trajectory.

---

[8] http://www.grc.nasa.gov/WWW/K-12/MarsV/rania.htm

Several meteors may also be aimed at the same point to increase the fracturing along the surface with their shock waves. Because the effects on Mars from the Argyre impact may have lasted millions of years it may only be necessary to create a smaller terraforming effect, a smaller impact and subsequent volcanism might make the planet stable for colonization in a shorter time.

The asteroid belt is close to Mars and large asteroids orbit close to Mars often impacting on it. The theory requires altering the trajectory of one or more of these asteroids sufficiently to create something similar to the Argyre impact. The Argyre meteor has been estimated[9] at around 900 kilometers in diameter but one much smaller may be all that is required for terraforming.

Asteroids that pass close to Mars would be surveyed and at least one selected. Then it would need to have its orbit altered so it would hit Mars like the Argyre impact, at a shallow angle. Proposals including launching rocks from such a meteor with rail guns or nuclear rockets would be used to nudge the meteor into the right impact angle, perhaps over many orbits around the Sun. The trajectory of the asteroid would be moved, much as we ourselves propose to move asteroids preventing one hitting Earth in the future[lxiii]. So again no more technology is required than we ourselves have proposed and planned for, though the cost of this may be well beyond budgets available at this time.

However if the theory is correct Mars might become habitable more quickly than with other methods[lxiv]. Once the water and CO2 from the Poles had melted from the impact this should reach equilibrium with the hot rock of the impact site and water returning to the cold trap of the Poles. Over time volcanism would further heat Mars, any danger should be unlikely as long as the volcano doesn't eject rocks and hit the colony. While some fallout from volcanoes might be dangerous the thicker atmosphere would cause more meteors to burn up in the atmosphere reducing the danger from them. The initial heat would come from the impact allowing a window of opportunity to use plants and algae to create Oxygen from Carbon Dioxide. Because Oxygen would never freeze at the current Martian temperatures the air pressure on Mars would become permanently much higher, or at least until it changed back into CO2 which would be prevented by plants adapted to the Martian climate. The

---

9 http://www.lpi.usra.edu/publications/books/CB-954/chapter1.pdf

volcanism would continue to add additional heat to the atmosphere, particularly around the Poles, preventing them from reforming quickly.

Summary

The advantages of this terraforming theory are then as follows:

Melting of ice

After the Argyre impact there is good evidence the Polar icecap melted and the floodwater poured down the future Valles Marineris into Margaritier Sinus [10], Xanthe Terra, and Chryse Planitia. This area has abundant evidence large amounts of water flowed there, probably from the Pole melting after the Argyre impact[11]. So we might expect the ice from the current North Pole[12] would do the same, melting and pouring into the Northern Lowlands around Arcadia Planitia and Acidalia Planitia. This should create an ocean[13] of ice and water[14]. The melting of much of the water ice on the Poles would tend to refill low areas on Mars near them. So the Northern Lowlands[15], comprising much of the Northern Hemisphere might be partially filled with water some of which might reform as ice, and other areas nearer the impact site remaining liquid if Mars is warm enough. The areas near the hot crater impact would vaporize water forming clouds, rain, etc which would likely condense in colder areas of Mars as rain or snow. Liquid water should form near the

---

[10] http://www.nasm.si.edu/ceps/research/grant/grant_marg2.pdf

[11] http://www.lpi.usra.edu/meetings/lpsc2001/pdf/1799.pdf

"GEOLOGY OF THE ARGYRE BASIN, MARS: NEW INSIGHTS FROM MOLA AND MOC" Lunar and Planetary Science XXXII (2001) 1799.pdf

[12] http://astrobiology.arc.nasa.gov/news/expandnews.cfm?id=9623

Life as we know it cannot exist without water, and the search for water on Mars is a vital part of determining whether or not the planet could have supported ancient life.

[13] http://www.psrd.hawaii.edu/July03/MartianSea.html

[14] Ibid.

[15] http://www.astro.virginia.edu/class/oconnell/astr121/test/mars-status-aaas-200.html

crater depending on the overall amount of heat release. Salts [16] would dissolve in the water and keep some of it liquid at lower temperatures, in other areas ice sheets might form. Since much of Mars is already warm enough to melt ice, these ice sheets might be confined to higher latitudes. This large amount of water and ice should form oceans, rivers [17] [18], aquifers [19], and lakes [20], probably in the existing low areas and ancient river systems [21]. This should create weather on Mars and rainfall in land areas, such as on most of higher ground in the Southern Hemisphere. For example in the Northern Summer there would be more water vaporized on the North Pole and more ice forming on the South Pole. In the Southern Summer the extra heat and air pressure should cause ice on the South Pole to partially melt rather than convert directly to vapor and so create a sea around the Pole such as in the Prometheus Impact Basin [22].

Thickening the atmosphere

The fracturing of the ground should cause gases to be expelled from volcanoes, and thicken the atmosphere. Also the impact heat on the Pole should stop CO2 from freezing there. The air pressure on Mars from melting the poles would increase greatly, especially lower areas [23] such as Hellas Planitia[lxv]. This is based on studies of increased axial tilt sublimating all the CO2 at the poles on Mars [24]. In addition volcanoes formed by this terraforming meteor impact could increase the air pressure

---

[16] http://www.spacedaily.com/news/mars-water-science-00a.html

[17] http://www.pnas.org/cgi/reprint/99/4/1780.pdf

[18] http://geology.geoscienceworld.org/cgi/content/abstract/31/9/757

[19] http://www.spaceref.com/news/viewpr.html?pid=6237

[20] http://www.centrofermi.it/download.php?doc=upload/doc/doc43d75bcfc73f5.pdf

[21] http://www.museum.hu-berlin.de/min/lehre/vorlesung/GeoDynMars/Mars7Fluviatile.pdf

[22] http://en.wikipedia.org/wiki/Promethei_Terra

[23] http://en.wikipedia.org/wiki/Colonization_of_Mars

[24] http://www.spacedaily.com/news/lunarplanet-2001-01a6.html

even more[25] [26] [27]. Initially this air pressure would be mostly CO2 but terraforming would include the use of plants to convert much of this to Oxygen. If this could be done then the atmosphere of Mars could not refreeze at the Poles because oxygen has a much lower freezing temperature than CO2. So the window of opportunity would be to convert this CO2 into Oxygen with various plants, lichen, algae[28], etc. So since most of our atmosphere is Nitrogen the higher percentage of Oxygen in the Martian atmosphere might make it easier to thicken later, perhaps one day to breathe it unaided or only wearing a respirator and no space suit. If one was breathing in mainly the Oxygen Martian atmosphere then each breath might contain similar amounts of Oxygen as on Earth. One analogy might be like a plane pilot flying at reduced air pressure with a respirator, walking on Mars might one day be similar to this experience.

Antipodal effects[29] [30]

Opposite the Argyre impact crater there is Elysium Mons. It is likely this formed from a shock wave going through the Martian core and fracturing the ground opposite, which created the volcano. If the same thing happened in the terraforming process a volcano or at least a hot spot[31] might form near the South Pole which would melt ice[32] and create a smaller sea[33] around the Pole.

---

[25] http://www.lpi.usra.edu/meetings/LPSC98/pdf/1125.pdf

[26] http://www.astro.virginia.edu/class/oconnell/astr121/test/mars-status-aaas-200.html

[27] http://www.lpi.usra.edu/meetings/sixthmars2003/pdf/3021.pdf

[28] http://www.physorg.com/pdf4156.pdf

[29] http://www.newgeology.us/presentation35.html

[30] http://adsabs.harvard.edu/abs/1994Icar..110..196W

[31] http://www.mantleplumes.org/WebDocuments/Antip_hot.pdf

[32] http://www.esa.int/SPECIALS/Mars_Express/SEMYKEX5WRD_0.html

[33] http://news.nationalgeographic.com/news/2007/03/070315-mars-water.html

Shock wave rock heating

Because the shock wave would extend along the surface then rock in its path would be heated. This heat would persist for a long time and also lead to many smaller volcanoes in its path from fractures in the ground. These are similar to those seen in Solis Planum as well as the larger volcanoes of Tharsis and Olympus Mons. So CO_2 and ice reforming on the poles should be heated and vaporized, creating a kind of weather pattern.

Weather patterns

The Poles would become much hotter, so any ice depositing on the poles would be vaporized for a long time. This might create weather patterns, Hadley Cells, etc on Mars since some parts would be much hotter than others and limited oceans should form. With a hotter landmass and colder ocean, this on Earth forms clouds and rain. The Poles then might be surrounded by storms, rain, snow, etc and this constant injection of water into the atmosphere should create weather all over Mars.

Moving ice and debris

It is likely that ice, CO_2 and rock from the impact would be sent up into the atmosphere, even into orbit and land in other areas. So this ice and CO_2 would melt and sublimate if it lands at lower latitudes on Mars.

Rifts

Fractures caused by the shockwave should cause rifts in the Pole area, into which melted water would fall and be heated by rock in them creating steam. So these rifts might continue for a long time to add humidity to the atmosphere and thus rain.

Buried ice

Ice buried in other areas such as ancient poles might also melt creating and replenishing the water table, lakes, etc. Examples include around Solis Planum[34], Meridiani Planum, Gusev Crater, etc.

Polar wander

The redistribution of water and CO2 would affect the weight distribution of Mars and likely lead to renewed polar wander[35]. So if this was relatively rapid then the old Poles would be moved closer to the Equator and thus fully melt, releasing all their ice as water. Since the Poles may have previously been near the current Equator the terraforming may cause them to move back in that direction.

Brine

It is likely the Martian soil would contain large amounts of salts[36] [37]which would dissolve into this relatively pure water from the Poles. This would lower its freezing temperature and result in water staying liquid at lower temperatures[38].

Triple Point

Potentially Mars might be covered perhaps 40% or more with water and ice if the Northern Lowlands is filled, and the higher air pressure should maintain this liquid phase of water for a time. Currently on Mars the low air pressure causes the triple point of water to only be reached rarely so water ice usually goes directly to water vapor without forming water. However it is believed that even now brines can survive for long periods

---

[34] http://epswww.unm.edu/iom/pubs/Elphic2011.pdf

[35] http://www.lpi.usra.edu/meetings/lpsc2000/pdf/1930.pdf

[36] http://www.physorg.com/pdf7981.pdf

[37] http://ndeaa.jpl.nasa.gov/nasa-nde/usdc/papers/Polar-Conf-Gopher-03-8019.pdf

[38] http://www.lpi.usra.edu/meetings/lpsc2001/pdf/1689.pdf

on Mars without evaporating, so with higher air pressure open water might be permanent[39].

Habitation

The North Pole impact might create a sea in the Northern Lowlands area particularly near the impact crater. If the crater is near enough to the Pole then in winter CO_2 and water ice would be vaporized creating more weather patterns.

The most stable area for weather would depend on the disturbances caused. If necessary people could settle far away or even on the opposite hemisphere to the crater to be shielded from the weather, for example there may be caves[40] and lava tubes around Elysium Mons. Also higher ground would be needed if widespread flooding occurs so the Southern hemisphere may be best. Hellas Crater might partially fill with water from the melting South Pole and the air pressure would be higher in it for settlements.

The water might be seeded with algae and there would be less radiation owing to the depth of water and thickened atmosphere. Snow[41] has been shown to reduce radiation enough for life to survive even without a thicker atmosphere. Some areas[42] may already have moisture in the soil, so the thickening of the atmosphere could only increase this. Also there are places on the Martian surface[43] which are more shielded from the solar wind by remnants of the magnetic field. These would have reduced radiation for human habitation.

---

[39] http://www.uark.edu/depts/cosmo/publications/pub%20by%20year/2004%20papers/sears%20et%20al%202004d.pdf

[40] http://www.astrobio.net/news/modules.php?op=modload&name=News&file=article&sid=2290

[41] http://www.astrobio.net/cgi-bin/h2p.cgi?sid=380&ext=.pdf

[42] http://mars.spherix.com/5555-14.PDF

[43] http://www.berkeley.edu/news/media/releases/2000/12/15_mars.html

Conclusions

This theory attempts to recreate an event in the Martian past to terraform Mars. If correct we can determine much of what is likely to happen from observing the geology of the Valles Marineris and Solis Planum areas. Such an impact would cause large amounts of heat to melt CO_2 and ice, but eventually this will reach equilibrium. People might then be able to inhabit Mars, particularly in caves, lava tubes[44] and areas with high levels of magnetism in the rocks soon after the impact. If the events depicted of the Argyre impact are accurate then such an impact would be a cost effective way to heat Mars, melt its Polar ice caps and thicken its atmosphere.

---

[44] http://www.norwebster.com/mars/lavatube.html

Chapter Five

Interstellar travel

This refers to the ability of someone, whether ourselves or hypothetical aliens, to travel from one star to another. It is important to establish whether it is feasible for aliens to travel to our own solar system, to do this it is best to look at whether we could travel to other stars one day, the problems involved, and how we would do it. Of course we cannot know what possible technology aliens might have that we have not even guessed at, but we can get at least some idea of whether being visited by aliens is plausible. Some of these ideas have been examined in science fiction while others are actively investigated by real scientists. A good explanation of the problems involved can be found here.

One reason why it seems unlikely that we have been visited by aliens is that SETI has not found alien transmissions from other stars, with a few possible exceptions[lxvi]. The recent discovery of a possibly habitable planet around another star[lxvii] is the first evidence that life might have evolved on other worlds, which then makes it possible for that evolved life to visit our own planet or for us to visit theirs. NASA has recently come to the conclusion that Earth sized planets may be common in our galaxy[lxviii] which again makes sentient life on other worlds more plausible.

One of the biggest problems in aliens getting to our solar system is the limitation that no one can travel faster than the speed of light[lxix]. If there is a secret to going faster than the speed of light then it's likely some aliens have discovered it, then it would be also likely that whoever did this first would colonize other worlds either vacant by terraforming or by force to build an Empire. It may be that faces and other artifacts on Mars imply that faster than light travel is possible because otherwise it would be too difficult to get here, especially since it is unlikely that the closest stars to us happened to have had this advanced alien race. They might have been 1,000 light years away for example and if it takes light 1,000 years to get here from there then even going ten times faster than the speed of light the journey is still a prohibitive hundred years.

However this theory runs into the problem that neither SETI nor anyone else has found any convincing evidence of other civilizations, if an alien race developed faster than light travel then we should expect to see some evidence of it in various electromagnetic frequencies such as laser light, radio waves, etc. So the lack of any signals implies that either these faster than light travellers emit no easily detectible signs perhaps to hide themselves from attack, that the communications they use emit no signals we are likely to detect, that there is no one else out there able to send signals we could detect[lxx], or else that no one can travel faster than the speed of light and stay on their own planets using relatively weak communications we are unlikely to detect. From our own scientific knowledge the last is most probable, that nothing can travel faster than the speed of light, but then there are huge logistical problems in getting to even one other star let alone finding one by trial and error that would be suitable for a colony. So then the prospect of aliens coming to our own solar system appears very unlikely. It may be possible to travel faster than light using wormholes[lxxi], an idea promoted for example by the physicist Michio Kaku. One problem is if aliens travelled to Mars in the distant past using a wormhole then why we haven't seen more recent evidence of them, though some might argue that UFOs could be evidence of this.

Another possibility is one that we have worked on ourselves, that of sending out artificially intelligent or Bracewell probes[lxxii] with instructions to start terraforming and genetic material to seed it with life. However even this would require such huge amounts of energy that if the probe was unsuccessful then it would be unlikely aliens could afford to keep sending them. One possibility is that a probe might have the ability to replicate other probes at its destination and perhaps build a power supply large enough to launch these probes. If so mutations in this artificial intelligence might cause it to start doing this on its own behalf rather than for its alien creators. We may then have a galaxy with artificial intelligence visiting other worlds for its own goals which may also explain the lack of signals from living aliens, they may have been overrun by whoever started off this seeding program with artificial intelligence. There is no evidence for any of these theories though; this is only an enumeration of some of the possibilities. However now that we have possible evidence of alien visits to our solar system it makes sense to work out how they got here.

The most likely way these hypothetical aliens could visit our solar system then is with a kind of Bracewell probe that either started a colony on Mars that failed or it was sent to leave some kind of message[lxxiii]. If so then it is possible the remains of this probe may still exist on Mars or orbiting the sun, or there may be some records on Mars of what it did. If faster than light travel exists then it leads to the possibility of aliens terraforming Mars long ago to create a colony and even that Von Daniken like contacts with extra-terrestrials and UFO sightings may have some basis in fact.

Chapter Six

Xenoarcheology

Xenoarcheology currently is a science without a subject to examine. It refers to the scientific study of alien ruins, much as archeology does on Earth with ruins made by humans. Arguably however this scientific field may have found its first subject on Mars with the various possible artifacts there. If so then this science will likely expand into the various related fields already existing on Earth, if not then eventually we may find alien ruins in our own solar system or beyond.

Some of these related fields are as follows, though most of these terms have not even been invented yet in regard to alien life and ruins. These are the names of real scientific disciplines on Earth to which I have added the prefix xeno. Other scientists favor the prefix exo, for example to make the word exoarcheology. Xeno is based on the Greek word stranger while exo is also from the Ancient Greek meaning outside or foreign. The actual names of these scientific branches will probably not be settled until we are actually on the ground examining these artifacts; however the names below will serve to illustrate how these fields might evolve.

Archaeoastronomy, Behavioural xenoarchaeology, xenobioarchaeolgy, Cognitive xenoarchaeology, Commercial xenoarchaeology, Environmental xenoarchaeology, Ethnoxenoarchaeology, Evolutionary xenoarchaeology, Field xenoarchaeology, Forensic xenoarchaeology, Gender xenoarchaeology, Geoxenoarchaeology, Historical xenoarchaeology, Industrial xenoarchaeology, Interpretive xenoarchaeology, Maritime xenoarchaeology, Processual xenoarchaeology, Settlement xenoarchaeology, Xenozooarchaeology.

It is logical to say that fields like this and many others will evolve upon the discovery of alien life, civilizations, ruins, etc. The point of this book is that some of this evidence may have already been found on Mars; however these possible artifacts can be analyzed by referencing these different scientific fields without positive proof of artificiality.

Archaeoastronomy

The possible ruins on Mars were likely built by either Martians that evolved there, aliens that visited there, or indigenous Martians bred by visiting aliens by genetic engineering. It isn't known which is more likely or even if these formations are artificial, this is just an enumeration of the different possibilities. Visiting aliens might leave records of where they came from, star maps accurate at the time of their arrival, images of their home world, and so on. A star map might not be enough to find their home world because stars might have drifted to completely different positions over hundreds of millions of years[lxxiv]. It would be necessary to infer where their home world is by other information such as a spectrographic record of the characteristics of their star, the number and type of planets around their sun assuming we could detect them using available technology, a record of how they anticipated their star to move so we could estimate its current position, etc. In the case of indigenous Martians we might expect to find ancient ways to predict the equinox and solstice, with the lack of a Martian magnetic field there might be novel ways to find a common direction such as the North Pole, this however would be more difficult with polar wander and a varying axial tilt on Mars. Also the seasons on Mars may have varied more than on Earth because it has an elliptical orbit and precession of the axis would cause the seasons to change in a long term cyclical way[lxxv], this could make the building of a Stonehenge much more complex. It may be possible to find an array of rocks used to work out the solstices, some of the mound configurations found so far may have been used for this purpose. The Crowned Faces shown later may also have been designed to show different shadows depending on the time of the day and the season. The Amphitheatre formation mentioned in this book may have been designed as a representation of our solar system, with concentric rings on the mesa and a large central mound for the sun. There has also been found on Mars a configuration of mounds closely matching the stars around Orion, this will be examined in an upcoming book.

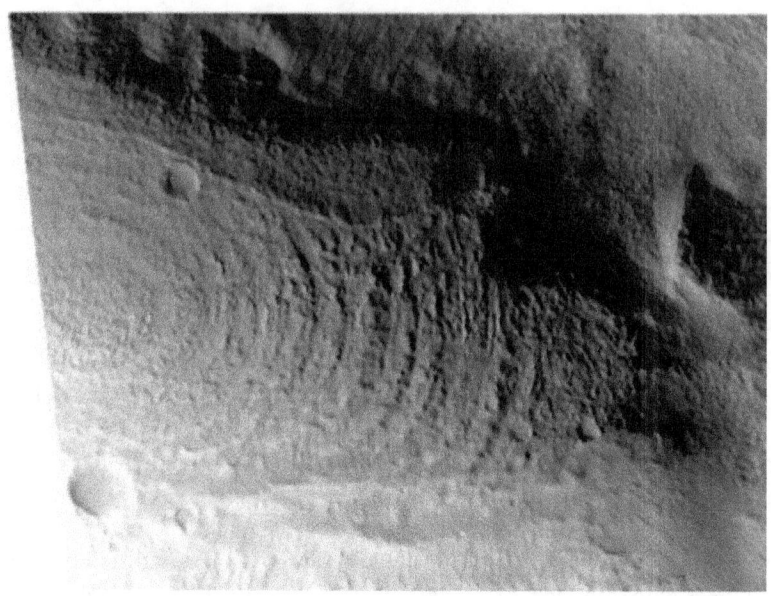

Figure 1

The Amphitheatre in photo M0303960 may be a representation of a solar system; the crater shapes may be from impacts or designed to represent planets. This formation will be more fully analyzed in an upcoming book.

Behavioral xenoarchaeology

The possible artifacts on Mars are a puzzling and incomplete record, we can work out some plausible theories based on common themes but at this stage this can be misleading. We don't know for example whether the Cydonia Face was damaged by erosion, defaced by warring races in a kind of iconoclasm, damaged by aliens who came later and removed traces of a previous Empire's claim, and so on. There is no real evidence for any of these alternatives except for various possible causes of some artifacts being degraded. There may be garbage dumps nearby where workers on these artifacts lived though there are no clear traces of settlements, quarries, roads, etc. around them. This may indicate the artifacts were built by a robotic technology for symbolic reasons and tidiness was part of this aesthetic aim. Indigenous Martians whether they evolved there or were created by visiting aliens should have left extensive traces of their lives unless these have all eroded away or are underground. Rubbish dumps should survive, though if they lived underground to avoid radiation we may not see any real evidence of this until there is a manned colony on Mars. We tend to assume these faces were highly important to these hypothetical colonists but this is only because they have survived, far more significant ruins may be buried under lava, mostly destroyed by meteor impacts, buried in dust, or have degraded so much they only be differentiated from natural rocks by spectrographic analysis.

Xenobioarchaeolgy

We have many different faces on Mars to try and understand their biology, if they were motivated to build large scale faces then it is likely there are many smaller monuments, statues, engravings, etc. of animals, plants, and aliens. There are also some interesting formations on the Moon of possible animals, this would imply visiting aliens rather than indigenous Martians because it is unlikely they could have evolved space flight and left so few signs of civilization on Mars.

Figure 2

This looks like a bird and was found on the Moon by one of the Apollo missions. The Moon artifacts will be analyzed in an upcoming book. That a particular artifact looks too much like an Earth animal is not necessarily a problem, there are so many different kinds of birds that a representation of a winged creature would be bound to look like one of them. The evolution of wings is an obvious competitive advantage and should often occur with alien life.

The bird sculpture above is not buried under micrometeorite dust like the other rocks around it; this would seem to be impossible to occur naturally unless it was ejected like this from a meteor impact. If all the other rocks are partially buried then the bird rock should also be partially buried but it sits on top of all the dust. It seems to be perched on legs or at least crouched down on them which if naturally formed was an unlikely way for it to land. The darker areas on it imply a piebald arrangement of feathers; it also has eyes and a beak. The eyes may be rocks that were put there by the Apollo astronauts for the photos; they may have also swept around it though there are no records of this. If artificial sculptures like this might be found on Mars as well, they may give an insight into alien flora and fauna or extinct life on Mars. If this life existed on Mars then there may be fossils of it there as well. The bird rock is only one of five possible animal sculptures on the Moon; these will be more fully analyzed in an upcoming book.

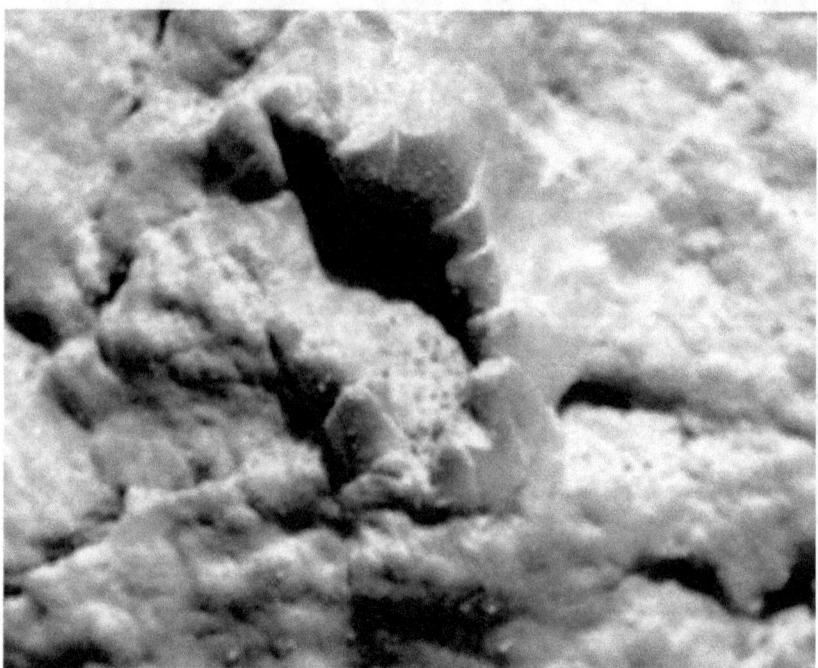

Figure 3

This was found by one of the Rovers on Mars[lxxvi] in 2004 and created quite a stir. It resembles crinoid life forms on Earth. There are several other possible fossils like this which will be analyzed in an upcoming book.

Fossils like in Figure 3 imply a relatively long period of life friendly climate on Mars; if crinoids could evolve there then perhaps indigenous aliens could as well. If life reached the level of crinoids and then was completely obliterated to the point where the discovery of this fossil was a surprise then perhaps higher forms of life may have also evolved. Before these fossil like rocks were found it was considered so unlikely that evidence of past or present life could exist that the Rovers had no way to detect life there, the new MSL Rover however will be able to look for signs of life[lxxvii]. So if fossils could exist despite our beliefs about Mars then it may be conceivable that higher forms of life also existed. This might mean indigenous Martians built the faces and other artifacts then died out before they could evolve high enough technology to save themselves from the cooling of the planet. If there was little fuel on Mars because of a lack of biotic coal and oil then a higher industry may have been impossible to develop. Life on Mars may also have evolved or been engineered during a warm period from terraforming with meteor impacts as described elsewhere in this book, this would be consistent with fossils and with possible artifacts on the Moon as aliens would have been to both places. If so then it is likely they also came to the Earth and had something to do with our own evolution. Searches have been made in Earth DNA looking for signs of alien intervention or messages but no evidence has been found.

Cognitive xenoarchaeology

We tend to think of the faces as Art, but we really don't know what the builders were thinking when they built them. One scenario could be like Ancient Egypt where large numbers of indigenous Martians may have slaved away building tombs for a royalty or religious elite. These Martians may have been much smaller than us, for example no bigger than squirrels which would make it possible for them to have evolved much earlier. There is little evidence these builders were humanoid, we see faces but not the rest of the body, they could have been animals of any kind we can imagine though they must have had a way to manipulate tools. We don't know whether there were sentient animals on Earth before we evolved because they may not have had the intelligence and ability to handle tools that leaving a record requires, some believe for example that Raptors were

85

intelligent[lxxviii]. Scientists now believe dolphins are very intelligent but they cannot work tools with their flippers and so could not leave archeological evidence of this intelligence, perhaps an aquatic animal evolved on Mars protected by radiation by staying underwater during the day. Eventually it may have become amphibious and evolved enough tool handling abilities to construct these artifacts. Another possibility is we see in dictatorships on Earth where the leader's face is portrayed ubiquitously in statues and portraits; there may have been a Martian leader who compelled his subjects to build monuments to him or her. This might also happen with an alien colony or an indigenous race created by aliens, it may have started out as or become a dictatorship and then the colonists were compelled to create monuments in the leader's likeness. The answers to these questions might only be found after years of research and a manned colony on Mars, their extreme age might mean nearly all of the artifacts are destroyed.

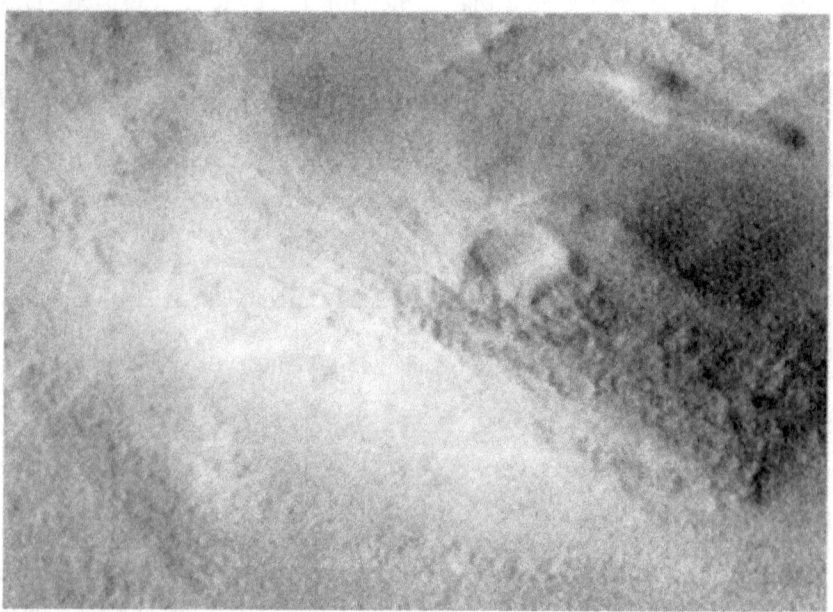

Figure 4

The angel formation appears to have a large left wing with the right wing either folded across the chest or less clear to the right of the head. This may be a burial mound. Nearby is a possible subterranean opening lending some credence to the idea that they may have lived underground. There are many possible artifacts in this area which for space reasons will be fully analyzed in an upcoming book. This may lend some credence to the notion of the faces being related to burial mounds like our own Egyptian pyramids and the Sphinx. If the aliens or Martians were able to fly then this may explain why they built faces more visible from above, they may have been landmarks for them while flying, for planes, balloons, etc.

Commercial xenoarchaeology

There is no real evidence of commercial activity in the possible artifacts to date, but it is very likely that a market based economy would have either evolved with indigenous Martians or with colonists. We might expect to find some records of business transactions in rubbish dumps like in the City of the Sharp Nosed Fish or in underground markets. If there were native Martians then judging from the low level of technology in the Faces the level of commerce may have been similar to Ancient Egypt, there could have been paper money, checks, renting houses and factories, posting letters, etc. but any paper made from local vegetation would probably be long gone. We would then rely on someone having made more durable records which are usually unnecessary in a primitive economy, of the paper records found in the City of the Sharp Nosed Fish virtually none of them had ever been preserved in stone or clay elsewhere in the Roman Empire. A more advanced civilization might store records in a more durable form such as optical media like CDs, but it is hard to imagine any of these lasting for hundreds of millions of years. The most likely chance for comprehensive records being found is if these were built by visiting aliens and they made records for us to find, they would have the capability of designing records that would last long enough. If these artifacts were originally part of cities then there was likely trade between them, such as fish caught from the sea in the Northern Lowlands might be traded for minerals mined in other areas.

Environmental xenoarchaeology

This would be an interesting subject because it is likely the Martian environment was only marginally suitable for life whether indigenous or alien. Radiation was likely high because of a lack of a Martian magnetic field though some areas still have a remnant magnetization[lxxix] in the rocks which would offer some protection from the solar wind. This magnetic field survives in stripes and it may be that they were used as roads, acting as a protection to colonists or indigenous Martians. If the atmosphere contained enough oxygen it may have developed an ozone shield against UV radiation.

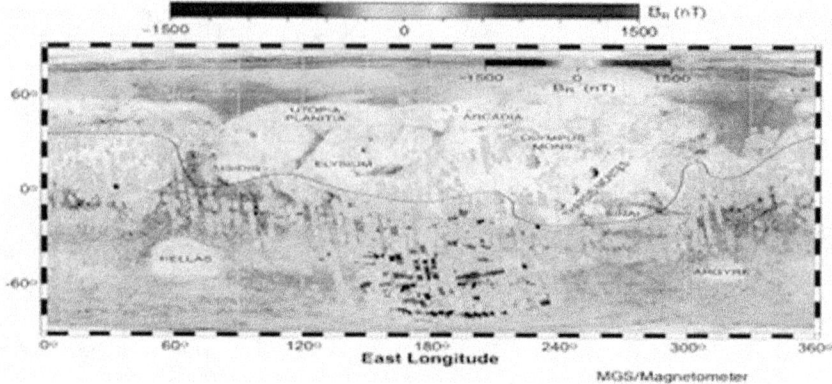

Figure 5

This shows the remains of the magnetic field on Mars. Logically inhabitants of Mars would have built in areas of high magnetization to protect them against radiation unless the magnetic field of Mars was much stronger at the time. This will be analyzed in an upcoming book but the City area for example is in an area of high magnetization. The King's Valley is in an area of medium magnetization as is Cydonia, this is important as no magnetization would be a serious drawback to the theory that these sites are artificial.

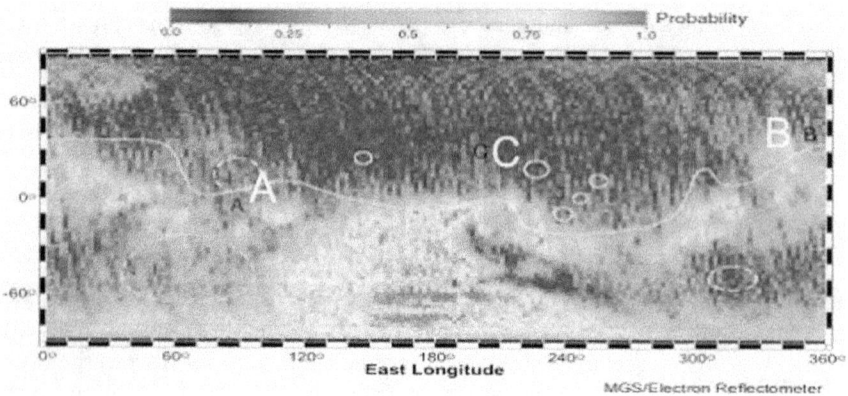

Figure 6

This graph[lxxx] measures the probability that Mars Global Surveyor will be in the ionosphere when orbiting at 400 kilometer's altitude. The dark areas represent a low probability, meaning the spacecraft was usually in the solar wind and the ionosphere was below the spacecraft. The lighter areas show where the ionosphere often protrudes *above 400 kilometers altitude. (Credit: David Mitchell, UC Berkeley.)The King's Valley is at **A**, the Amphitheatre formation at **B** and Cydonia at **C**. There are 15 candidate formations for artificiality and these will be analyzed in comparison to this magnetic field in an upcoming book.*

Ethnoxenoarchaeology

The material remains of these artifacts are easier to analyze because we know virtually nothing about any associated culture. There are clues about the methods used to build them, most seem to rely on natural foundations which are then altered to the desired appearance. This would save a lot of time and energy, the same technique was used in Mount Rushmore for example. It also leaves many of these artifacts open to the charge of being natural formations because parts of them are natural rock, to prove otherwise it is necessary to differentiate parts from this natural geology and prove that natural forces could not have made these alterations.

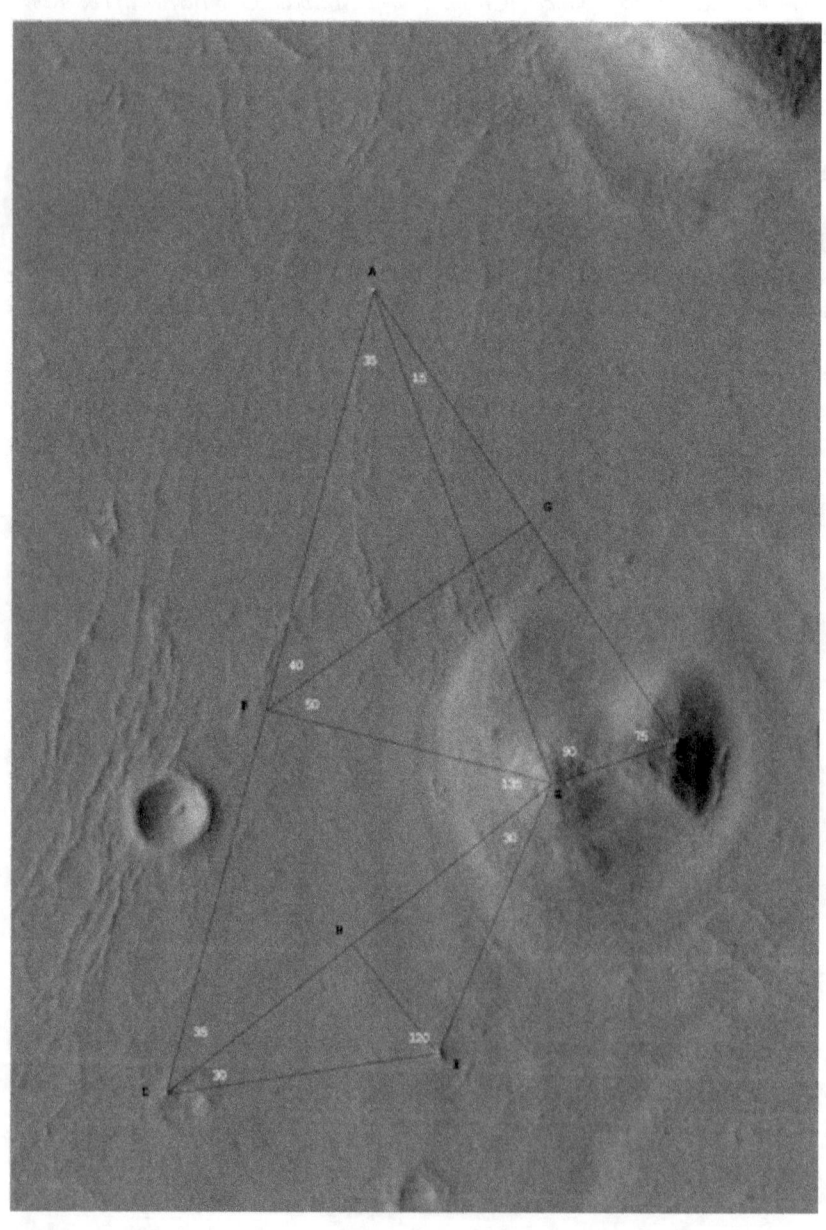

Figure 7

Near the Angel formation at SP243304, the angles between the mounds are highly unlikely to occur by random chance.

The letters in Figure 7 are placed on mounds easily visible in the original image. When these are connected with lines the geometric relationships as shown are found. Joining together the mounds on this plain gives 2 isosceles triangles, one with the significant 60 degrees (**BDE**), and the other with 35 degrees (**DAB**), often found in other photos. **DEH** and **BEH** are 30,60, 90 degrees is formed when the isosceles triangle is bisected. The right angled triangle of 15,75, and 90 degrees (**ABC**) is also common in other areas. The right angle at **F** is divided into 40 and 50 degrees, 2 other common angles in other areas. The most likely reason for this is some kind of surveying, there may have been buildings here made of non-durable materials that have eroded away. The raised oval area on the right appears to have two ruins on it. There are many sets of mounds like this on Mars associated with possible artifacts; all will be analyzed in an upcoming book.

Evolutionary xenoarchaeology

It may be highly likely that we would see an evolutionary change in these artifacts on Mars. The ones that survived are probably early parts of a civilization, whether it was indigenous or a colony just like the Egyptian pyramids have survived when many modern buildings made of more flimsy materials have disappeared. We might expect to find a substantial difference in the ages of these artifacts, the mistakes made and the lessons learned from building one artifact would have been applied to the others. The King's Valley has many Faces which would probably have been built sequentially, they may be of different but contemporary people, be all of the same alien or Martian at one time or at different stages of their life, or be of various historical figures.

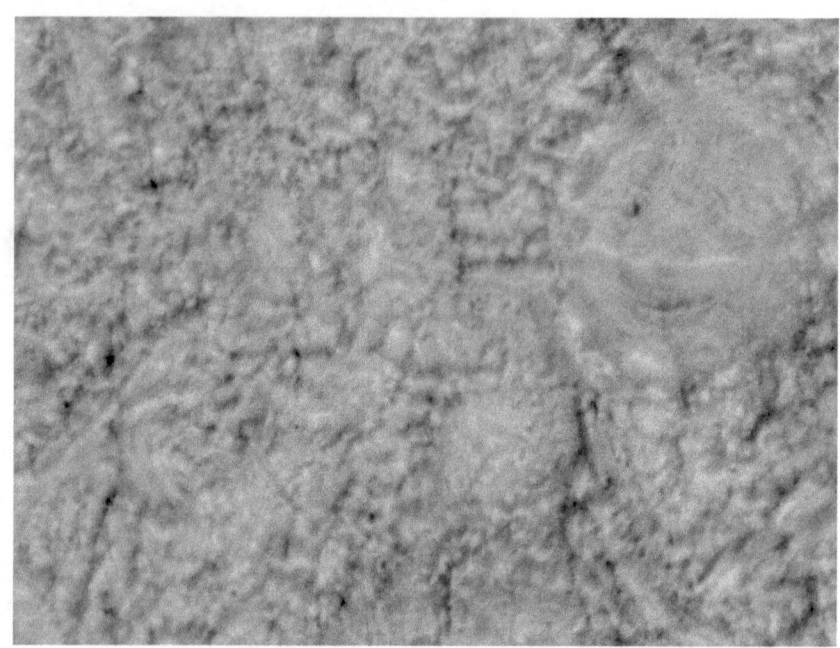

Figure 8

Figure 9

This is the Spaceman formation in photo FHA00533, Figure 8 shows *the photos and Figure 9 some possible humanoid Faces outlined.*

The Spaceman formation is one of the strangest on Mars because if artificial it shows an alien distinctly non humanoid, near to it there is a more humanoid head and neck. This might imply an evolution of one from the other, two races coexisting on Mars, or that one race genetically engineered or evolved from the other. Since we are humanoid in that case it is likely that the Spaceman on the right created the humanoid face on the left, we don't have any animals on Earth that look like the Spaceman. The images are extremely old, the many cracks and ravines are probably caused by glaciation and ice moving over this area over many hundreds of millions of years. There is also a second Spaceman formation that looks similar to the alien on the right; this will be analyzed in an upcoming book. Both appear to have a bubble around their heads consistent with wearing a space suit like our astronauts might wear but the figure on the left is not wearing one, this implies a toxic atmosphere for the alien that the humanoid on the left could breathe.

Field xenoarchaeology

This is the main process currently being used around Mars, mainly through remote sensing using satellite imagery. GIS[lxxxi] is an area where much can be done before a manned or robotic expedition to these possible artifacts. An example was shown above with the comparison of the remnant magnetic field and artifact sites, more precise measurement of the magnetic field around these areas would test the theory that they were built in areas with lower radiation. Data modeling also has a great potential, for example these areas could be modeled on computers or in materials such as plaster to test how they look from different sun angles. It's likely that they were built to use shadows to stand out from space or at least from above for planes or flying inhabitants. This is because when photos are taken of these without shadows the Face like aspects are much harder to see, this was shown with the King's Valley and in Cydonia with the HiRise images. THEMIS has a similar problem where infra-red images also show no shadows. Hydrological modeling in the King's Valley could help to understand how erosion occurred there and how the dunes were formed, whether from floods or a long term flowing river. There is a possible dam on the King's River show elsewhere in this book, modeling this could determine whether it held water for the inhabitants or to prevent flooding the river and destroying the faces. This remote sensing can also include looking for likely ice deposits or artesian water to be used in an expedition to help make methane fuel using the Sabatier process[lxxxii] in the Mars Direct and other mission plans, the King's River may contain ice deposits from when water flowed there which would make a colony much easier to create. There may also be a cave further up on the King's River next to the dam; this is analyzed elsewhere in this book. Analysis of the minerals on and near these artifacts may yield insights where they came from, whether made locally or transported from elsewhere with more suitable building materials like marble was transported on Earth during the Roman Empire. The artifacts may have been built in areas where a city near them could access certain minerals or there were certain kinds of raw materials needed for their construction, otherwise the positions may have been symbolic for some reason. Three faces appear on a great circle which was probably the Martian equator, this implies a symbolic reason for their locations and so the materials used in their construction may have come from elsewhere. This should be easier to determine with a manned colony or robotic expedition.

Figure 10

An artist's impression of a mission to Mars. A similar vehicle could land near the King's Valley and make methane for a return journey if there is ice nearby.

Forensic xenoarchaeology

While there is no evidence of any crimes associated with these artifacts there might be some evidence of wars and conflicts with a colony or indigenous Martians. Like humans any life form tends to fight with others to some degree, so if Mars was highly populated it would be logical for some conflicts between those in a better location and those with fewer resources. Also if the faces relate to an alien visitation then aliens capable of travelling here from another star would be likely to be able to form a star Empire. If so then there might arise conflicts between the different colonies and even other star Empires or alien races, a manned colony could look for signs that some artifacts were destroyed or damaged, and what weapons were used. For example lasers might create molten rocks around an artifact, atomic bombs and the like might be differentiated from impact craters by radioactive materials left from the explosion. A more primitive conflict might use axes, shovels, etc. and these should leave marks in the rock to work out what kind of weapons or tools they were. This might be the real reason the colony failed or the indigenous Martians disappeared, for example the two alien races shown in Figure 8 might have warred with each other. The humanoid race like the humanoid face on the left might have created a colony and the faces and then the alien race on the right destroyed the colony. It might also have been abandoned if under attack or the Empire found it too expensive to maintain, a similar scenario has happened many times with Earth's Empires. The Hellas meteor impact probably occurred towards the end of this colony or indigenous race's time, it might have been launched from the Asteroid Belt to be a planet killer as an act of war. There is no evidence at this point to favor one theory over another; this is only to enumerate the various possibilities which would be examined when we go to Mars.

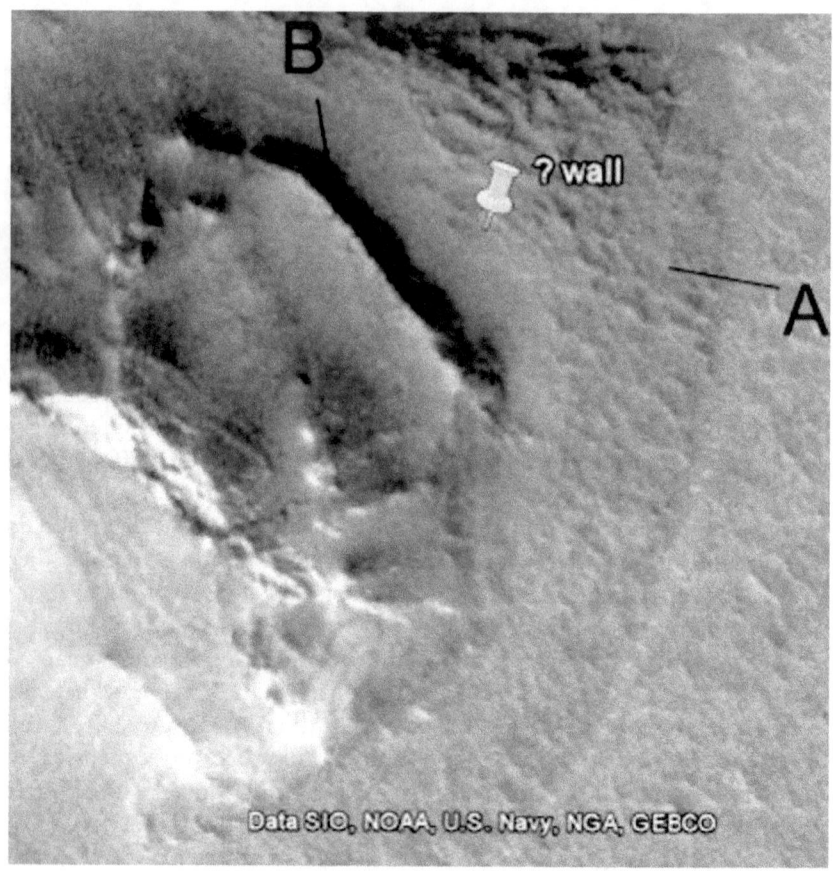

Figure 11

*I call this the Part Face because it has a similar shape to the Cydonia Face at **A** but only part of it has been imaged so far. In this raised area there is a wall at **B** which could have been a defensive formation against animals, flooding or other humanoids. It may also be collapsed rooms that were built under the face; this may indicate that the other faces are hollow as well with dwellings, temples, meeting places, etc. inside them. This photo gives an a priori prediction that the face shape will continue to the left and below, a random formation is unlikely to have a shape connected to this that looked like the Cydonia Face.*

Gender xenoarchaeology

Trying to understand the genders of the indigenous Martians or alien colonists would be an important goal, many say the Crowned Face looks feminine as perhaps the leader of a matriarchal society. Hypothetical aliens might not have sexes or have more than two sexes; all of these factors would critically affect how a society evolves and has been explored in science fiction. Dual Faces such as with the Crowned Face might also indicate a male female duality.

Geoxenoarchaeology

This might be done mainly by geologists who are exploring Martian geology. Faults are found in the King's Valley running near the faces, dating these faults might give some insight into the age of the faces. Trace element geochemistry can try to ascertain where the materials used to build these artifacts come from, whether nearby or from a common location which would be strong evidence for artificiality. For example on Earth marble is often shipped from specific areas for statues, it may be the faces had a particular kind of outer layer for decorative purposes. If the faces used a kind of paint or coloring for different parts, such as dark colors in the eyes, then there might be some remnants of this dark paint, clay, dye, etc. still near the eyes. The crowns might have been colored differently, the simplest way to give different colors and shades would be different kinds of mud or clay, these would stand out on a visual inspection as not being from the King's Valley. The faces seem the same shade as the rocks around them but this might be from a layer of dust, we might look for different shades around areas on steep slopes such as around the eyes or on the nose. The crown also appears to be darker than on the face so this may be from a lack of dust or more wind higher up.

This is a commentary on the geology of the King's Valley by Peter K. Ness; he has contributed several papers on the subject in the Appendix.

The Crowned Face represents a large slump structure with multiple circular faults. Fault movements are marked by arrows below, and the direction of the major forces shown with larger (like grey) block arrows. On earth we would call this a "type area" due to the pristine nature of the structures. The fault plane surfaces are exposed in most cases giving both direction and rotation of that fault block. With geologic processes if you reverse the faults the original pre-fault (pre-displacement) layering is restored.

The entire Crowned Face region is controlled by the two dark fault lines. The left side is rotating left to right and the right side is collapsing to the left. This has the effect of having the appearance of compressing the Crowned Faces. It also has the uncanny effect of improving some features if the faults are reversed. For example, the right eye socket of the most well defined Crowned Face has slumped. A stone has slid down the nose, partly flattening it. The area below the mouth has slumped, distorting the mouth. The left ear of this same Crowned Face is missing, truncated by the fault. The Face on the right has draped over the one on the left.

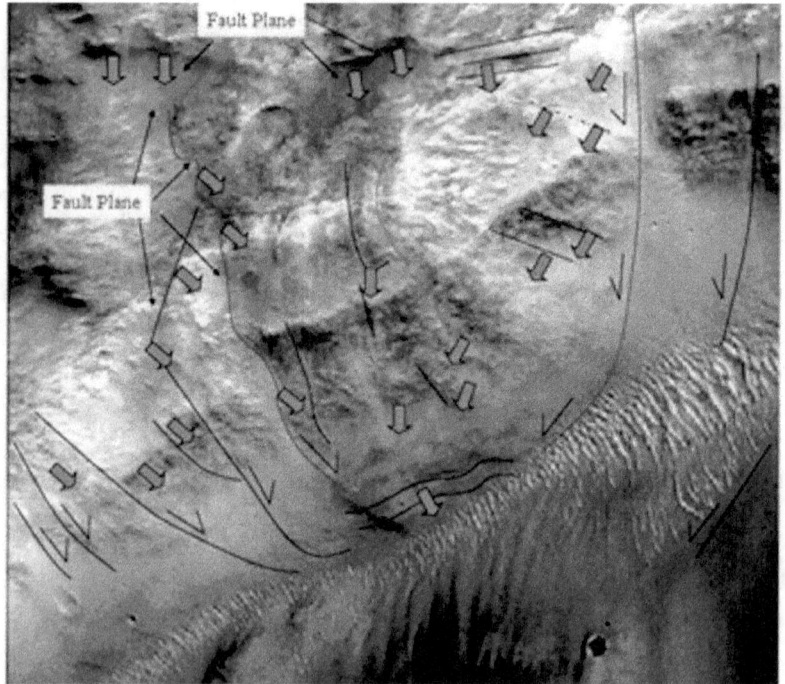

Figure 12

This shows an analysis of faults in the King's Valley.

Historical xenoarchaeology

There may be some written records found at these sites though the problem then would be how to translate them and how much would be eroded away. We may have some of their language with the symbols mentioned earlier found near the Cydonia Face, these may describe the area in some way.

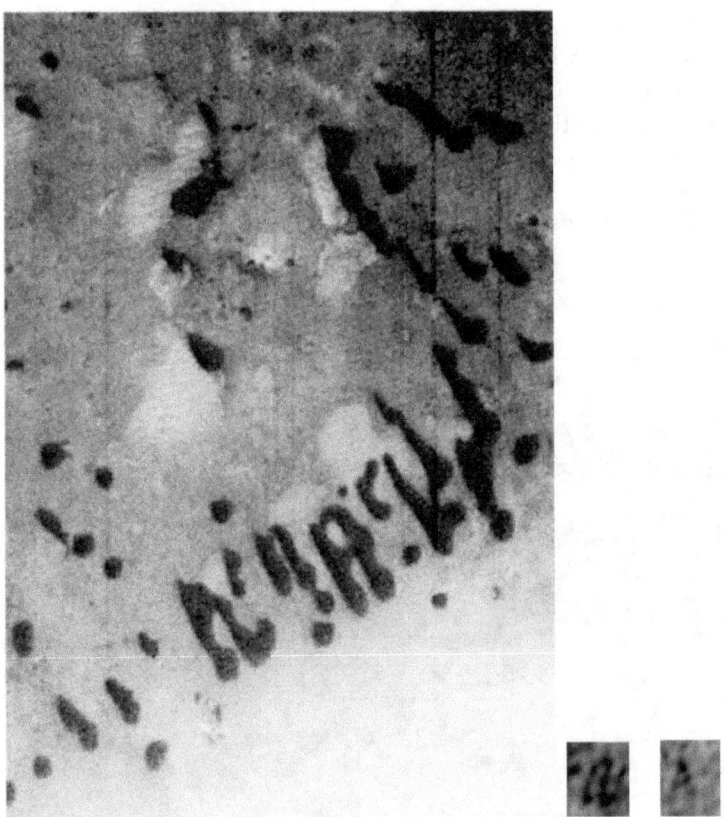

Figure 13

The image on the left is some dark dunes with shapes like letters, the two on the right are letter like shapes in Cydonia. More examples of these will need to be examined to see if they occur by random chance or are a real language. They also run into the problem of being made from dunes and probably too recent to be associated with the Faces unless the dark soil is sitting in ditches carved in these shapes. These letters and others will be analyzed in an upcoming book.

Industrial xenoarchaeology

There do not seem to be any signs of industry surrounding these artifacts, which is strange as a substantial amount of work would have been involved in building them. The same however could be said of the Egyptian pyramids and others in South America where the areas were tidied up for aesthetic reasons. One example is Machu Pichu where few signs of its construction are found even with its remote location. The construction of these faces seems different from the Great Pyramid for example; there are no indications of blocks cemented together. Instead they may have either been carved out of the cliff face with the Crowned Faces or a mesa with the Cydonia Face. Another possibility is a form of concrete, mud, or clay which would have been molded into the correct shape. Some of this may have disintegrated in large pieces leaving substantial deformations of the original shape.

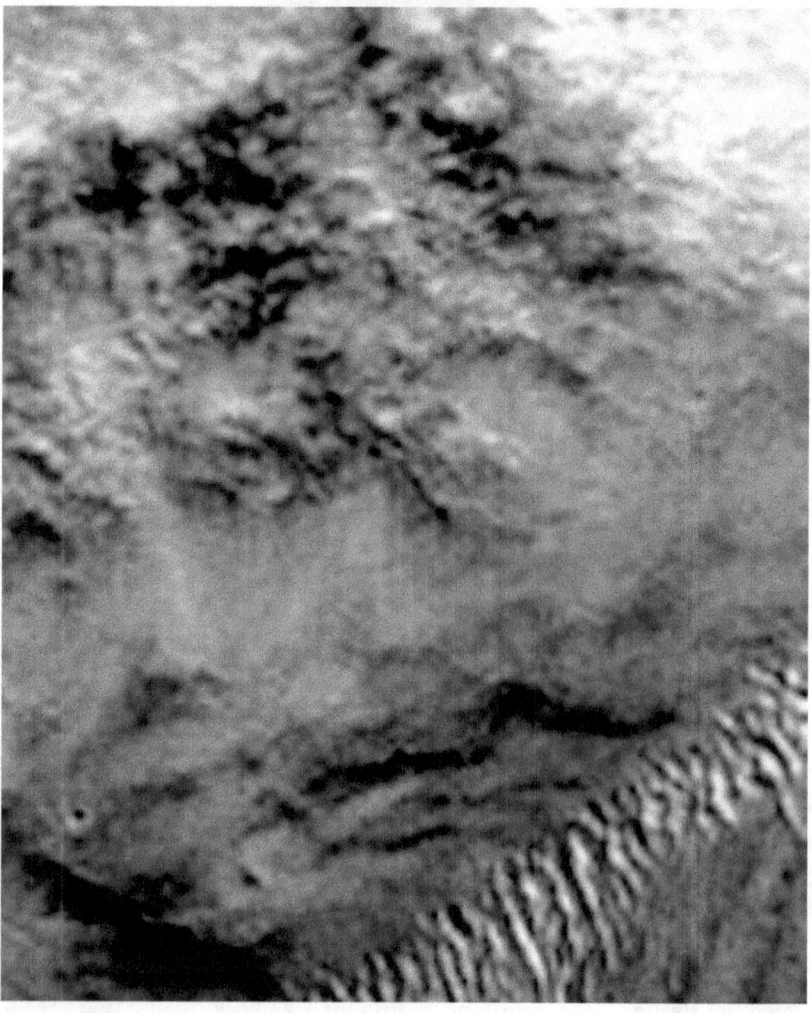

Figure 14

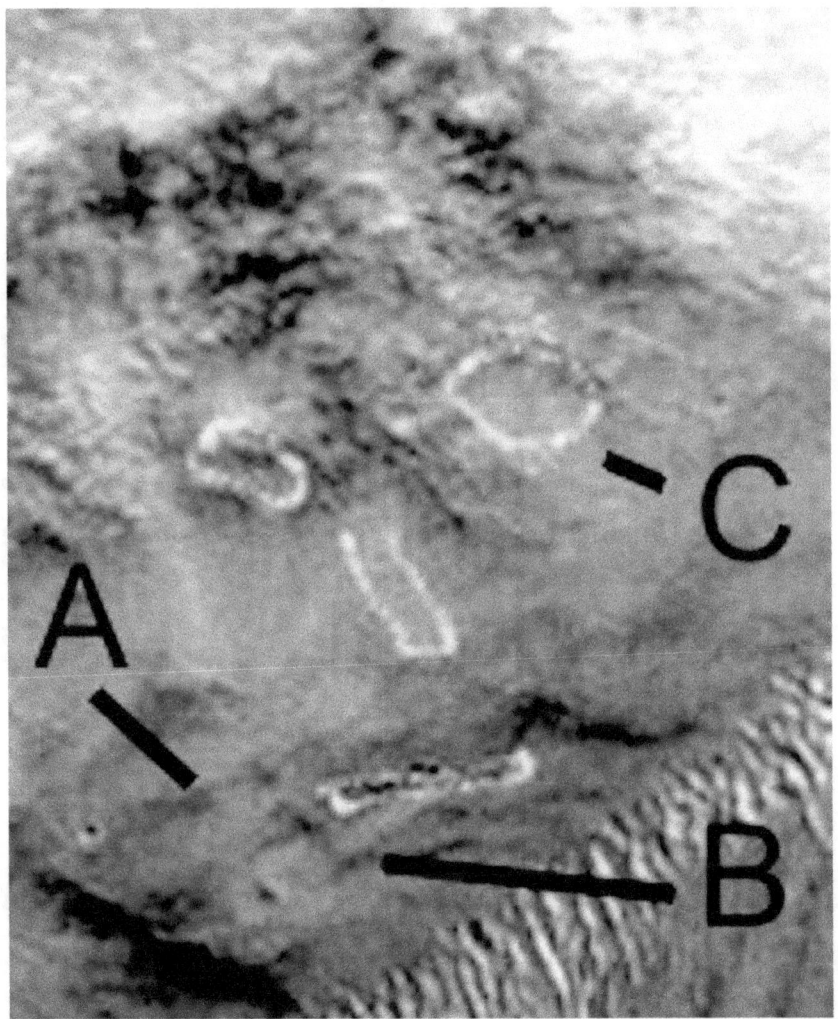

Figure 15

*Face Three is shown here outlined in white. The right eye is partially damaged by the large circled depression at **C**; this may be where a large piece of solid concrete or mud slid down the slope and disintegrated though there is no remaining debris below it. The eye shape remains in great detail and was reimaged with HiRise, shown later in this book. The mouths of this face and the main Crowned Face or noon face to the left also appear to have some damage at **A** and **B**. There are some near vertical ridges extending down from the nose into the mouth area, these may have been braces designed to hold together a layer of mud or concrete. They are however very similar to ridges on the Cydonia Face and on the Meridiani Face and it is possible that the original Face was* *designed to look like this around the mouth. It implies that these Faces are intended to be of one individual, it seems unlikely that nature would make an alien or indigenous Martian always with lines like this around the mouth. They might symbolize a wound or the right vertical ridge at **B** is next to the edge of the evening face and might be designed to give a shadow and define the edge of that face at certain times of the day.*

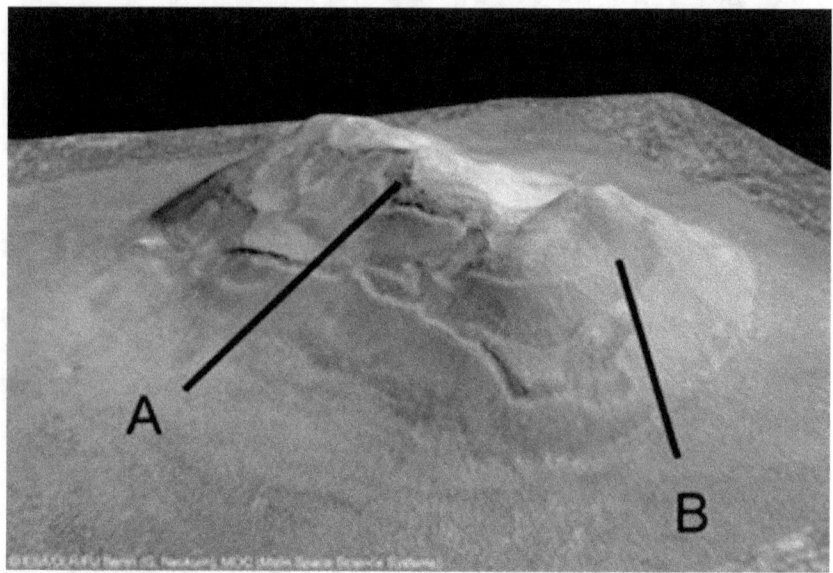

Figure 16

*This is a 3D estimation of the shape of the Cydonia Face done by the European Space Agency; it has the problem of being too tall compared to its width but is otherwise fairy accurate. This distortion of being too tall can be proven by Mars Orbiter Laser Altimeter measurements which run over the face and will be analyzed in an upcoming book. **A** and **B** may indicate areas where the outer layer has worn out through erosion or broken off leaving a more rough carved area showing instead. This would explain the uneven appearance of the face if the more well shaped areas such as on the Crowned Face were missing.*

Interpretive xenoarchaeology

It may be we can never be certain of many aspects of these artifacts, especially if records are either indecipherable or too eroded to read. Currently there are many theories held by different researchers about them, these ideas should become better defined with a robotic or manned expedition to the areas but the artifacts may remain enigmatic. For example if created by visiting aliens they may be designed to show their presence without revealing where they came from or give information to make them vulnerable to attack. They might even give a deceptive appearance such as giving an impression they were very powerful or unfriendly to dissuade others from trying to contact them, much as Empires on Earth used bluff and deception in their foreign policy. If indigenous Martians built these artifacts than most of their other creations may have eroded away leaving mysteries of how they could have evolved in such an inhospitable place. We may never know much more about these artifacts than we already do by observing them from space.

Maritime xenoarchaeology

The various artifacts seem to be clustered around the Northern Lowlands or have features related to water such as dams. It may be that their civilization or colony was greatly influenced by maritime commerce, fishing, living underwater if they were amphibious, etc. There may be ancient boats buried under mud and dust from this former ocean, perhaps fossilized if they were made of wood.

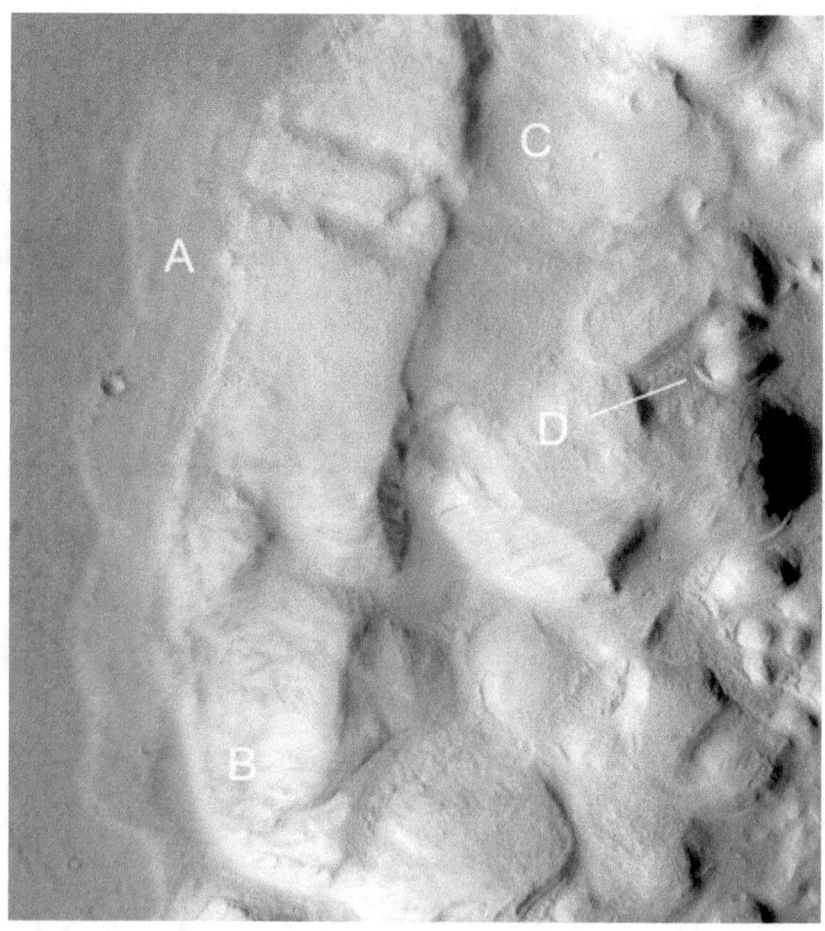

Figure 17

*An area near the Hangar formation in SP240503 may have been on the edge of a sea, it appears to have a sandbar offshore at **A** and there are signs water may have sometimes come in at **B** and **C**, **D** shows a possible small wall built around a hill, this may be hollow like the Hangar formation nearby. This area will be fully analyzed in an upcoming book.*

Processual xenoarchaeology

Understanding the culture of the builders of these artifacts, assuming of course they are artificial, may be possible by looking at the similarities between the various formations. For example Figure 17 implies living next to the shoreline and this then implies a culture of fishing or harvesting food from the sea, not relying on farming on land as much because this would have been an island. Trade might have been similar at some stage to that of the South Sea Islands on Earth where boats may have relied on wind created by the interaction between the one large sea in the Northern Lowlands and the Southern large landmass. This would have been similar to Pangaea for example where on Earth there was one large continent surrounded by one sea, however there doesn't seem to be signs of tectonic plates on Mars breaking up this land mass. The prevailing winds might then have been mainly north south but there could have been Easterly or Westerly trade winds with Hadley Cells forming between the warmer equator and the colder North Pole. Winds would also tend to blow onto the hotter land mass during the day and out to sea at night as on Earth. There would have been no Moon for tides so possible dwellings as in Figure 17 could be built closer to the sea's edge, only vulnerable to waves from storms. Cydonia may have been another group of islands in this sea so sailing to there for trade was likely though we don't know what level of technology the inhabitants evolved. The City area in the Southern Hemisphere is a long way from the sea and could not be reached by boat so that implies trade like on the Silk Route on Earth. They may have followed strips of the remnant magnetic field arrayed North-South for protection though this is pure speculation.

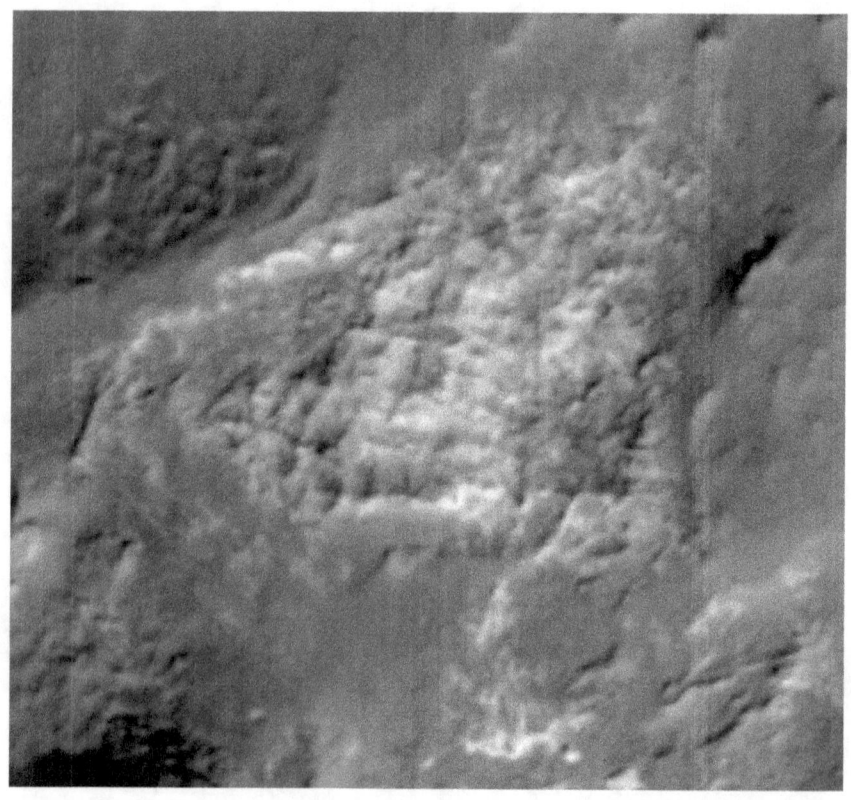

Figure 18

Unusual shapes like a city in photo FHA01046, to get here from islands in the Northern Sea such as Cydonia there would have been at some stage a culture based around trading sea goods with those on land. The City area will be analyzed in an upcoming book.

Settlement xenoarchaeology

This would be mainly around areas with the possibility of having been settlements. The layout of these structures would be examined such as the purpose of walls, whether to keep out water, intruders, animals, etc. They may also be the remaining parts of buildings where the roofs have disappeared. Some may be laid out in symbolic shapes, like early churches were laid out in the shape of a cross[lxxxiii] or parts of the Taj Mahal as a representation of Paradise[lxxxiv]. Parts of the city like structure in Figure 17 might resemble letters of an alphabet or it may have been intended to appear like a portrait or Face but too many parts have eroded away to see this.

Xenozooarchaeology

Representations of animals have also been found on the Moon like the bird formation in Figure 2. These animals might have evolved on an alien home world or have been bred on Mars by visiting aliens as part of terraforming the planet. The similarity between these animal sculptures and our own animals on Earth might indicate similar DNA or that life evolves in this way because it is often the most fit for survival in Darwinian selection.

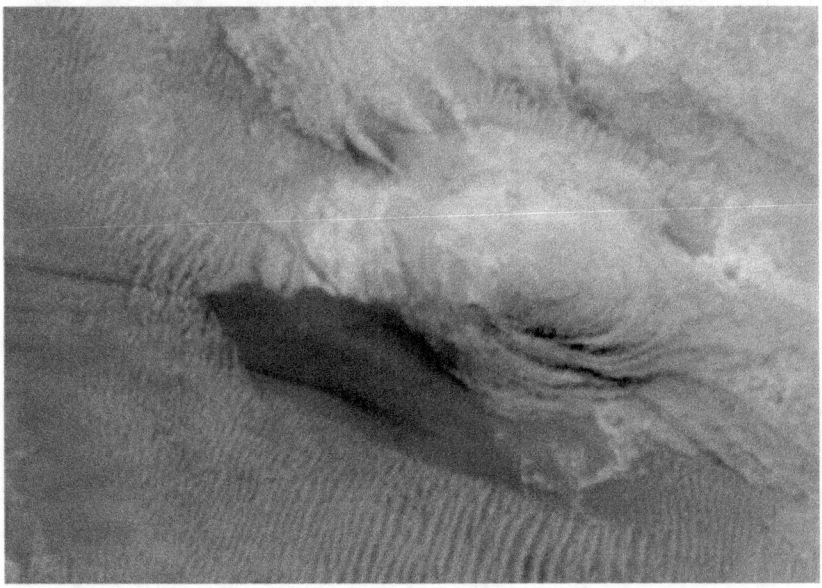

Figure 19

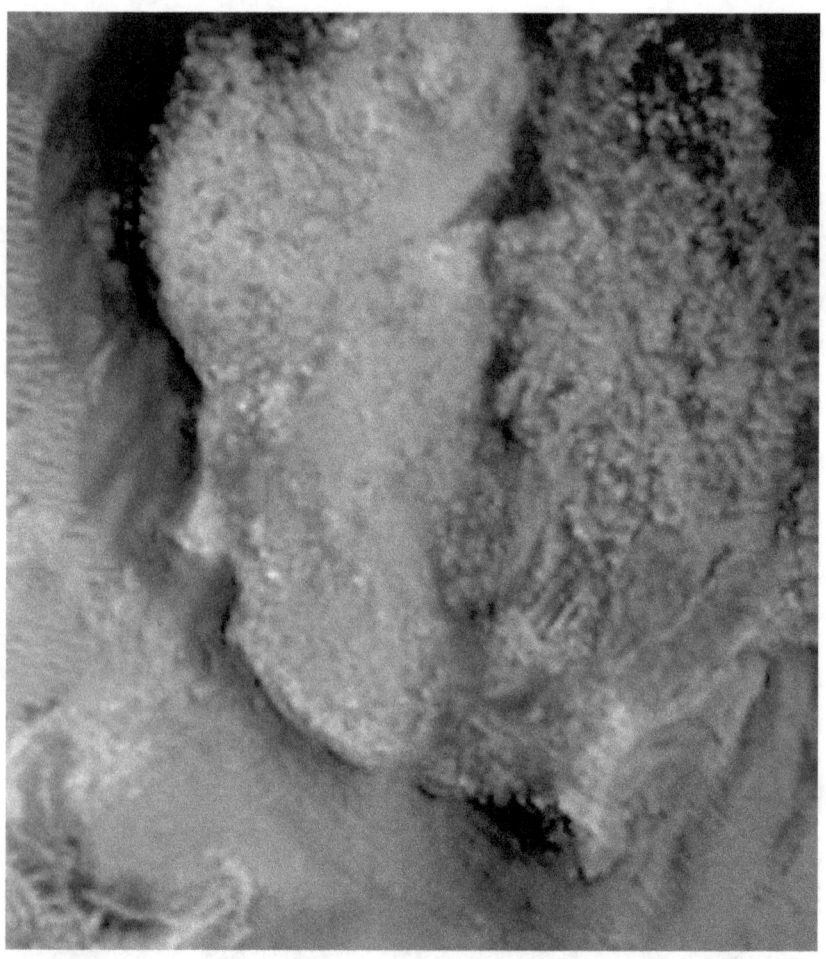

Figure 20

The appearance of a bird in Figure 19 is compelling; this formation called Parrotopia and the Face were discovered by Wil Faust on photo M1402185. The Face shape in looks up at the Parrot like formation, they have been separated here into two images. Parrotopia has much more evidence bolstering its claim to artificiality, for example it lies on several degenerate triangles with other anomalous sites; these will be examined in an upcoming book. There are so many different kinds of birds on Earth that a winged creature was likely to look like one of them, in this case a parrot.

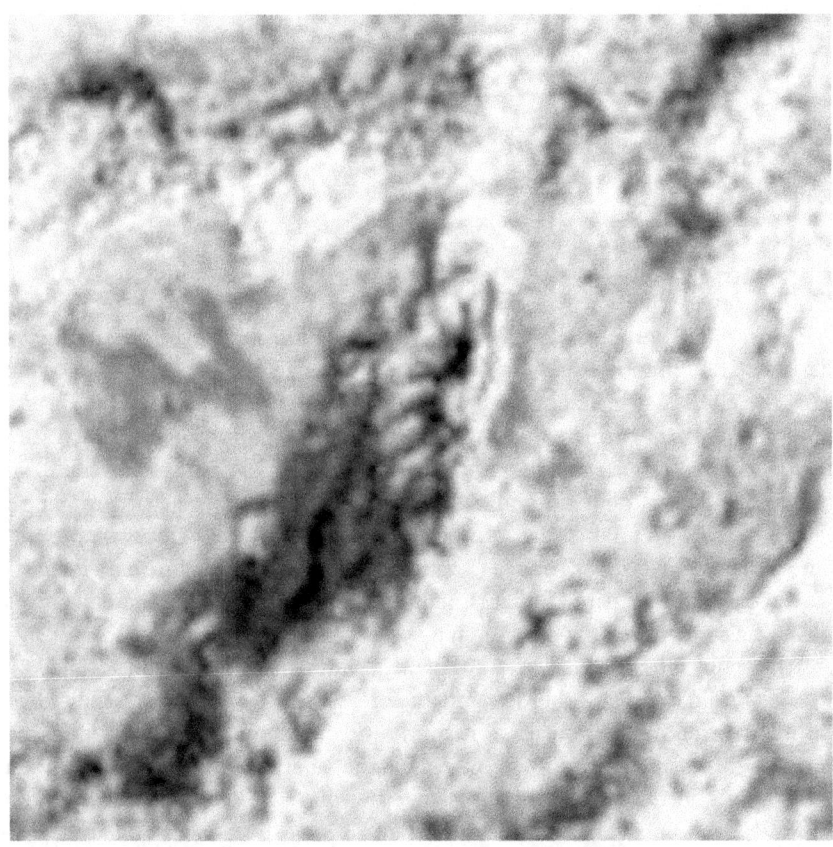

Figure 21

Another fossil like structure photographed by a Martian Rover[lxxxv], these lends some credence to the claim that more advanced life may have evolved on Mars though it appears unlikely. Another possibility is when aliens terraformed Mars this involved seeding it with life, some of this became fossilized. Compare the Martian images to Earth Crinoids[lxxxvi].

Chapter Seven

Archeological plans

At some point the amount of evidence for artifacts on Mars may reach a critical level which will then cause mainstream science to take it more seriously and perhaps plan to visit Mars to examine these artifacts more closely. If so then the path of this process can itself be analyzed as to what may be the most efficient way to proceed. We already have good quality imaging probes around Mars; if particular formations deserve further study then they can blanket those areas with image taking looking for more detail and other potential artifacts nearby. In the case of the King's Valley each potential face could be reimaged at different times of the day and year because this produces shadows that accentuate different features and would allow a three dimensional model to be made of the area. This is done below where shape from shading is used to construct a 3D version of the main Crowned Face. As these shadows moved across the faces in the different times of the day and different seasons each Face may have looked different like a variation of the Stonehenge idea. There is some reason to believe this was the intent of the builders of these faces, it is a logical reason for joining the faces together like this. It is difficult to know how the shading varied in this area because the Pole wandered to many different positions each of which would give differing shadow patterns on these faces. In a previous chapter the Pole position was estimated by assuming that three of these faces, namely the Crowned Face, the Cydonia Face, and Nefertiti were on a great circle because it was the equator. This can give certain shadows in the King's Valley but other factors such as continued polar wander and changing axial tilt also need to be estimated to validate this theory.

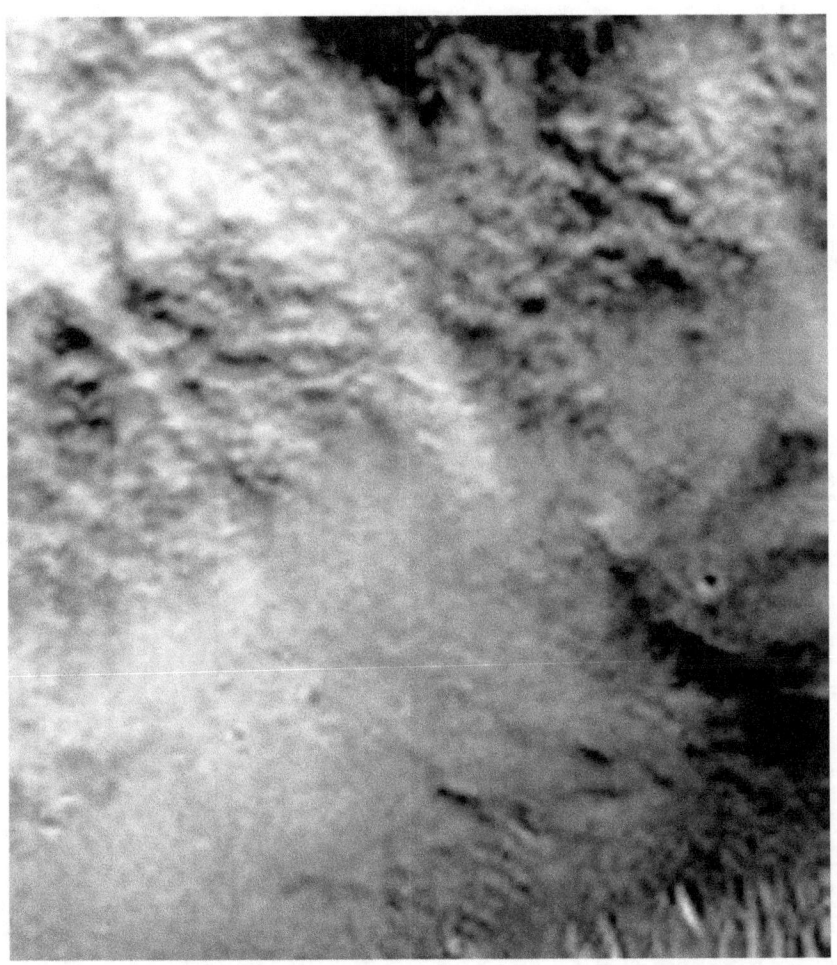

Figure 1

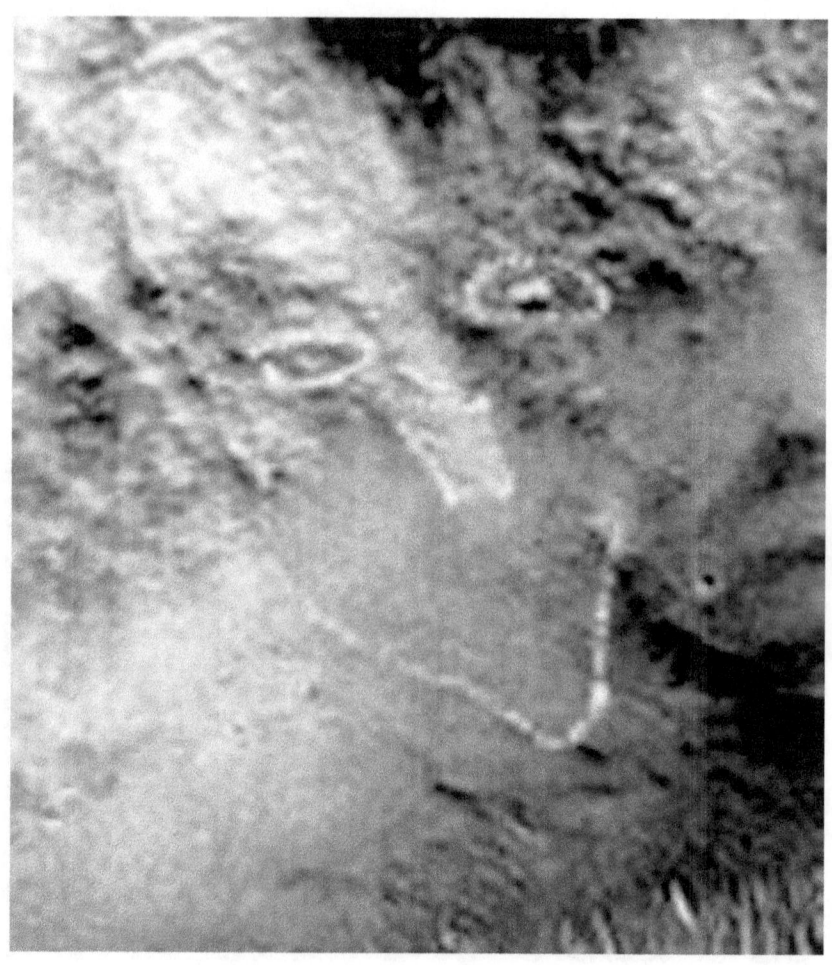

Figure 2

I call this Face One; shadows may have made this stand out in the morning or afternoon or perhaps on the Summer or Winter Solstice. More images of this face are needed to tell if it is plausibly artificial, the latest HiRise images didn't have enough shadows to see its features clearly.

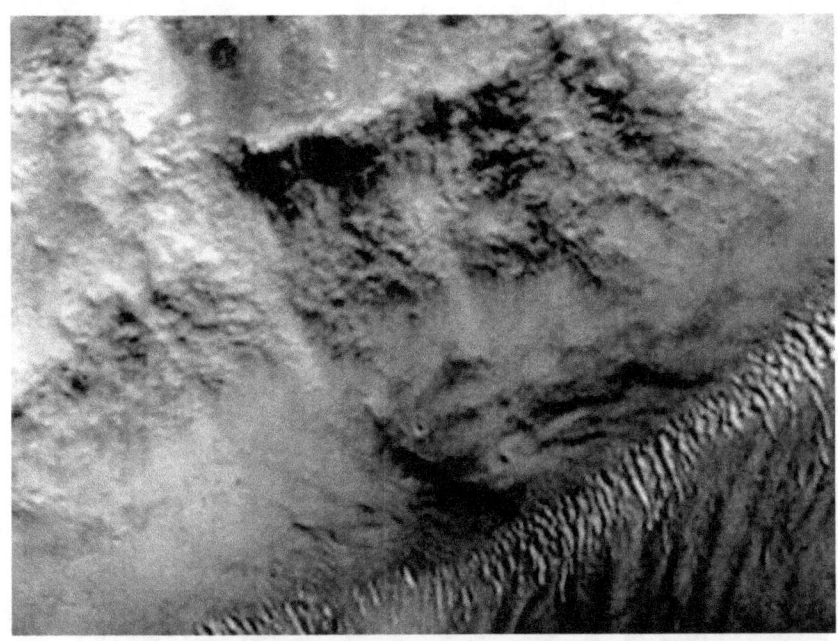

Figure 3

This is the main Crowned Face or Face Two.

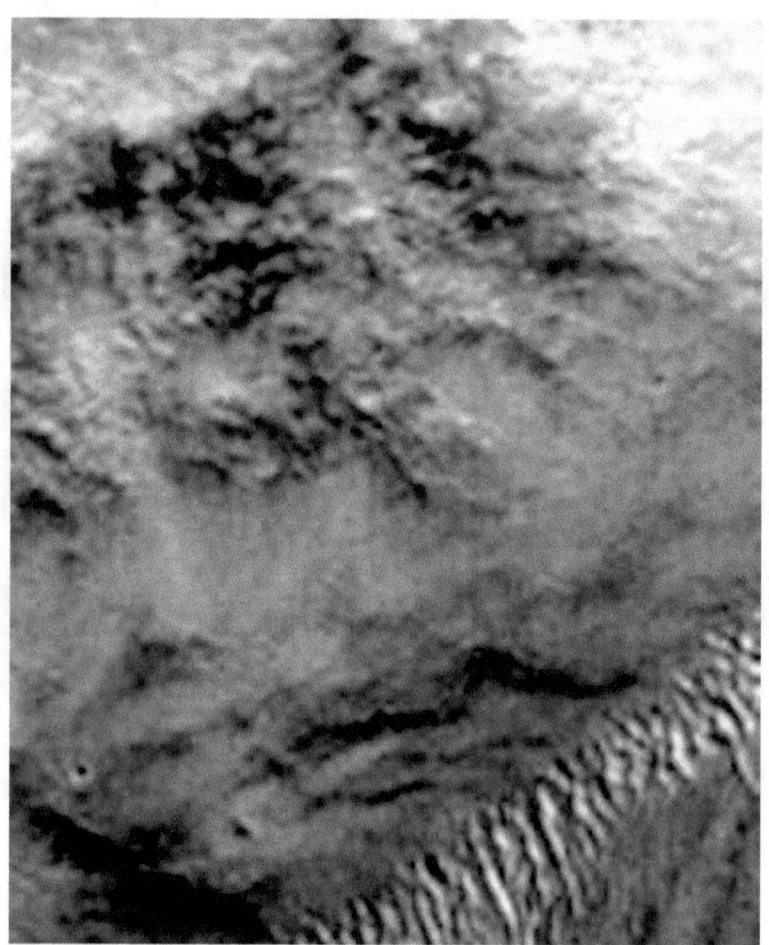

Figure 4

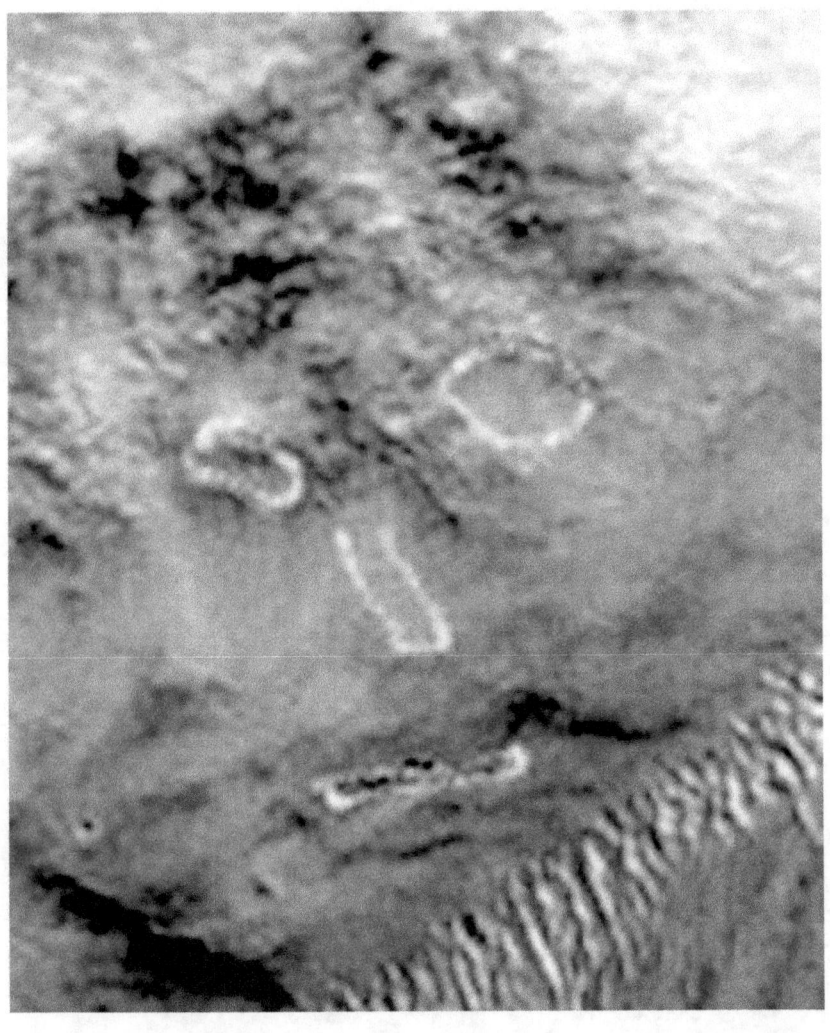

Figure 5

This is Face Three, more images are needed of the right eye to see whether it is damaged or intended to be part of the Profile Crowned Face to its right.

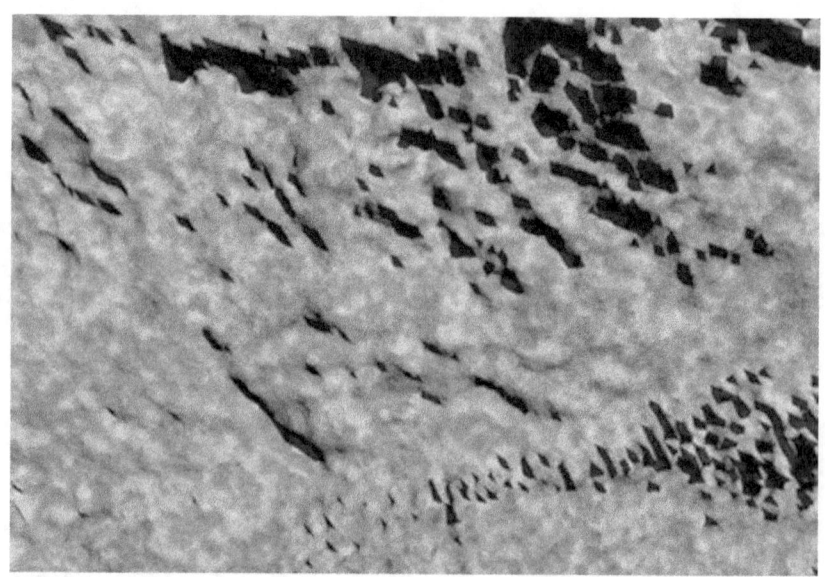

Figure 6 Sun azimuth 18.9, altitude 54.6.

This simulates the sun coming from above the Crowned Face. The shape from shading is done using the program Bryce 4.0. HiRise should be able to take images of the King's Valley at different times of the day to show some details more clearly with shadows.

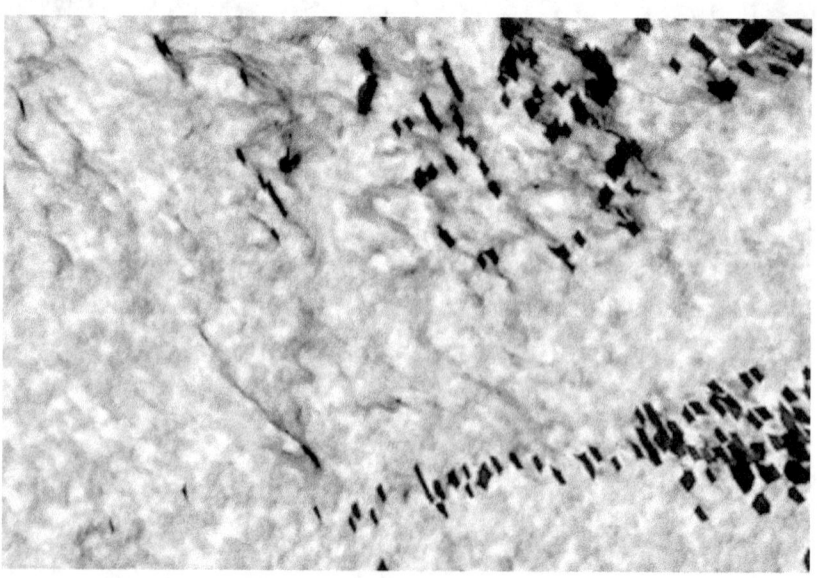

Figure 7 Sun azimuth 90.6 Altitude 54.4

Here the light is simulated to come from the right, note how the main Crowned Face stands out in the eyes, the nose and the left jaw line.

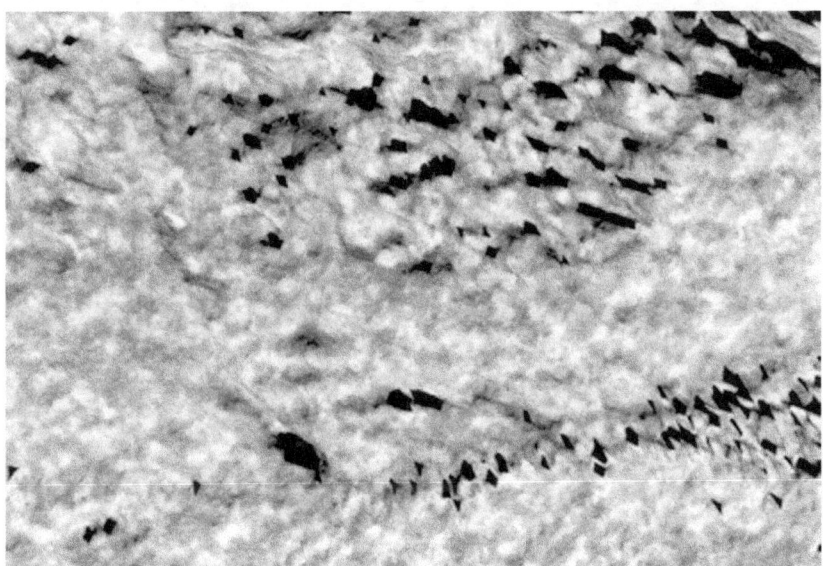

Figure 8 Sun azimuth 145.3 Altitude 51.1

The sense of a face persists through the different shape from shadings; here the shadows would not be accentuating the left jawline of the Crowned Face. A left ear stands out however.

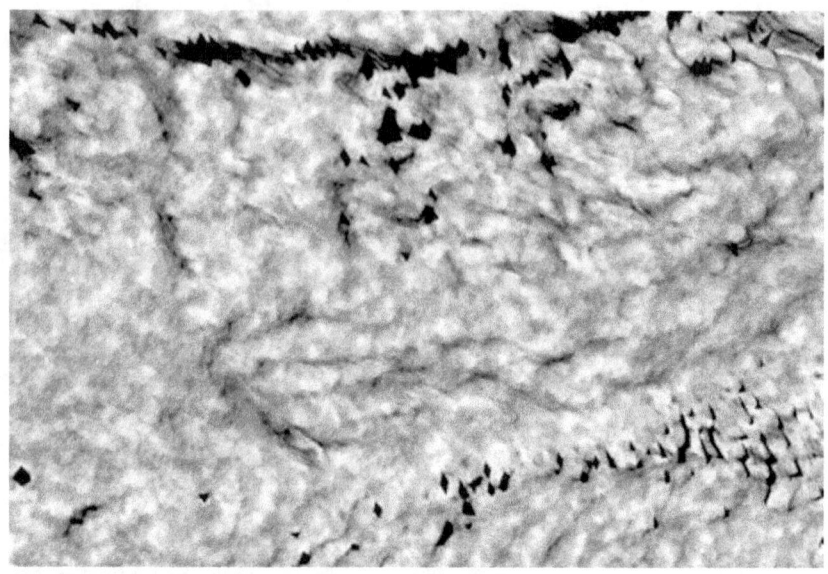

Figure 9 Sun azimuth 287.5 Altitude 52.4

The shadows tend to make the main crown of Face Two stand out as separate from the cliff, with a gap on both sides of it. The different shadows might also convey different emotions; figure 7 for example looks angrier but figure 9 more pensive and perhaps older. Of course we have no basis to say that facial expressions here would equate to our own, however it's likely that a Face which could convey different expressions would not use this ability. It may be that different seasons could also change the facial expressions, for example in summer the solstice would have a different sun angle at various times of the day than at the equinox.

Figure 10 Sun azimuth 49.6 Altitude 46.2

This is looking from ground level under the chin of the Crowned Face, it gives an idea of what a robotic rover or manned expedition would see. If HiRise images with enough shadows were taken this this ground level view could be simulated with much higher resolution. A robotic Rover for example could have its path calculated through the King's Valley in this way.

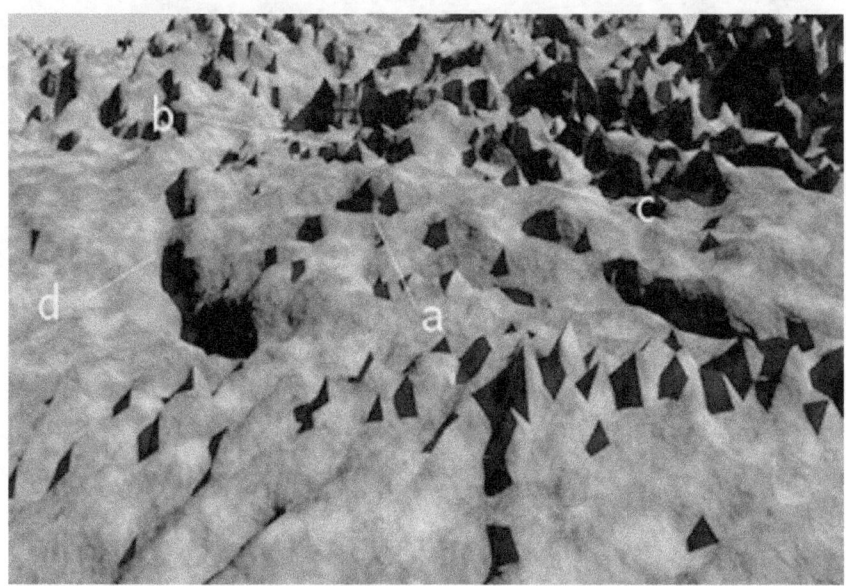

Figure 11 Sun azimuth 49.6 Altitude 46.2

Here "a" shows the nose. "b" and "c" are pixelated but do appear to show 2 raised eyeballs. "d" shows a clear jawline as a ditch. An expedition would examine this ditch for signs of having been carved out, given the low amount of erosion there should be signs of tool marks if these faces are artificial. The eyes do not be appear to be craters even though they are depressions, a raised eyeball could not form if they were impact craters. There would also appear to be no reason for the wind to carve two nostril depressions without wearing away the rest of the nose.

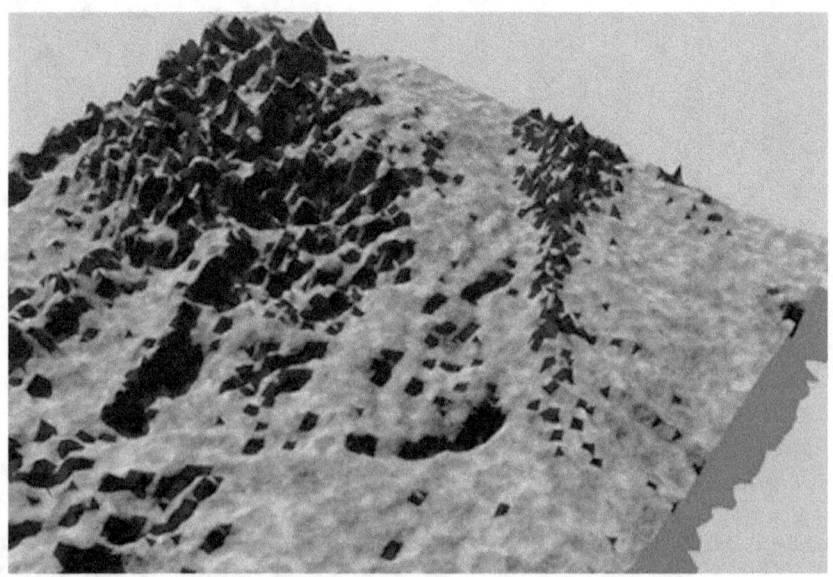

Figure 12 Sun azimuth 17 Altitude 47.3

This would be an approximate view of the noon face from the opposite cliff edge. Note how the crown appears to be separate to the rest of the cliff from this angle. A robotic probe or expedition would get a good view from this position but it would be more difficult to get to the other side of the valley.

Particular features can be reimaged at the highest resolution available from the HiRise camera to maximize the chance to seeing if they are artificial, for example the eyes of the Crowned Face look highly artificial but high resolution images with different shading could bring out details hidden because of a lack of shadows. Also if multiple images are taken of this area then extra data can be extrapolated out of them to simulate a much higher resolution. The nose seems to have two nostrils which stand out more in the original MOC image but with less shading in the HiRise images they are less clear. Because nostrils are hard to create from geological processes then additional images may show more details of them, if they are very clear then they would be almost impossible to be naturally formed. For example with the area so highly eroded the nose might have been expected to have eroded away, and erosion under the nose should be forming more random shapes than two caves that happen to be where nostrils would be on a humanoid. Generally then erosion should randomize the features so those shapes that are small and well preserved may be made of materials harder than the surrounding rocks such as a kind of concrete or clay. Some face like areas appear to be buried under the sand, some kinds of radar from space or on the surface may be able to look under this sand to see the shape of the rock underneath.

Many of the techniques used on the Egyptian pyramid exploration would be useful here. Datable materials such as charcoal[lxxxvii] would be hard to find after so long, Carbon 14 methods for any organic material found might work if the Faces are comparable in age to fossils here on Earth. Magnetometry and resistance tomography have been used on the Great pyramids[lxxxviii], topographic mapping, drill coring, seismic sounding, microgravity[lxxxix], Ground Penetrating Radar[xc] and laser scanning are also used. Radar has been used to look for the tomb of Cleopatra[xci], there might be hollow areas inside these faces perhaps as tombs or receptacles for a message or historical records. The Bosnian Pyramid of the Sun may offer a guide as to how the area can be excavated[xcii]. The Akapana Pyramid Mound used ground penetrating radar to construct a 3D model of the mound both above and below ground[xciii]. This would be useful to work out exactly how the shadows changed around the faces with different pole positions. If this is a burial mound this implies someone may be buried there, and fragments of DNA might possibly survive. On Earth burial mounds like these usually contained art work and clues about the life of the person, this may be one of the best ways to find out more about them.

These areas can also be analyzed geologically for faults, where the sand and dunes in the King's Valley came from and how old they might be. For example if the dunes were formed by erosion from the wind and show little sign of having moved then that would imply the erosion came from the cliffs and the faces there should have been eroded away or were formed more recently. The dunes may have come from debris washed down the valley from flooding or the valley may have contained a river for a long time. Regular flooding might be expected to damage these faces;

some of the faces may be buried or eroded closer to the bottom of the valley. The Cydonia Face is in a lower area in the Northern Lowlands and being submerged or covered by moving ice may have removed some of its features. In that case the faces with the jaw and mouth buried may have been eroded away from the water so we could never reconstruct them. One example is the erosion on the Sphinx; it makes it difficult to reconstruct exactly what it looked like originally. The number of small craters on these faces and features might indicate how old they are as well, generally the older they are the more likely large craters would have destroyed them. The craters on these faces are usually small and rare which may indicate they are not as old as in the terraforming theory. If they were built at a time when hypothetical aliens were terraforming Mars then much of the erosion would be very old from that time because the atmosphere would be thicker and hence the wind stronger and likely with some rain or snow. Later this erosion would have been much less as the atmosphere froze so the main erosion after that period may have come from micrometeorites as on the Moon[xciv]. Water erosion signs on the faces then might date when they were built, while signs of more massive damage such as in and below the mouth on the Crowned Face might indicate a cataclysm after their construction perhaps from the Hellas meteor impact[xcv]. The debris on the faces could be analyzed for its chemical signature to see whether it came from somewhere else with dust storms or eroded off the faces.

A logical next step in investigating these formations would be an unmanned lander perhaps like Spirit and Opportunity. This would likely happen if the evidence was sufficiently compelling but the problem would be in getting a lander into this area. There are three main ways of doing this, one is to land it where it could move up the valley to where the faces are, the second is to land on the opposing high area and move to the edge of the cliff to see the faces from there, and the third to land on the side of the valley where the main faces are and try to navigate down the slope. There is also a suggested Mars Rover landing site very closer to the King's Valley.

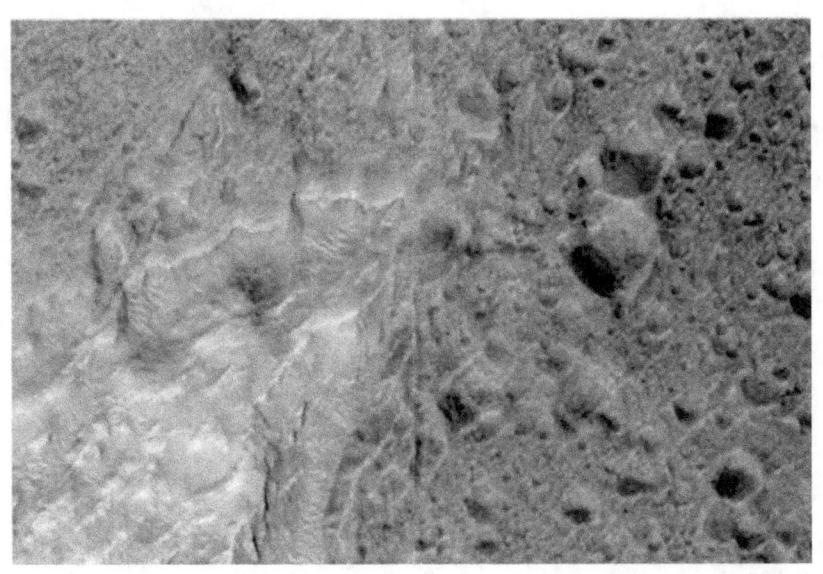

Figure 13

This shows the terrain in the MSL proposed Landing Zone near the King's Valley.

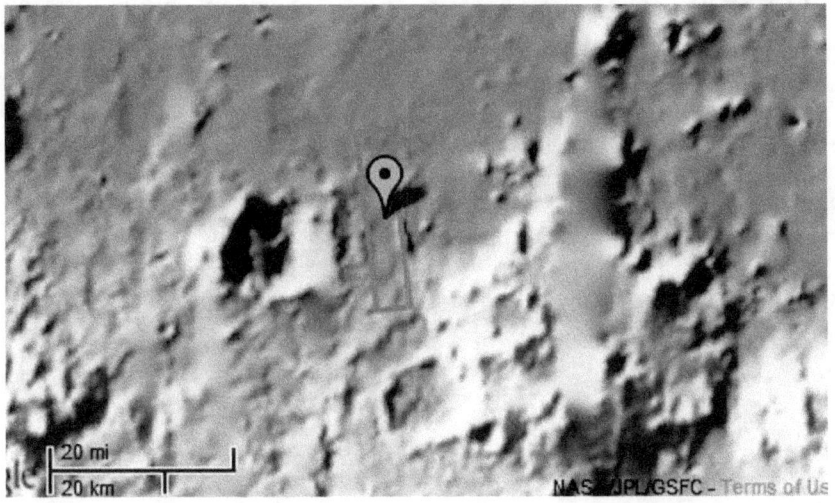

Figure 14

This is a suggested MSL Mars Rover landing site just near the King's Valley[xcvi]. From this position a Rover might be able to try all three suggested routes. The King's

Valley is in the dark line heading downwards from the dark rectangle. As can be seen from the photo the ground is very pitted which can cause difficulties in landing here.

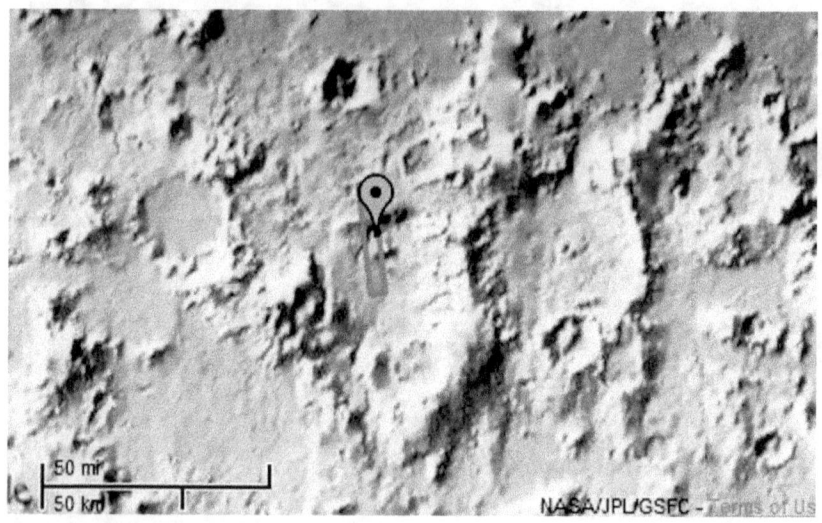

Figure 15

This dark rectangle is the boundary of the most recent HiRise image taken of the King's Valley which is just below the Rover suggested landing site. The Rover landing site is in the smoother terrain of Isidis Planitia to the North. Since Rovers have a limited range it may be difficult for one to travel to the King's Valley from a safer landing zone. As will be analyzed elsewhere in this book it may be better to approach the King's Valley from the south because there may be a cave there. This is seen in the image below.

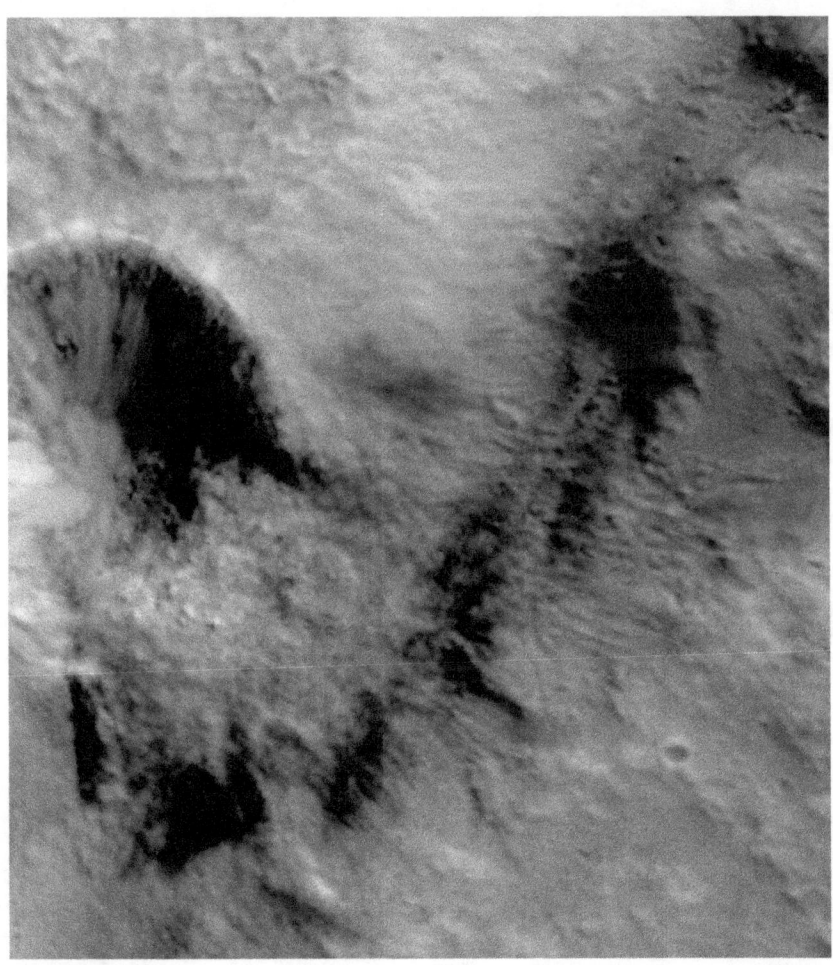

Figure 16

The King's Valley may have been dammed here to prevent flood waters damaging the faces, this might make it more difficult for a Rover to get down into the valley but it could approach the Crowned Face side and climb down there. The dam and possible cave are analyzed later, but it would be a priority to found out with reimaging if this is really a cave as it would coincidentally make the best known place for a Martian colony right next to the King's Valley. If this is an artificial cave then many others may exist on Mars making it much easier for colonies to spread protected from radiation.

A major problem in landing on Mars is to avoid large rocks which can damage it[xcvii], so a lander might have to drive a large distance from a safer landing zone to get to the King's Valley. If it could climb down the slope next to the Crowned Face it would have the best view of its shape to be able to determine if it was artificially carved or naturally formed. If artificial then the materials should be different from the opposing rocks such as having used a form of concrete or clay. A trail might be possible to follow by going down near the left ear as shown in the picture below.

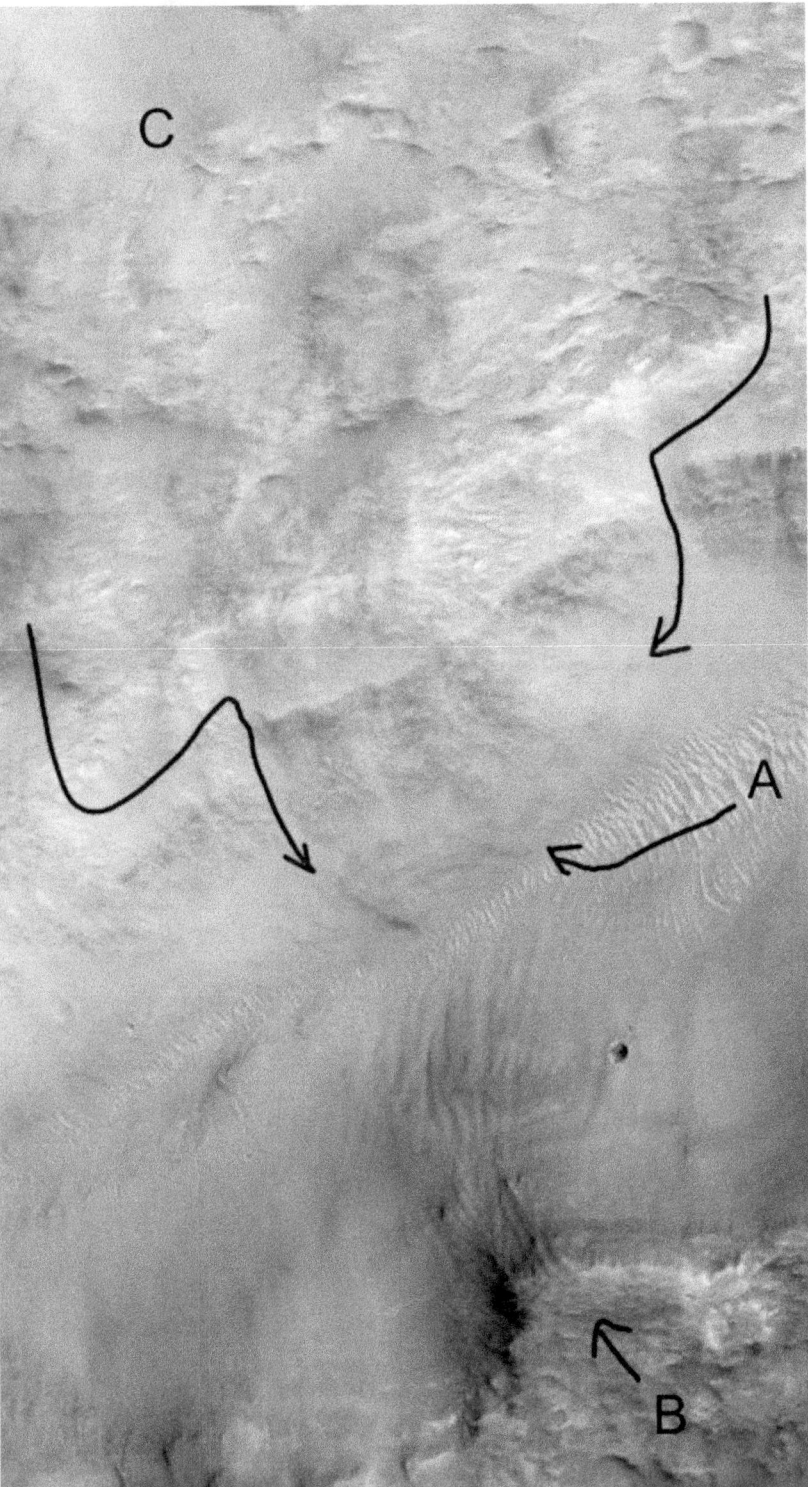

Figure 17

*A shows an approach going up the valley from Isidis Planitia and then onto the faces, care should be taken into driving into an area that with so much archeological significance. The route **B** has the problem that it is much steeper to get down to the valley, a Rover would have a lot of problems here after having seen the grades of slopes Spirit and Opportunity could traverse. It may also be impossible to move up this valley in a Rover because the sand may be too soft, like the sand that the Spirit Rover got stuck in[xcviii]. **C** is a potentially flat area that would need to be carefully examined for a suitable place to land, more likely the whole side of the valley would have to be imaged carefully for rocks and a bigger area needed to give a margin of error in landing off course. Landing on the wrong side of the King's Valley could be a bad mistake as it may be very difficult to cross to the other side. Once landed there are two possible ways down the slope to the faces as shown, avoiding the more steep edges. The arrow on the right hand side may be more hazardous because of the dust and sand there. The left hand arrow may be the best path as it has less debris to get stuck in and it comes directly to the edge of the noon face around the left ear. This area should be less significant archeologically because it should be designed to not be part of the faces.*

Figure 18

In this image the King's Valley is seen as the dark line crossing the white rectangle which is Mars Orbiter Camera photo M0203051. It may be possible for a lander to travel up the valley as shown. There may also be other artifacts along this valley, if the whole valley can be reimaged by HiRise it will help to determine the full extent of this site and perhaps give some clues as to why this valley was chosen instead of other valleys in the area.

The tools used by a lander exploring the King's Valley should be similar to those used on Spirit and Opportunity. Abrading rocks[xcix] and using a Mössbauer spectrometer and an alpha particle X-ray spectrometer could determine the composition of rocks there looking for clay, concrete, paint, or traces of former joiners such as metal nails and screws. A Microscopic Imager might show where rocks had been cemented together and take samples of this cement. There might be writing engraved into the rocks like our Earthly statues would be expected to have a plaque to commemorate its building or explain its purpose, the images taken by Spirit and Opportunity might have sufficient resolution to examine this if found[c]. This writing might also be buried underneath the sand at the bottom of the faces so a digging tool or an ability to brush off dust from these formations may be a critical advantage. The area might also have some hollow areas like caves or buildings now filled with sand, it would be unlikely for a robotic lander to be able to open or enter these because of the potentially large amounts of sand to be moved but perhaps such rooms could be examined with radar. If significant artifacts are seen inside with radar then more effort could be made to remove the sand and dust. Another useful tool would be like a leaf blower used in gardening, this could blow dust out of cavities and off surfaces though the low air pressure would make this more difficult to do. The dust on Mars is very fine like talcum powder; because it is fine enough to be blown around in dust storms it should be movable by a blower. Some areas may be more highly preserved if there has been insufficient wind to move the soil for millions of years, rocks just partially buried may be able to be examined in a more protected state if the lander could remove a few centimeters or more of soil. Organic material could be looked for in buried areas though it is virtually impossible for it to have survived in direct sunlight because of the high radiation[ci]. The soil might also contain levels of moisture from ice freezing at night[cii], this might tend to decompose materials buried near the surface.

Pottery may be found if the faces were formed by an indigenous race, but if aliens created the faces it is less likely low technology artifacts would be found. Some kind of message would be highly valuable though it is hard to image how such a language could be decoded without knowing the names of various objects referred to. The problem may be similar to some languages which cannot be deciphered today on Earth[ciii], there are symbols near the Cydonia Face which may represent a language. The more examples of these symbols that are found the more likely they represent a real language and can be decoded.

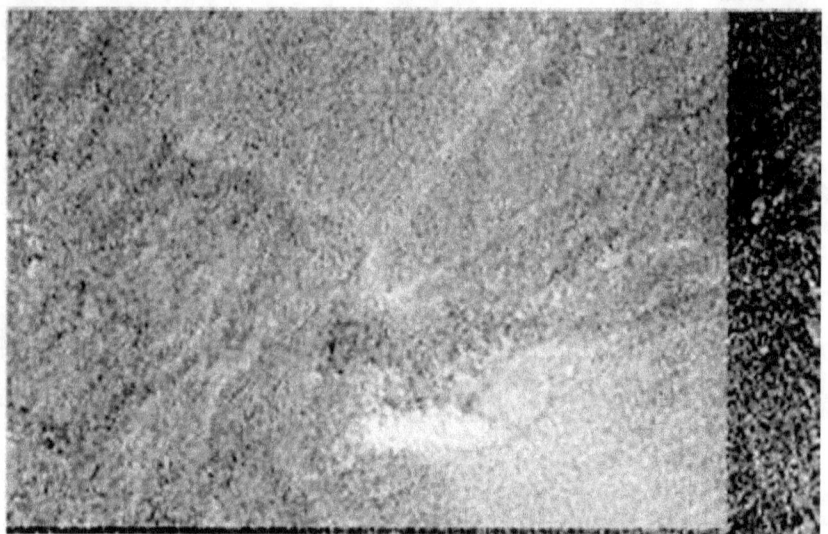

Figure 19

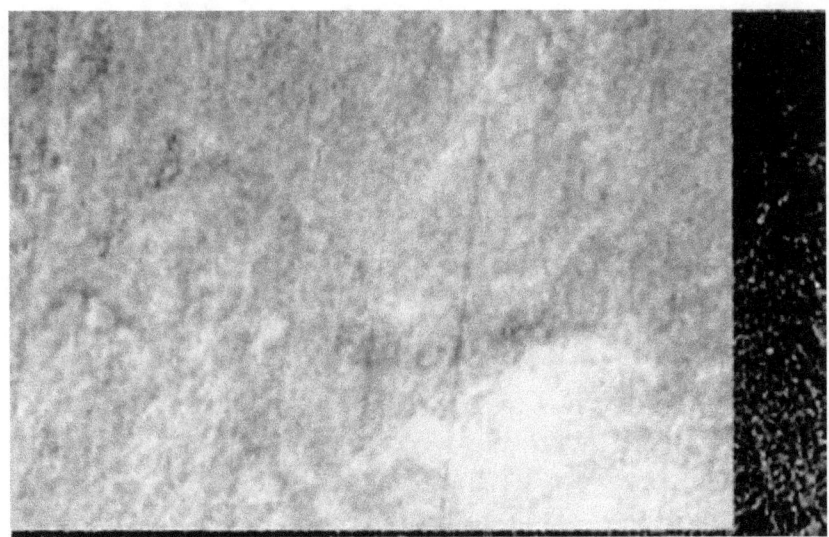

Figure 20

Figures 19 and 20 are overlays of two different Mars Orbiter Camera images taken from different sun angles. In Figure 19 the markings are seen as raised ridges, and in Figure 20 the possible letters in the first MOC image are clearer as dark shadows. This shows the letters stand out more with shadows and implies the builders may have used shadows to have some formations be seen more easily from space or from planes. More of this area will be analyzed in an upcoming book.

Features of faces might be more easily associated with names in a recorded language such as engravings, also the names of planets and stars, valleys, rivers, rocks, etc. might also be possible to decipher if they have pictures or coordinates associated with them. Mathematics would be an ideal record to find because it would indicate the level of understanding of the creators of these formations; if indigenous then this mathematics might be very simple by our standards. Euclid would be an example of the level of mathematics available to the ancient Egyptians, there are indications Pythagoras' Theorem have been understood by the Martians as well as the angles in a Tetrahedron from Horace Crater's research on the Pentad formation, explained in more detail in an upcoming book.

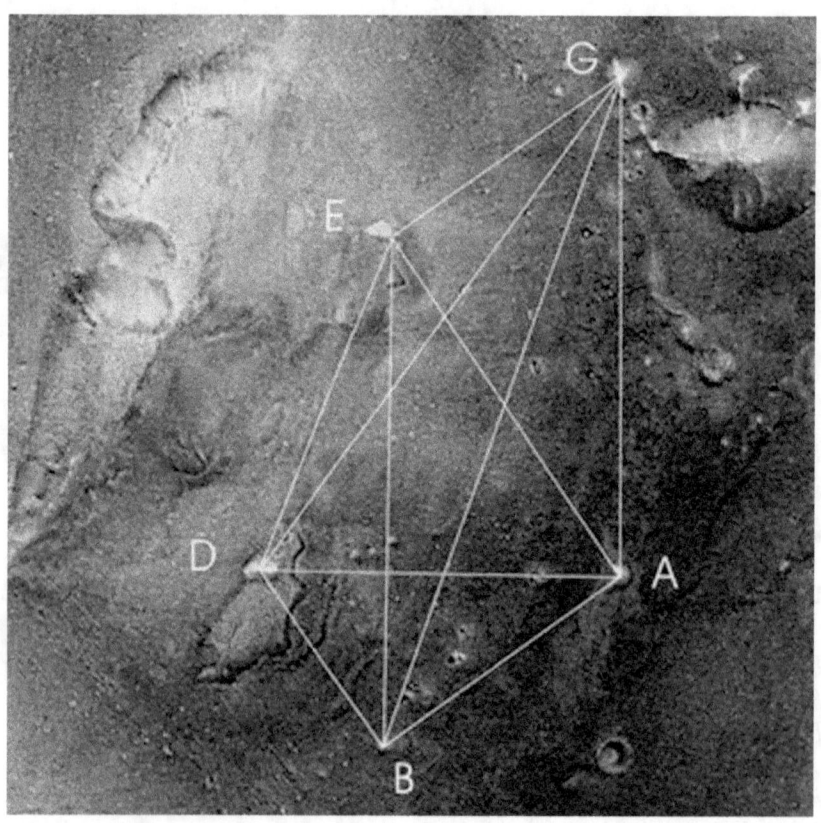

Figure 21

This is the Pentad formation discovered by Horace Crater of SPSR[civ]. It contains complex mathematical relationships related to the tetrahedron and possibly quantum mechanics. If artificially created then this would give an estimate of the mathematical knowledge of the visiting aliens or indigenous Martians. Finding such patterns engraved in rocks in the King's Valley might indicate technological knowledge or important theorems, for example relativity and time dilation can be shown in a geometrical diagram[cv]. This will be analyzed fully in an upcoming book.

If such an exploration was successful and sufficient artifacts were found then the next logical step would be a manned exploration of the area and preferably a permanent manned base to explore other sites. Because of the amount of work involved it is unlikely an expedition could accomplish much without staying in a permanent colony, the best ways to do this seem to have been worked out already by the Mars Society with Mars Direct and some NASA projects. It's likely the curiosity from Earth once artificiality was proven would mean such an expedition and colony would be made a priority, perhaps within fifteen years of the decision to go people would be walking on these sites and preparing for a permanent colony. The simplest way to do this would be to make a base in lava tubes, natural caverns created by lava flows from nearby volcanoes such as Elysium Mons or Nili Patera, or in the possible cave near the King's Valley. Once these caves were found then protection from radiation would be much cheaper underground as well as providing a more constant temperature. If there was still some methane escaping from volcanic deposits or life forms[cvi] then this might be collected for fuel for cars and ships to return to Earth, otherwise methane can be readily created from the CO2 atmosphere and water ice using the Sabatier reaction as explained in the Mars Direct plan[cvii]. Radiation may also be avoided by travelling more at night and using lights to examine artifacts, then getting underground during the day. This would avoid an accumulation of radiation exposure that might become serious over a permanent stay[cviii]. Part of the King's Valley could be made into a makeshift cave to protect from radiation, a dome constructed and then soil heaped over it to give additional radiation protection. This could connect to the cliff face of the King's Valley; alternatively a cave might be dug into the side of the cliff for a local shelter. Craters could provide another way to make a cheaper shelter, a geodesic dome constructed in one could use the sides if steep enough and then soil put over the top would create a hollow structure. As shown in an upcoming book hollow structures on Mars used as dwellings may exist there already.

Figure 22

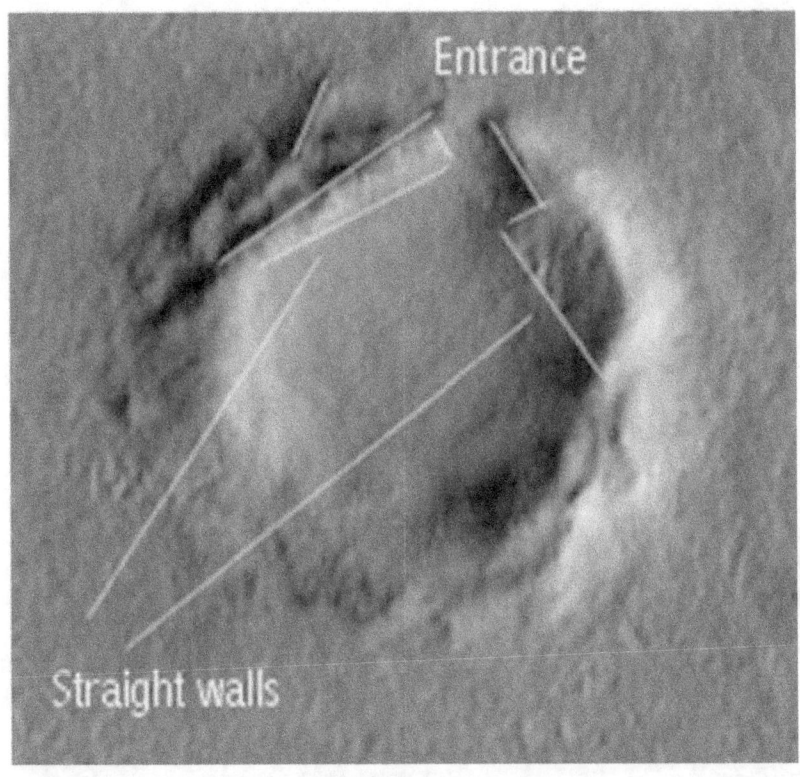

Figure 23

This crater in photo SP253104 seems to have been altered in shape, the walls made more angular and an entrance made. The floor of the crater also seems flat rather than being curved from an impact. Placing a dome over an altered crater would make a good shelter against radiation.

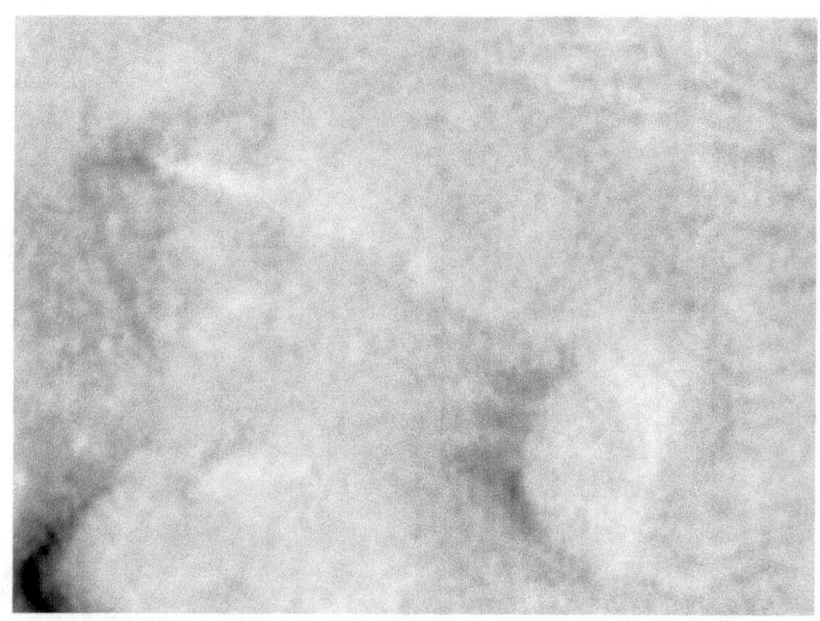

Figure 24

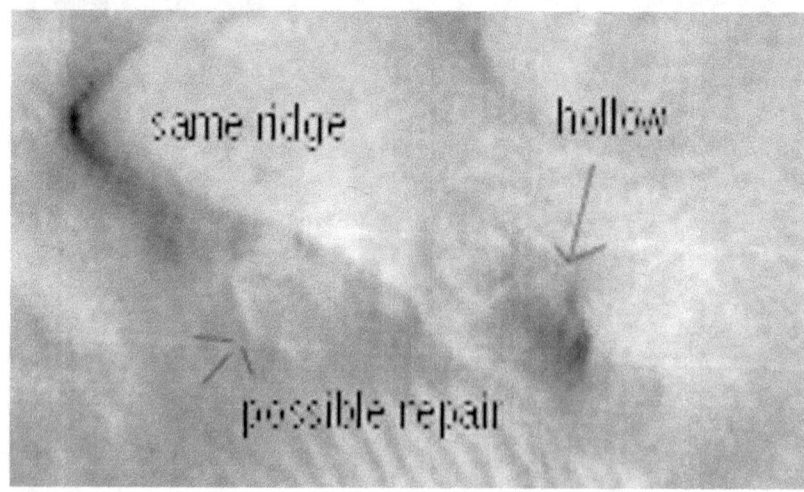

Figure 25

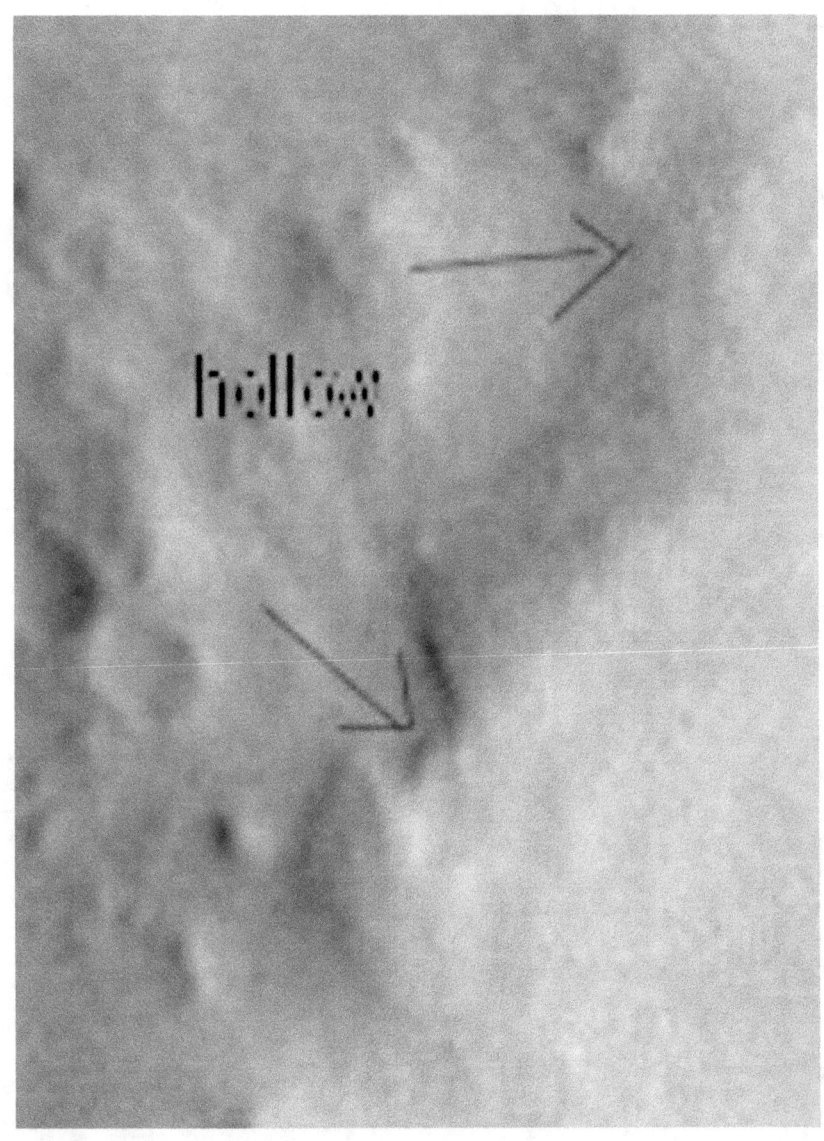

Figure 26

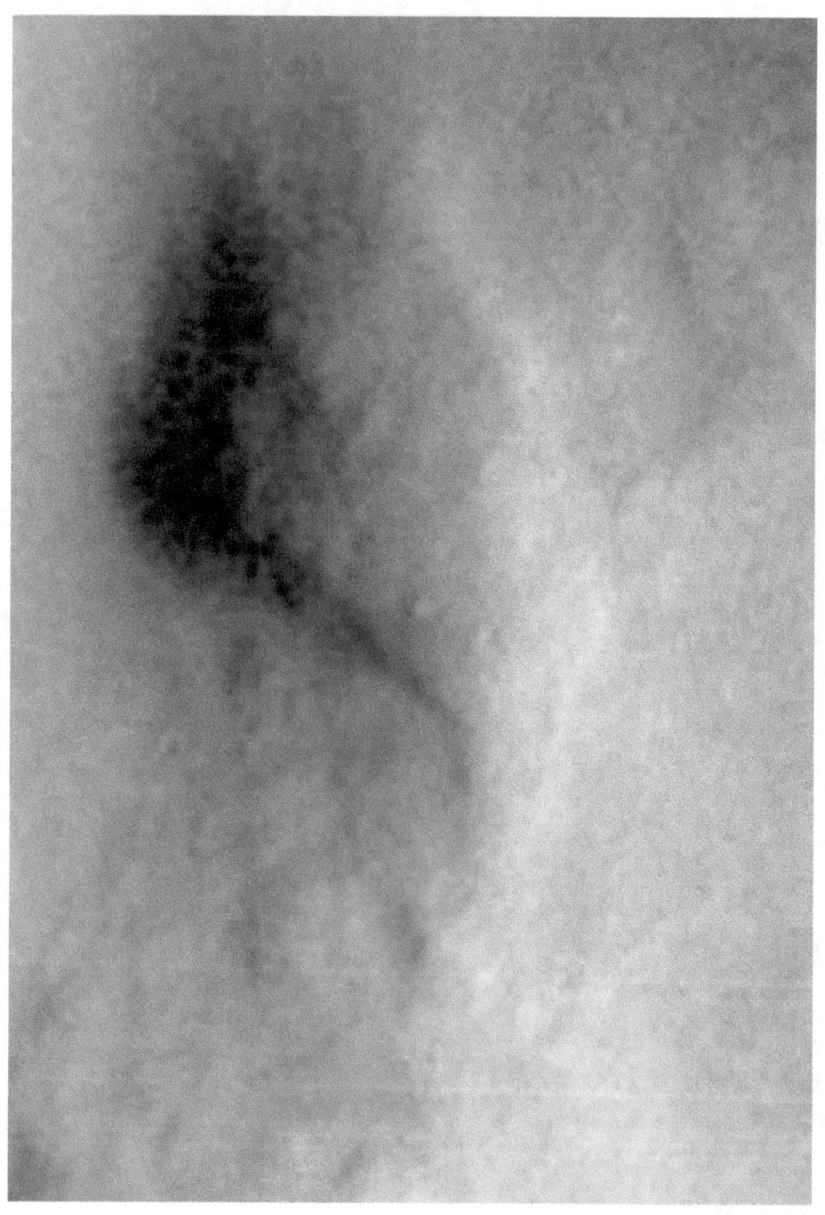

Figure 27

These formations in photo SP240505 may have been hollow dwellings that collapsed or were swept away by flooding from the Hellas impact. In Figure 24 the higher area on the right in the oval may have been a support along with a central ridge to hold up the roof. There are many of these on Mars; they have no obvious geological explanation if

138

they are natural. They are similar to what we might build ourselves to protect against radiation, perhaps inflate a balloon or build a roof and walls, cover it with mud or cement to give its shape and then build supports inside to make it strong. Figure 25 shows a possible hollow still remaining after the roof collapsed, also in Figure 26. Figure 27 shows part of the hill missing like it was hollow. This theory of their being buildings will be explored in an upcoming book.

Chapter Eight

Methodology

I try to use a consistent methodology in examining these formations, relying on proving artificiality rather than trying to accumulate a large amount of suggestive evidence that might all turn out to be natural formations. To do this I try and disprove the natural hypothesis, that these formations are natural by ruling out the various kinds of geological processes that could have formed them. I also look for small parts of formations that are least likely to be naturally formed because it only takes one proven artificial part of a formation to prove the whole case. For example the tip of the Crowned Face nose may be hard to explain by natural processes with the two nostril indentations and the lack of erosion compared to other rocks nearby. Small details like this might be the most valuable to determining whether any of these structures are ultimately artificial or natural. Often this kind of evidence is ignored by both sides. Those that already believe these formations are artificial think it unnecessary to look at small details and those very skeptical tend to dismiss all the evidence with hand waving arguments and also then avoid looking at the smaller features of a formation.

It is not necessary to prove that every part of every formation I discuss there is artificial, no doubt many of them were just naturally created by the wind and rain. The point is that no part of any of them should be artificial, once we prove some parts are artificial then we know for a fact that either aliens visited Mars or there were indigenous Martians. This book then often concentrates on small parts of these formations while ignoring the more vague impression of looking artificial because this often cannot be proved. This book then sometimes skips over what appears to be good evidence because it is impossible to prove, for example the Crowned Face obviously looks artificial to many people but it is impossible to prove this is not just a coincidence caused by erosion. There is also the obvious objection that an alien or Martian Face should not look like us so its convincing appearance is a reason to some people why it cannot be artificial. This leads to a circular argument where the better the evidence is the more likely it is not artificial, but then the worse the evidence is i.e. not looking like a clear face the better the reason to dismiss it again. The idea of a face is subjective to most people and so is of little use in proving artificiality, many people for example see the lack of symmetry in the Crowned Face as evidence for it being formed naturally. To counter this idea it is necessary to detail how these asymmetries are accounted for in making dual faces, for example the right eye of the Crowned Face also needs to be the left eye of the face to its right so this explains why that eye cannot be anatomically correct for both faces simultaneously. It also helps to disprove that this is a natural formation; it is less likely that random chance will not only make two faces on Mars but make them a complex duality that shares the features like this. The more a facial feature is shared by another face, like the nose of the Crowned Face acting as the edge of the face to its right, the less likely this occurred by chance because there are many more opportunities for chance to have marred the perfect way these faces meld together.

I also try to overcome this problem of subjectivity by using statistical arguments. For example I compare the three Crowned Faces with the Meridiani Face and Cydonia Face on the theory that they are all highly eroded versions of the same face. This allows a stronger statistical argument, a natural formation might appear face like from wind erosion but generally hypothetical aliens visiting Mars would tend to look like each other. We should then find that faces and other formations resemble each other and the more similarities that are found the more they point to a consistent history of their construction and what their builders were like. This avoids what I call the Star Wars Café problem where in the movie there were so many different looking aliens in the café. It is unlikely Mars was visited by ships with hundreds of different kinds of aliens on board because there is hardly any evidence of artifacts there. More likely there was either one alien race or one indigenous Martian race, or perhaps both. Because of this I try and concentrate on faces that are similar to each other because this is less likely to occur by chance. A person might see a face in a cloud but it is much less likely they will see many faces in clouds that resemble each other closely.

It is also important to use the best quality data possible because distortions in photos can make natural objects look blocky and artificial and also obscure the artificial looking formations. For example the resolution of the Crowned Face can be artificially lowered by resampling it in an art program. First is the original which comes from the NASA photo.

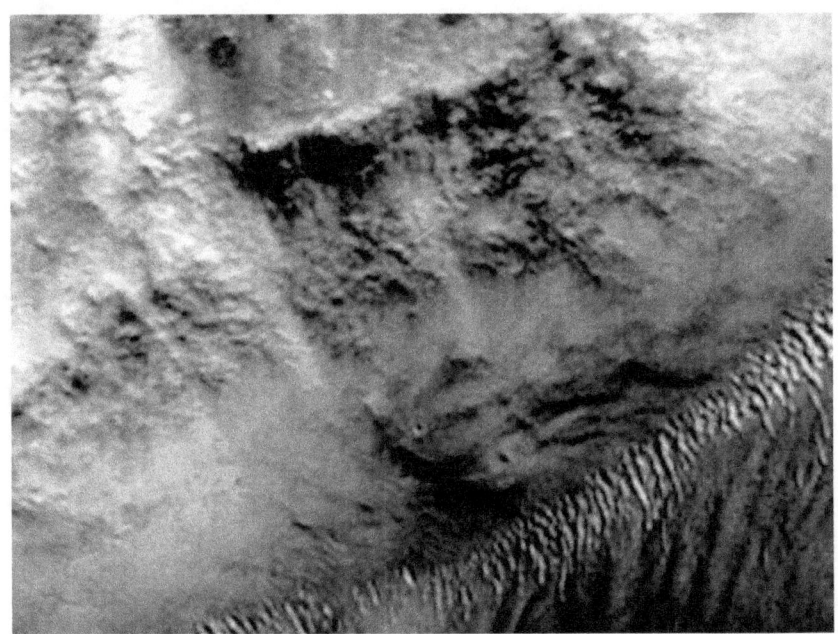

Figure 1

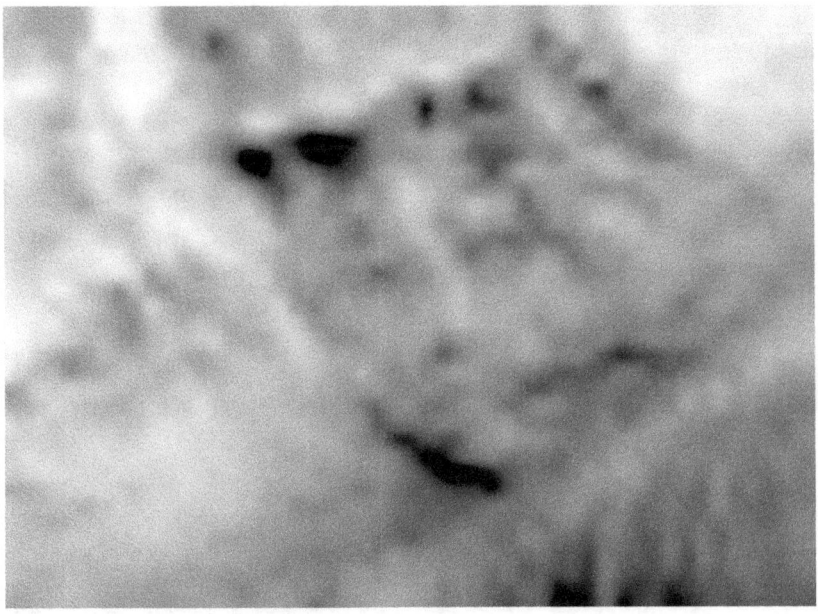

Figure 2

If the Mars Orbiter Camera had taken the image in Figure 1 at a lower resolution, which it often did to take larger photos, then we might have had a photo more like in Figure 2. The Crowned Face still looks compelling but the faces to its left and right have vanished. With no follow up photos this might be written off as a coincidence like seeing faces in clouds.

Researchers would have taken this inferior image and tried to extract more information out of it such as by resampling to a higher resolution and manipulating the brightness, contrast, and gamma. It still looks face like above at 12 dpi instead of the original 96 dpi (dots per inch). If I now resample this back to 96 dpi the lost detail is not regained, instead this adds even more distortion seen below. It appears clearer however until you compare it to the original above. Note how a large grey area has appeared above and to the left of its left eye, we might then conclude that this feature was real and we had found this by resampling the original. This might then be used to prove the formation is natural because a face should not have this grey area there. Compare this to Figure 1 to see how far we are getting from the real image. It is certain that some of what we see there has the same kind of distortion; fortunately we can compare this to the latest HiRise photo of the Crowned Face to see how misleading this image was.

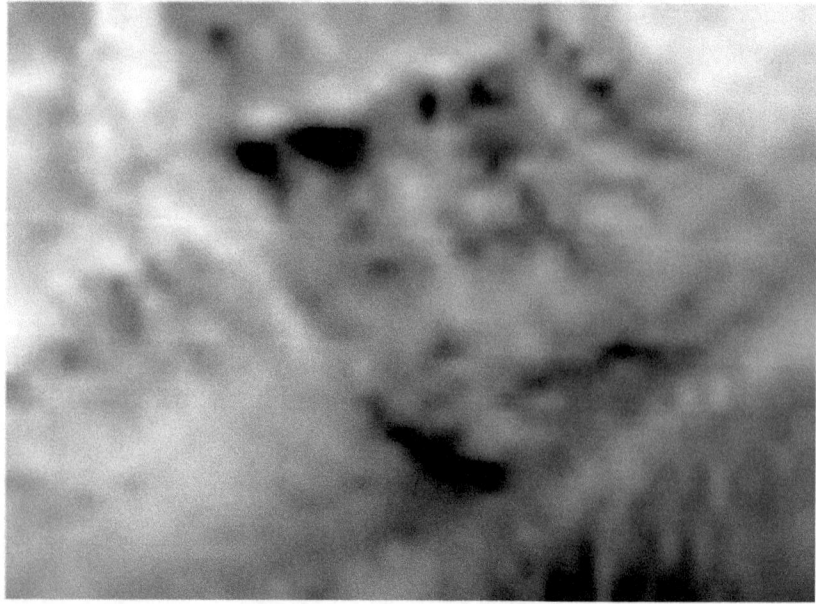

Figure 3

Generally then it is better to avoid low resolution images of anomalies because this can leave out crucial information or make the formation appear to be something it isn't. Imagine for example the resampled image above was the one originally taken, this leads to an a priori prediction that future images will appear more rather than less face like. Comparing this to Figure 1 we can see that the Crowned Face passes this test, when its resolution is made lower it still looks face like and this only improves when the resolution is restored. It would be a reasonable prediction that a reimaging would appear even more face like and this succeeded with HiRise.

Figure 4

The HiRise image shown above has even more detail in it, but this needs to be left out because the photo would be too big for this book. The full photo of the King's Valley is over 500 megabytes in size compressed in JP2 format. Note the area to the left and above the left eye which became featureless and grey with the resampling earlier; this has disappeared and it looks more like it did in Figure 1. Other details however appear more artificial such as in the eyes. The nostril cavities are still there though they disappeared with the resampling earlier. We might have expected them to be clearer in this photo but they are still distinct but a lack of shadows prevents them standing out like in Figure 1. There is much more evidence of erosion hidden by the smoothing effect of lower resolution in the Mars Orbiter Camera photo.

In a future book I will be examining how this happened with the Cydonia Face, when it was reimaged many features thought to exist were found to be an optical illusion caused by the low resolution of the Viking orbiter's camera. However many other features were found that appeared even more artificial than the impression of a face in the first image.

Later the Crowned Face HiRise images are compared to the MOC images in much more detail; this can show how many features were distorted by the low resolution in the MOC image. Generally these features appear more artificial in the new image which fulfills an important prediction as they could have appeared more natural looking. The new HiRise image has a resolution of 55 centimeters a pixel compared to the older MOC image of 5.78 meters a pixel in M0203051. There is then roughly 100 times more information in the new images which is a powerful tool to validate the hypothesis of artificiality.

An important tool to determine artificiality is the a priori prediction; this means that predictions should be made at different steps of the artificiality investigation so that a successful prediction is less likely to occur by chance. It is a reasonable prediction that similar formations should be found close to the Crowned Face so requests for reimaging the whole long valley have been lodged with the HiRise team. The more similar any formations that might be found are to the Crowned Face the less likely this can occur by chance.

Another important point is that some geological features are innately random, such as sand dunes, shapes of rocks weathered by the wind, rocks that were moved in the past by ice flows, etc. If there enough of these random varying shapes it becomes very likely some will resemble almost anything, including faces. This can be seen in exaggerated shapes such as faces with missing features, too large a nose, etc. and gives rise to the Star Wars Café syndrome where too many different kinds of alien faces are found. It is generally better to avoid random geology like this; more plausible anomalies seem to be formed against the common shapes of erosion so this explanation doesn't work.

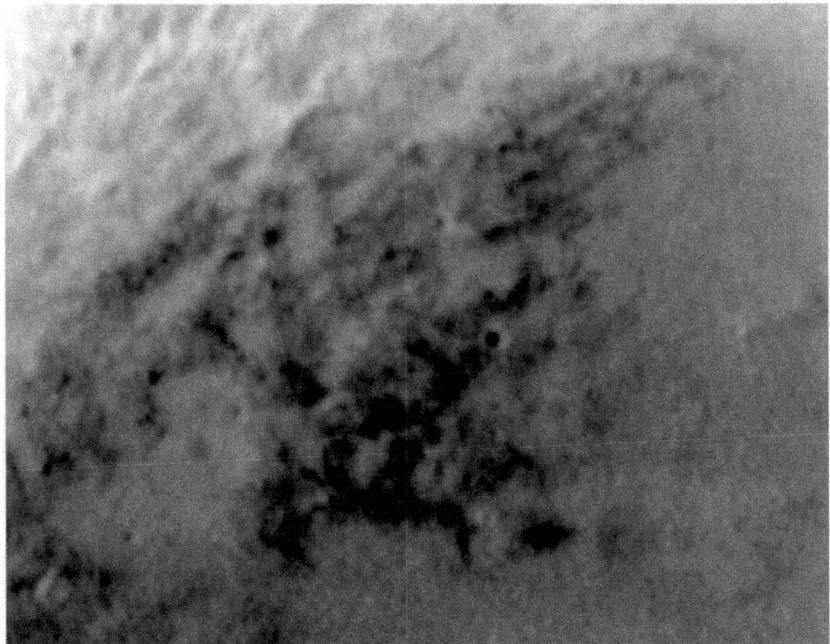

Figure 5

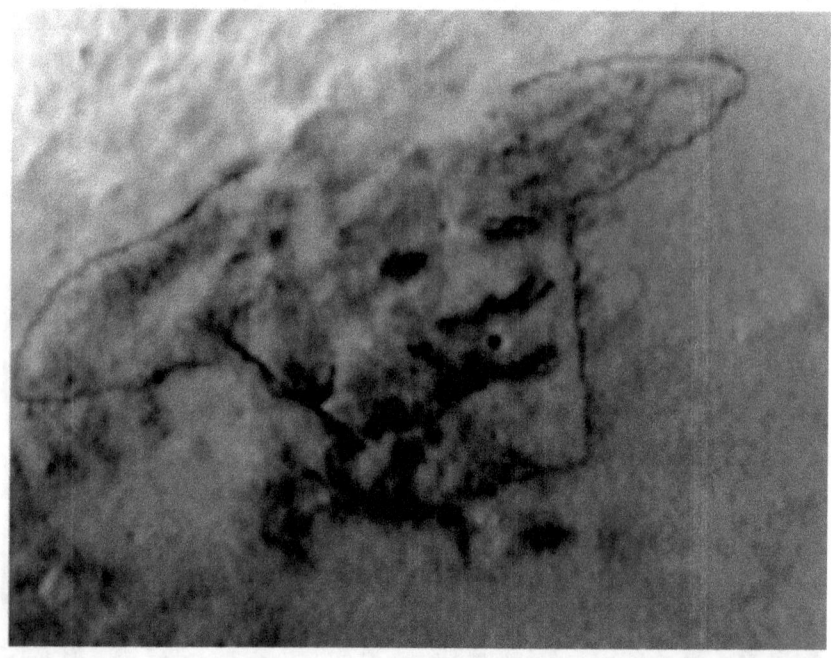

Figure 6

I called this face Pancho for somewhat obvious reasons. The original image is at this reference[ax].

The image above shows the face outlined. The problem with this face is the ground has random variations in it and dark soil can be caught against these undulations. Eventually drifts of soil like this could produce any shape. There is then no way to falsify the face because it can be completely explained by random geological processes. It may be artificial or natural; it only relies in a nebulous feeling of improbability which is of little use in proving artificiality. It is close to the old Martian equator that many of these other anomalies are clustered around which is a point in its favor but again this cannot be falsified. It has another improbable aspect in that it falls on a direct line with three other faces, namely Nefertiti, the Crowned face (the subject of this book), and the Spaceman. The face also looks nothing like the other faces which is the Star Wars café problem of how many different alien races could there have been on Mars. It may be however that the hat is different which would be a natural variation on Earth. Since it's likely Martians or aliens would have worn hats to shade themselves from the sun rather than just crowns the shape of the hat is

148

not a problem, just that there is only one face with this kind of hat. This and many other faces will be analyzed in an upcoming book.

Another anomaly I called the Martian Sphinx:

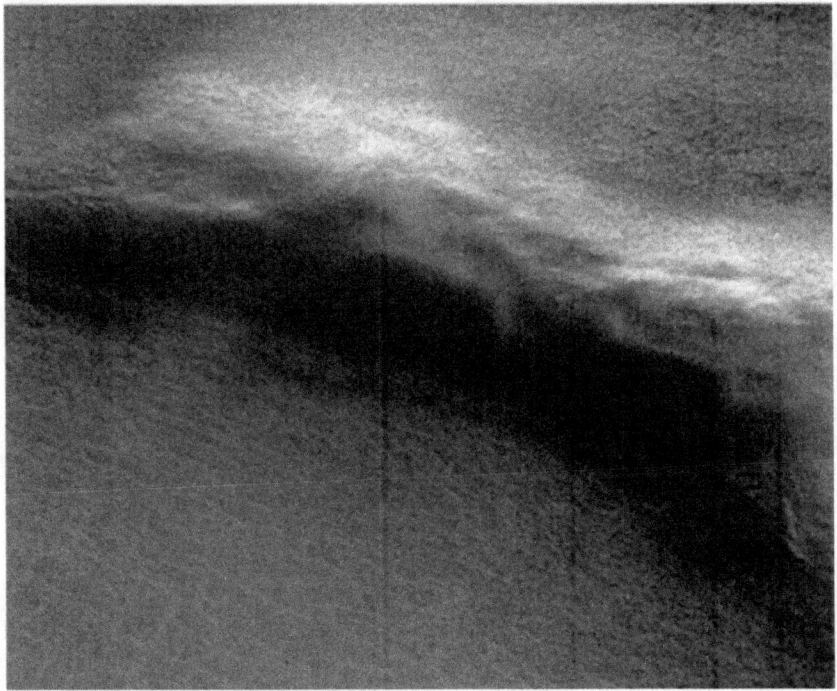

Figure 7

It appears to have a facial shape and a symmetrical body. It forms part of a crater wall in photos M0200090ᶜˣ. It is part of a crater wall, this is some evidence for it being natural but it is also common for these possible artifacts to be built on natural formations. The shape of the crater wall can account for some of its features and some random variations of so many crater walls on Mars make it hard to differentiate it from natural processes. Unusual angular shapes in the formation are shown below:

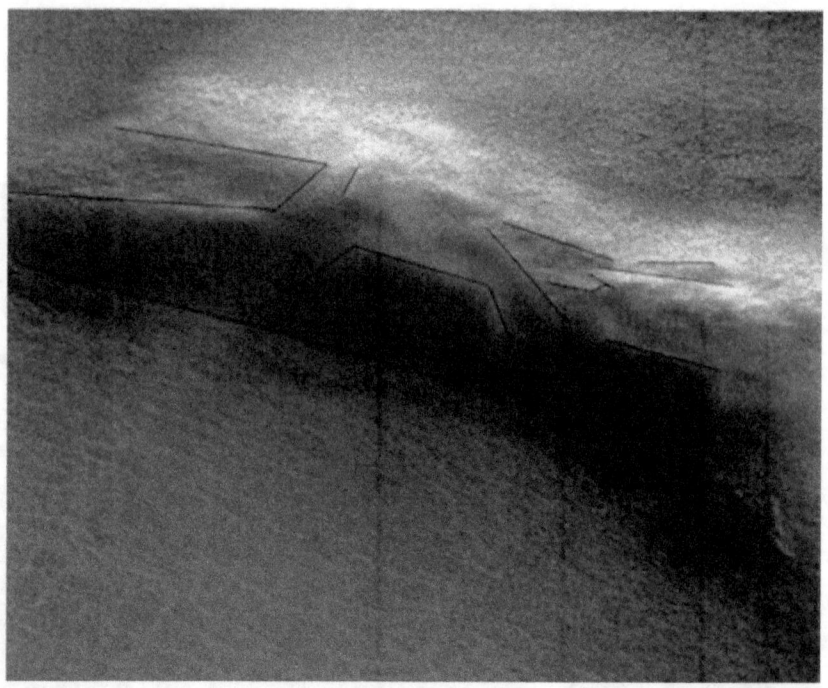

Figure 8

Arguably it should be difficult for the shock wave of a crater impact to create a symmetrical shape like this, the shock wave should have a different force on the interior of the crater wall compared to the exterior. However the randomizing force of a crater impact combined with random wind erosion might be capable of creating this. Wind erosion should however make these features more amorphous rather than accentuating sharp edges like this, as erosion changed the crater wall it perhaps should have made it less rather than more symmetrical in shape. Crater walls are generally random in shape without symmetry. Another problem is it doesn't look much like the other faces, if artificial it is the only one except for the Spacemen formations which show the body. The Spacemen are much less humanoid looking, it is possible this is related to them rather than the more humanoid ones like the Crowned Face. There is an impression on the side of the face like a helmet; this might be part of a crown that eroded away at the top. More evidence may be found if it is reimaged by HiRise.

Generally research like this should begin to find similar anomalies that start to form a coherent theory and exclude other possible theories. This has happened consistently with formations like in the King's Valley and in an upcoming book I will explain more of this emerging model. Another reason for this is in the nature of probability. There are two kinds of events in probability, independent and dependent events. An independent event is where one event has no influence on the other, for example if the wind carves a face in one area of Mars it has nothing to do with the wind carving a face in another area because the individual air molecules move randomly. In the same way clouds can form faces and all kinds of shapes but they do this from the random movements of water molecules, one face in a cloud cannot help to form another face in a different cloud elsewhere. Probability can determine to some degree with separate events are independent or have some kind of connection to each other where they are dependent events. For example if hypothetical aliens carved one face on Mars then this makes it more likely that other faces on Mars might also be artificially made because it is more likely that the creator made two or more faces than the huge odds against any alien getting to Mars in the first place. If this creator made one face or monument then it is likely that other creations would have similar designs because this is what we see with artists on Earth. Also a symbolic reason for creating one face is likely to be related to the reason for a second face. Because of this it is important to look at small parts of these features for evidence of artificiality rather than just an overall view, it may be that one small rock is in a place that could not occur naturally and from a small piece of evidence artificiality is then proven. Similarities between parts of candidate artifacts would point towards their being artificial because randomness would tend to make them different. For example the King's Valley has many faces with similarities to one another, this implies dependent events where a creator depended on a common reason to shape them the way they are.

This can be illustrated by throwing a pair of dice, throwing two sixes will occur one out every 36 throws because independent odds are multiplied together and 1/6 times 1/6 equals 1/36. In the same way the odds of dependent events often rapidly diverge from independent ones because multiplying odds together can quickly give huge odds against chance causing some events. In the case of artifacts on Mars the odds are difficult to calculate because we don't really know how likely it is for the wind and ancient rains to carve a face. For similar faces to appear means that the odds against chance multiply together and can become astronomical in size. There are 24 faces in the King's Valley more or less depending on where the line is drawn and how complete a face is, one main face in Meridiani, one in Cydonia, and 4 in the area of the Nefertiti image. This gives 30 faces with some similarities to each other, there are other faces which will be analyzed in an upcoming book. Depending on the similarities between them in style and construction the odds against these common themes multiply together to diverge rapidly from random events if indeed some are artificial. If each Face was 10 to 1 against occurring randomly then the odds against chance would be multiplied together to give 10^{30} or 1,000,000,000,000,000,000,000,000,000,000 to 1 which should be enough to take seriously as a candidate for artificiality. If each face had for example 5 features in common with the others and each was also 10 to 1 against occurring ransom then this would be 10 the power of 31 times 5 or 10^{165}, a 1 followed by 165 zeroes to 1 against chance. So by examining the features of these faces carefully for common features a strong argument for artificiality can be made. These odds need to be weighed against the total number of images; usually the Mars Orbiter Camera images are sufficient for this analysis. There are more than 100,000 of these narrow angle images. This however is a much smaller number than the other odds mentioned. With so many faces to compare it is still likely that the odds against chance will be highly significant even though there are 100,000 images for something odd to occur in. For example if you looked at 100,000 clouds the chances are you would see some strange shapes but it is unlikely you would see 30 faces with common features over that time, though it is very difficult to defend this idea without precise mathematics. It is better to make some attempt at working out the odds against chance rather than just relying on the familiarity of a face compared to our own or the potential impossibility for a feature to be formed naturally. The odds above are just an example but more detailed analysis of the odds will be made later in this book.

Chapter Nine

Controls

These are some images from other areas near the King's Valley for comparison, to see what formations that are not artificial look like. To prove a formation is artificial it is necessary to eliminate the known geological processes that might have created it. To see what these processes typically do it is useful to look at areas nearby and see how the Aeolian and fluvial erosion changed the formations there.

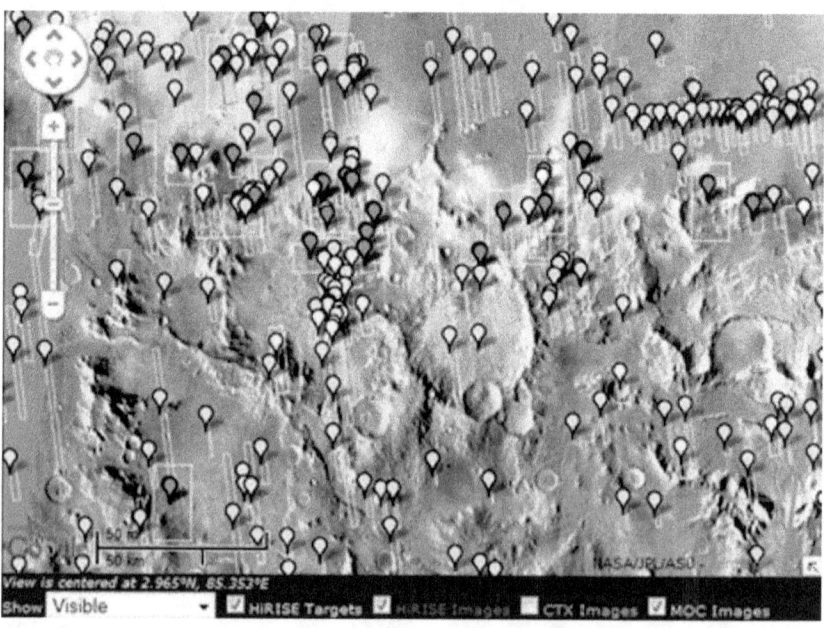

Figure 9

These are photos in the King's Valley area, the white pins are suggestions made by various researchers to examine Martian geology while some are to look for artifacts. The dark pins are HiRise and Mars Orbiter Camera images. The smoother terrain at the top is Isidis Planitia, an old impact crater.

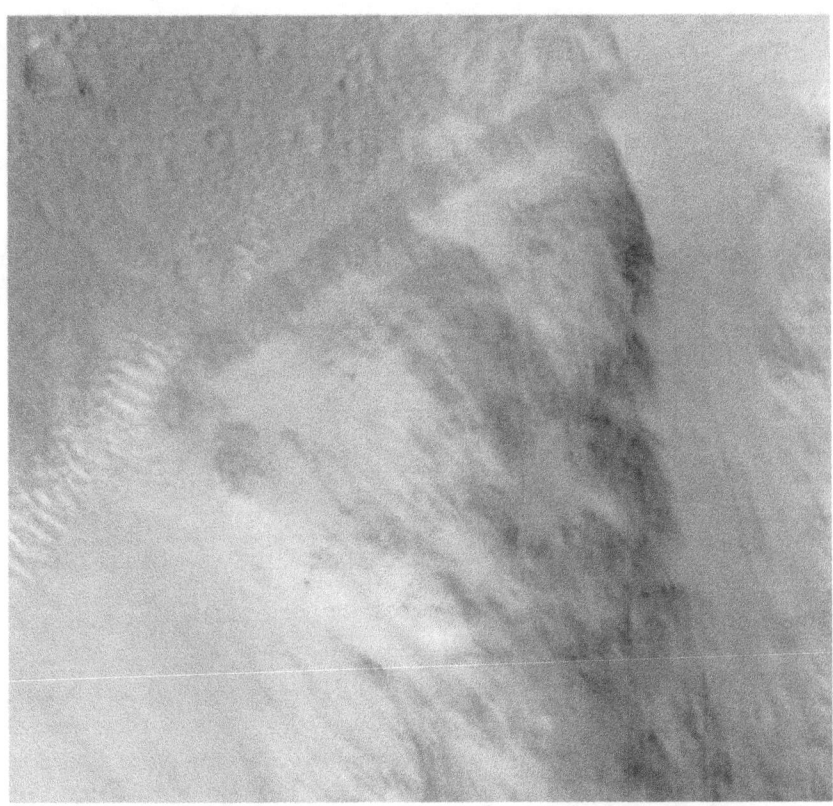

Figure 10

This image is of *Terraces at Base of Massifs and Talus in Southern Isidis Region Rim (ESP_012289_1840). This has an approximate shape like a face and the edge of the cliff looks slightly like a crown, but the features seem completely random like a natural formation.*

Figure 11

Libya Montes (ESP_016522_1835) This is a color image from HiRise, it shows a cliff face similar to where the faces are in the King's Valley but there are no face like features except for the top of the Crowned Face where it meets the cliff. It would have been relatively easy then to create the Crown on the faces using the natural terrain.

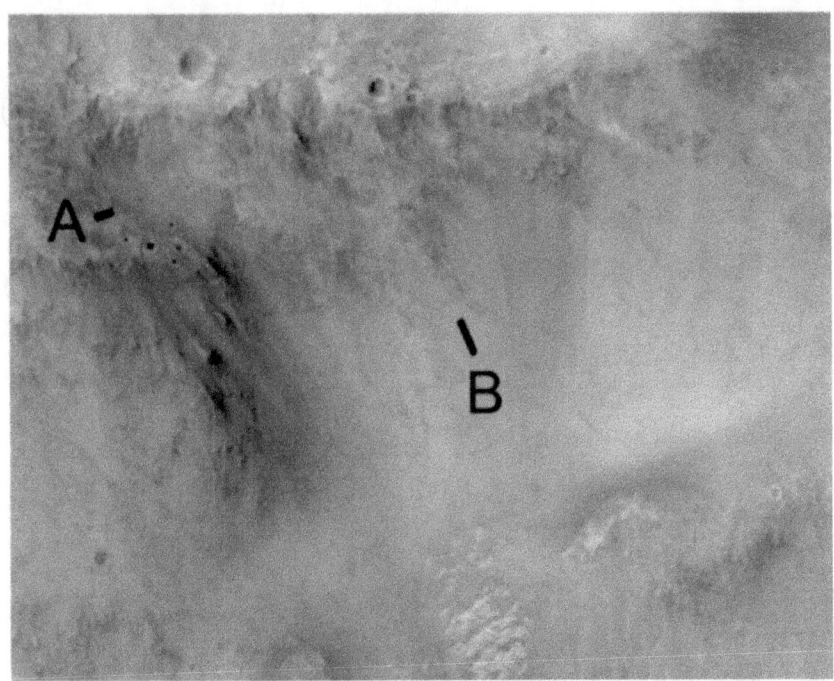

Figure 12

*Carbonates in Libya Montes (ESP_016034_1835) **A** shows a small face shape, it is little more than a triangle partly formed by the outcrop under **A**, it appears to be natural. **B** has a ridge which could be used for a jawline and would have been a good candidate for alteration to make a face.*

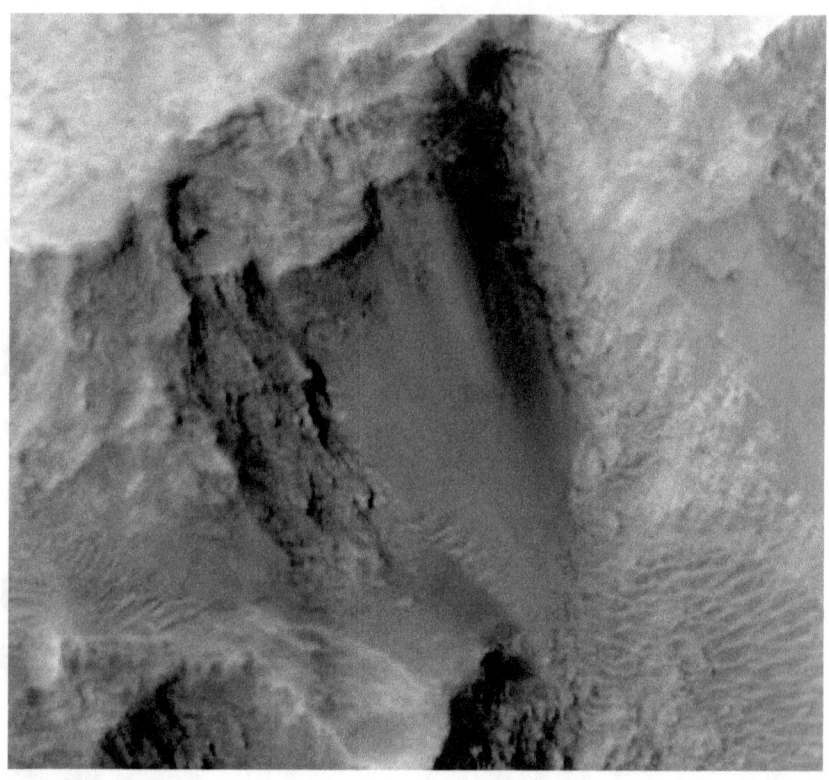

Figure 13

Possible Olivine-Rich Terrain (ESP_017445_1835) Another triangular shape formed by an outcrop of rock, this also has no facial features. Interestingly this area near the King's Valley has both olivine and carbonate deposits which may be associated with previous life on Mars.

Figure 14

Valley in Libya Montes (ESP_018223_1830) This is in the King's Valley near to the faces but contains nothing that looks artificial, unless something is buried under the sand.

Chapter Ten

The King's Valley, Mount Rushmore on Mars

The idea of Mount Rushmore is a fairly obvious comparison because on Earth we created monuments of faces on a cliff. The notion of aliens or indigenous Martians doing the same thing, while unproven becomes easier to understand.

Figure 1

Mount Rushmore. If the Crowned Faces were constructed then the difference in materials in their makeup should be apparent when visited by a robotic lander or manned expedition.

We can imagine many reasons why aliens might want to visit our solar system; they would be similar to why mankind explored other continents on Earth by ship, to look for places to colonize, to look for raw materials, to compete against other explorers or Empires, to expand an Empire by creating military bases, etc. As part of this exploration icons of various people were constructed as statues such as of religious figures with Jesus, military figures such as George Washington, royalty such as Queen Victoria, or of explorers themselves such as Sir Francis Drake.

Figure 2

Sir Francis Drake, the faces on Mars may represent an explorer, leader, or Deity.

It seems logical then that hypothetical alien explorers would build Faces on Mars for similar reasons, there may be statues on Mars as well or they could have been eroded away because they would be standing upright into the wind for perhaps hundreds of millions of years. They may also have been similar to what we used on the Pioneer 10 probe and not meant to be of any particular individual.

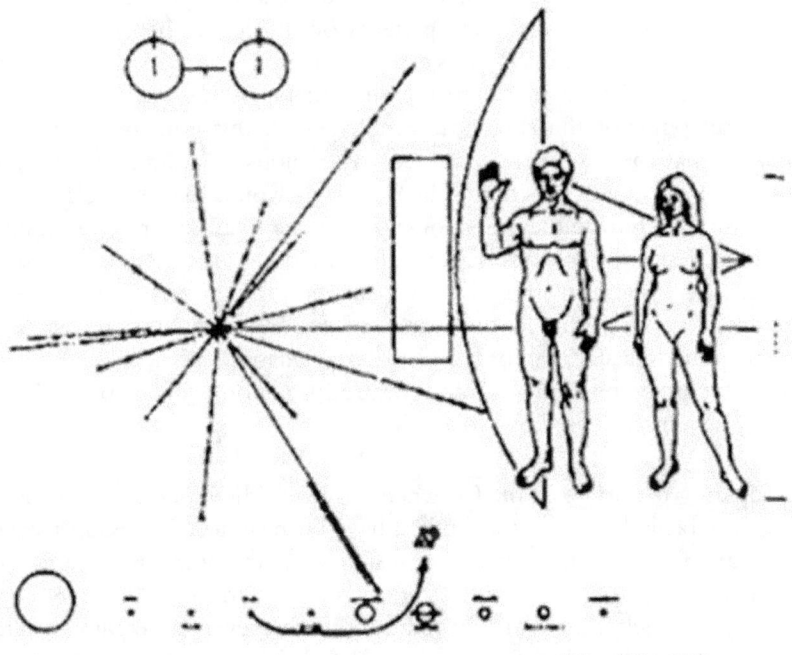

Figure 3

The faces on Mars may be intended to be a generic version of the visitors and not a particular individual, just as we used on Pioneer 10 to send a message if aliens found it. If we considered it likely enough that aliens might find Pioneer 10 when it could not possibly get far from our own solar system then it is also possible aliens might come closer and visit our own solar system.

This image is a good starting point because this is what we actually constructed in case any aliens found the probe. There was then a consensus of scientific experts at the time as to what information could be conveyed, one part of which was the human form and faces. It is not too surprising for hypothetical aliens or indigenous Martians to follow the same reasoning and use faces on Mars or the Moon to deliver a message or greetings. Also included on the Pioneer 10 Plaque were mathematical shapes based on the idea that these would form a kind of universal language or at least be recognized by anyone likely to find the probe. As we will see in an upcoming book there are many examples on Mars of possible mathematical symbols, configurations of mounds in complex mathematical relationships perhaps designed to convey a message. To prove these mounds are not just randomly arrayed in mathematical theorems is difficult to do but various statistical methods can be used to calculate the odds against such patterns occurring by chance. The other part of the Pioneer message was a representation of our solar system and the distances of each planet to the sun shown by the lines joining to a common point of the sun. One problem with this kind of analysis on Mars is that the stars from hundreds of millions of years ago may have moved from their positions then and it would be difficult if not impossible to retrace their positions to compare them to arrays of mounds to see if they are star maps. Even if this could be done there are so many patterns of stars in the sky that almost any pattern might occur somewhere. Interestingly one array of mounds was found on Mars conforming to a star map of the Orion constellation, this will be discussed in an upcoming book. The Amphitheatre formation will also be analyzed; it may have represented our solar system.

Comparing the photo of the Crowned Face with the Pioneer image above, the ideas behind each message might have been similar. The King's Valley has many faces in it in various states of erosion and with various amounts of plausibility. At some point this analysis can go too far and be trying to tease too much detail out of random rocks, even a legitimate alien monument could still have coincidental face like arrangements of rocks near it so some or even all of these formations may eventually turn out to be naturally formed and not artificial. Nothing is certain however there are so many interesting formations in this Valley that they all deserve a fair explanation and analysis.

Some artifacts can be small which raises the problem of pixilation distorting their shape. When a formation is small in a photo the pixels on the screen and the pixels in the original photo are large by comparison and these change its shape to make it appear to be something it may not really be. For example if small enough the right eye of the Crowned Face might only have a few pixels in it and so there is not enough information to examine it accurately.

Figure 4

This is an array of pixels which makes up the right eye of the Crowned Face. What appears so eye like at smaller resolution looks very blocky and is seen to be made up of very few pixels. The other images on this page are also made of pixels like this but because there are more pixels the images are clearer. The higher the resolution of the image the more pixels there are, and therefore the more detail can be seen. When I say the right eye in an image I mean the right hand side of the face according to our viewpoint, not the right eye of the face from its own viewpoint.

The main iris in this image is a group of about seven grey pixels which in the image we interpret as a smooth oval shape but as we can see here the actual iris is quite angular. This is even more of a problem when looking for buildings on Mars because right angled shapes can be caused by the angular shape of the pixels themselves. A sign of this can be where lines and shapes in a Mars photo are parallel to or at right angles to these pixel lines; this might mean these pixel lines are creating the impression of lines and right angles in the photo.

Another problem that occurs is that compression using the JPEG standard creates shapes in the photo which can look like alien artifacts or even parts of a face. For example if I save this part of the King's Valley at high compression it makes it look quite different:

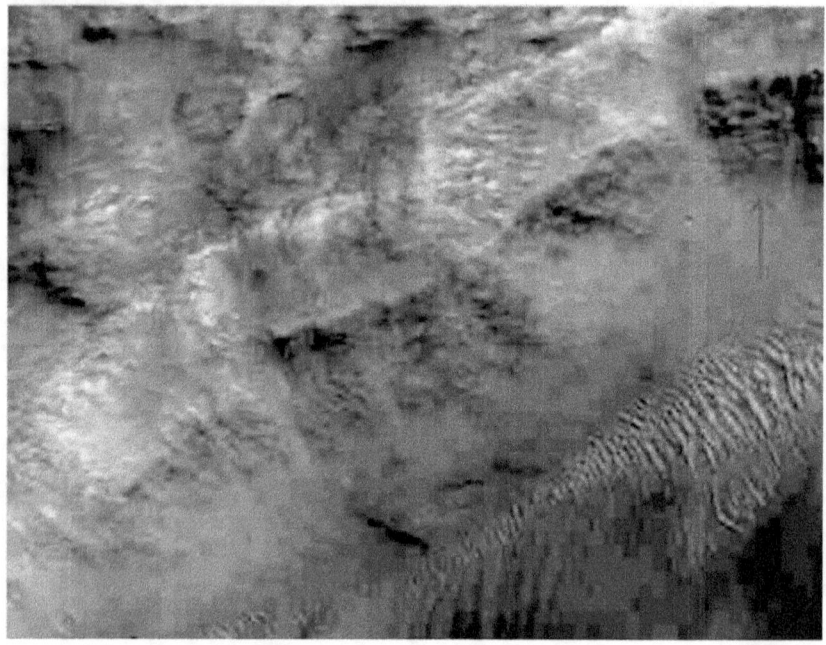

Figure 5

If the original MOC image of this area had been highly compressed like this then the Faces might not have seemed as credible. Compression makes images smaller and easier to transmit from Mars back to Earth. The kind of image compression is then important to evaluate in this kind of research.

The blockiness in this image is caused by the JPEG algorithm trying to make the picture smaller to save space; many people would have seen this effect when taking photos at low resolution on a mobile phone or digital camera. Malin Space Science Systems also used JPEG compression in their images so when downloading a JPEG version there are subtle distortions like this. They also provide a GIF version of the image which uses lossless compression; this means that no features are distorted by this compression. The pixels themselves however, like the squares in the eye image in Figure 4 already introduce some distortion. Someone seeing the image above might think these squares are signs of an artificial construction because we expect to see engineers on Earth making buildings with right angled corners. Note also the left eye is obscured in this image and the right eye looks different in Figure 5, so generally the higher resolution the smaller the pixels are and the less compression used the more accurate the features we see are. The new HiRise images use wavelet compression in the JP2 format, this introduces less distortion to the images but it is still there.

This can also be tested with the Mars Orbiter Camera images first taken of the King's Valley because the area has been partially reimaged with much higher resolution by HiRise. We can then tell with smaller details in much of this image whether the faces we see are real or caused by the low quality of the images. For example in Figure 5 the compression made some parts of the photo look more angular and artificial but it made the faces less face like.

Chapter Eleven

Twenty Four Faces in the King's Valley

In this chapter some of the main formations in the King's Valley are shown, and a classification scheme is defined so eventually each possible artifact or part of one will have a designation. The three main Crowned Faces are shown below as, Face One, Face Two, and Face Three.

Face One

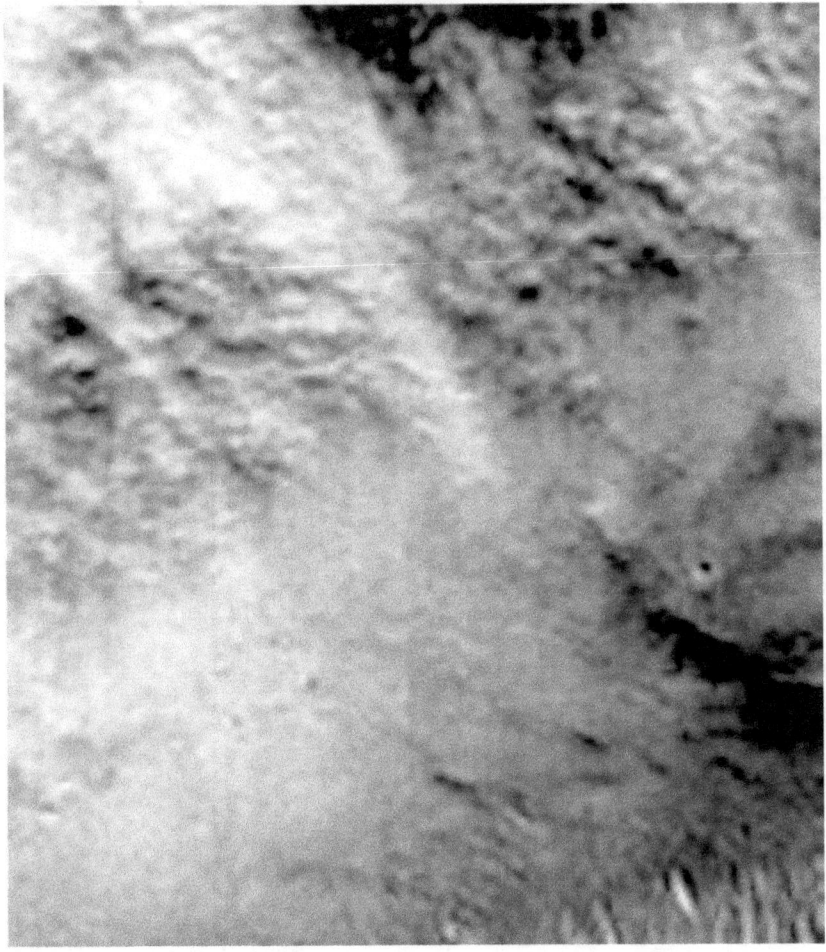

Figure 6

This is Face One. It is highly eroded but as will be shown later in this book it has many similar features to the Crowned Face. The latest HiRise image managed to get part of this Face.

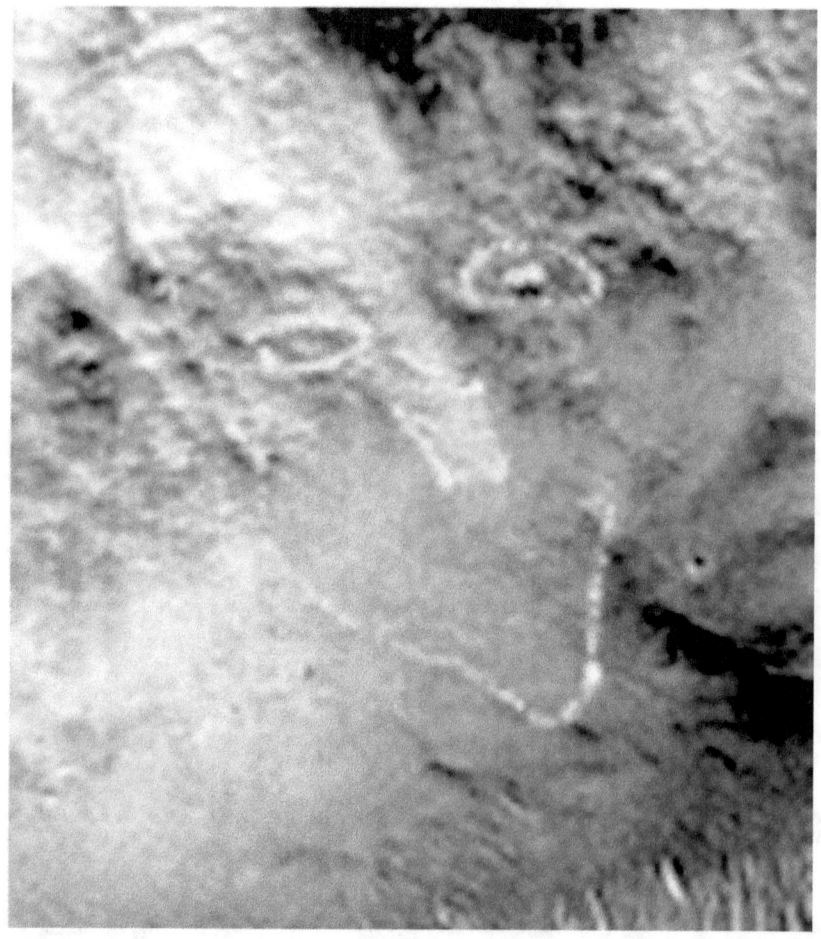

Figure 7

This is an outline of Face One.

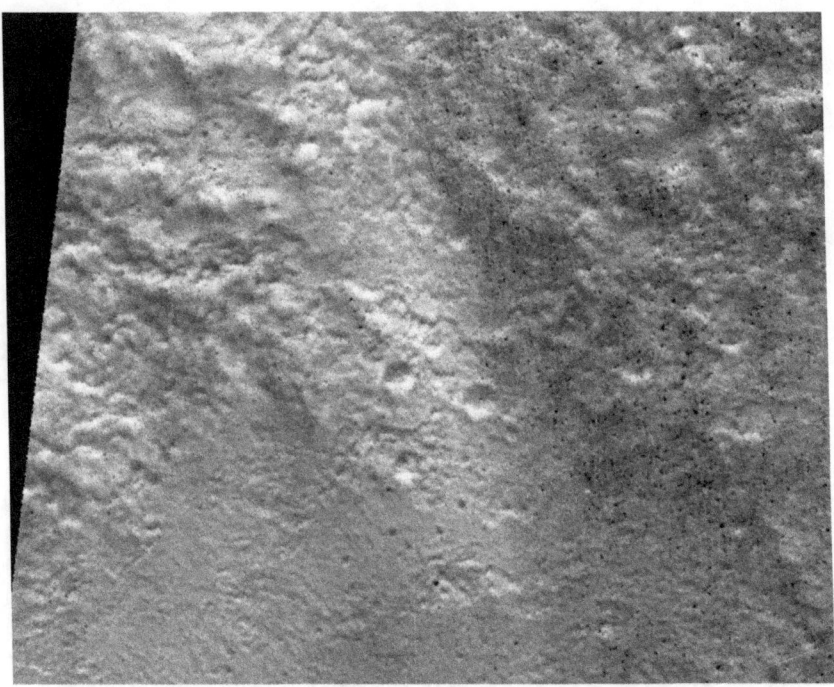

Figure 8

The extra detail shows much more erosion and damage from meteor impacts but the lack of shadows also makes the features harder to see. This indicates the Faces are very old because most of the features have meteor impacts on them.

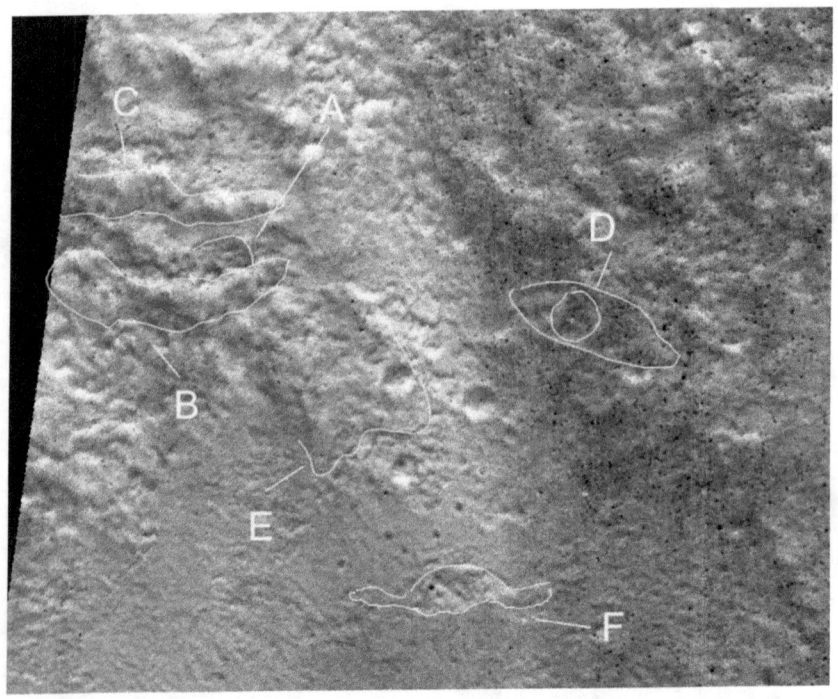

Figure 9

*There is a rounded iris shape at **A** which represents a successful a priori prediction. **B** and **C** show a similar kind of eyelid to the right eye of Face Three just to the right of the Crowned Face. This is also a successful prediction because the more similarities there are between the Faces the less likely this can occur through random chance. **D** shows an eye shape with an iris, like with the Crowned Face it seems slightly out of place. This is another successful prediction because this was not seen in the Mars Orbiter Camera image. **E** shows a nose shape which seems broken off, another successful prediction because the lack of a nose tip can only be explained by it being buried or broken. It also implies the nose is of a material like clay or concrete added onto the rock, and which broke off. An expedition could test the edge of this break to see if it is added onto the rock underneath. **F** is a shape like the nose tip in the right position and the same shape as the nose tip of the Crowned Face. There is also a similar shape just to the left of the nose tip which may be the real tip, this would give the nose more of a turn to the left like with the Crowned Face.*

Face Two

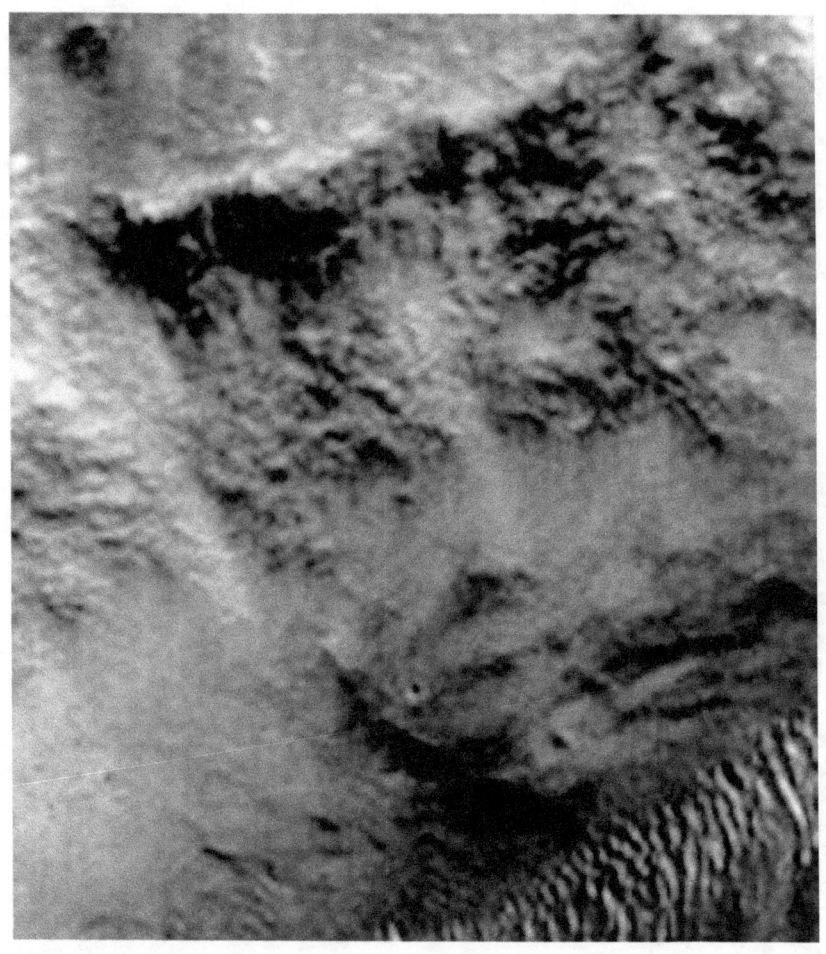

Figure 10

Face Two, the Crowned Face. It was recently reimaged by HiRise, a lot of features are now much clearer, looking more artificial.

Below some of the features and defects of the Crowned Face are shown.

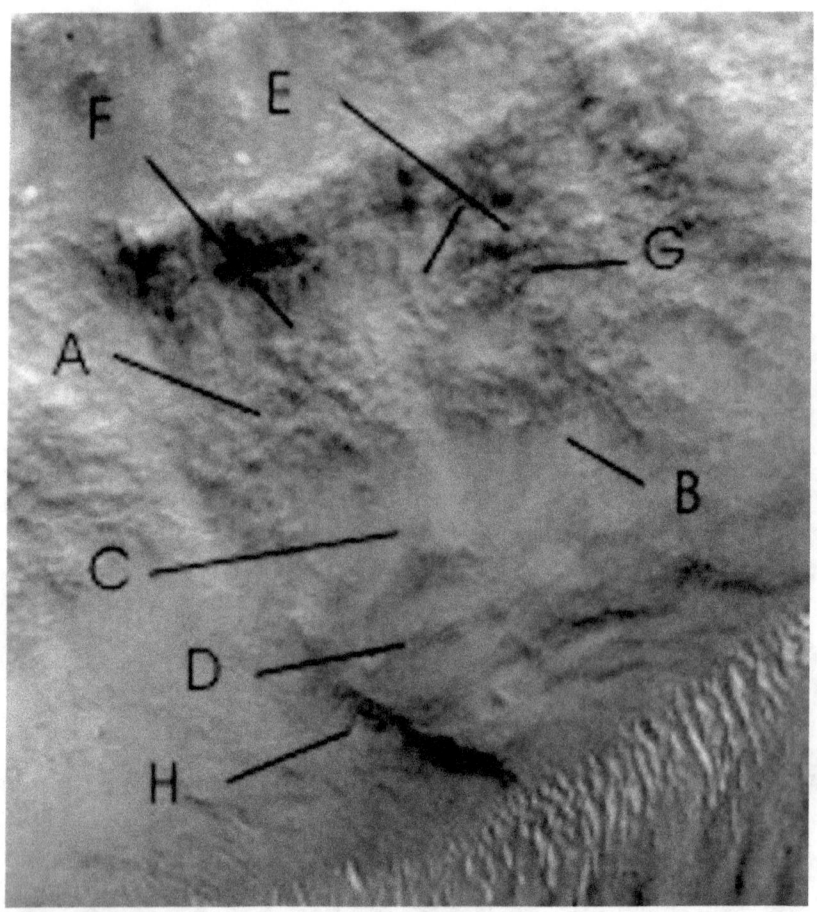

Figure 11

A shows the left eye of the Crowned Face (also called Face Two), this eye is quite well shaped. **B** shows the right eye which is more oddly shaped. The left eye appears to have an iris in the eye but the right eye has an empty shape in the iris. Also the shape of the right eye seems more similar to a left eye in orientation, and as we will see later there is a second face to the right this eye can also be used for. It may be then that this eye was made to be appearing as one eye shape with certain shadows and another eye shape with other shadows.

If so then one face may have appeared more in summer or perhaps in the morning for example. **C** shows the end of the nose in which a tip and two nostril shapes are clearly seen. There is also a nose like ridge connecting the nose tip to half way between the eyes, though the nose seems to curve

174

to the left side rather than being symmetrical. **D** shows the mouth shape which is more indistinct but is very close to the correct anatomical position. **E** shows a line consistent with the bottom edge of a crown, which fits in with the idea of a hat. The crown shape is reasonably symmetrical. **F** shows an eyebrow like shape that is not very accurate anatomically. **G** shows the right eyebrow which is also not very accurate. **H** shows the left side jaw line which is quite clear and accurate except it goes more to a point, though this could be a beard shape. The right side has no clear chin shape, though this could be from erosion or having to connect to the face to its right.

Overall then the main Crowned Face has some features which give it a striking humanoid appearance, while others are out of shape or missing altogether. Since it is likely a builder was skilled enough and acquainted with humanoid faces there seems little excuse for anatomical defects. There should then either be a reason for these defects, they could have occurred through erosion, or the subject of the design was not fully humanoid as we would define it.

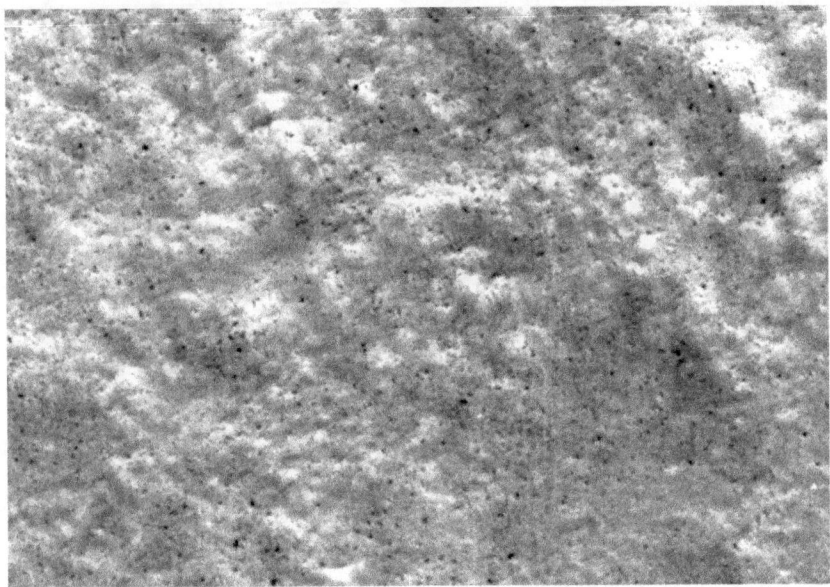

Figure 12

This is the left eye of the Crowned Face, Face Two. The high amount of erosion makes seeing details more difficult.

Figure 13

I have drawn an outline of the eye; there is little more detail than from the Mars orbiter Camera image. There is a shape like a tear gland in the left and right corners of the eye. This is not a human feature as we only have one tear gland per eye.

Figure 14

This is the right eye of the Crowned Face; it has much less erosion perhaps because the sand which damaged part of Face One also damaged the left eye of Face Two.

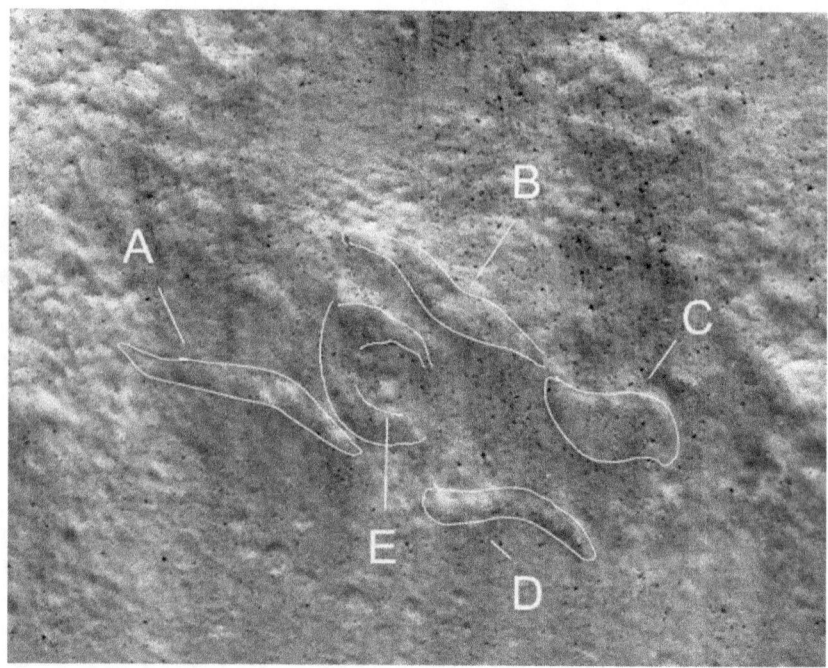

Figure 15

A, B and D show similar folds around the eye to that of Face One, this is not a human eye but may be amphibious if these folds would assist in closing the eye against water. C shows another possible tear gland because this eye also acts as the left eye for Face Three, it doesn't look human. E shows a round iris with a rounded pupil perhaps designed to stand out with shadows.

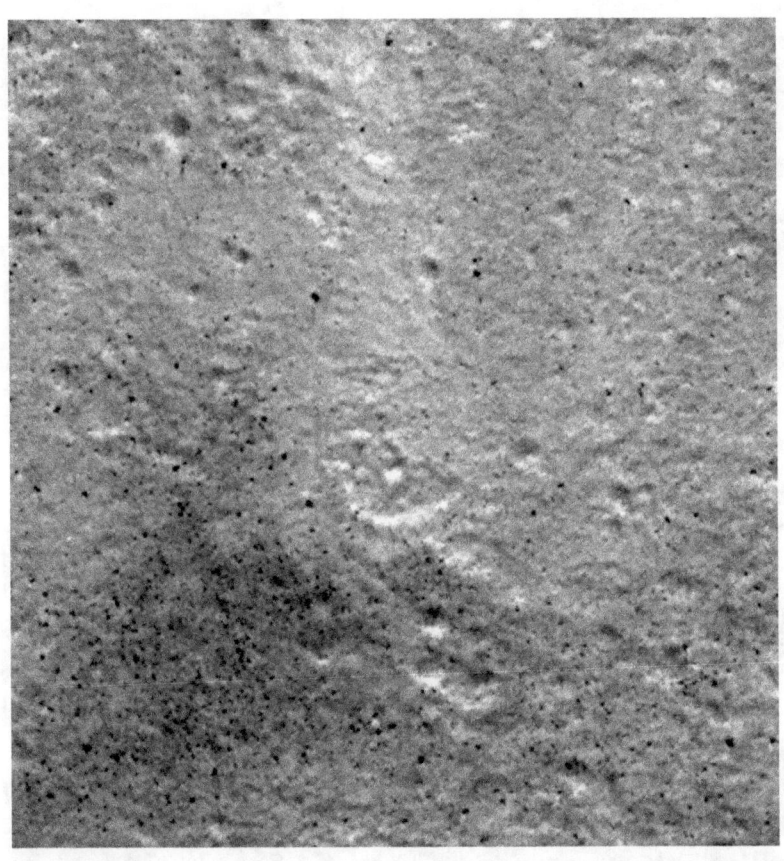

Figure 16

This is the nose of the Crowned Face, Face Two.

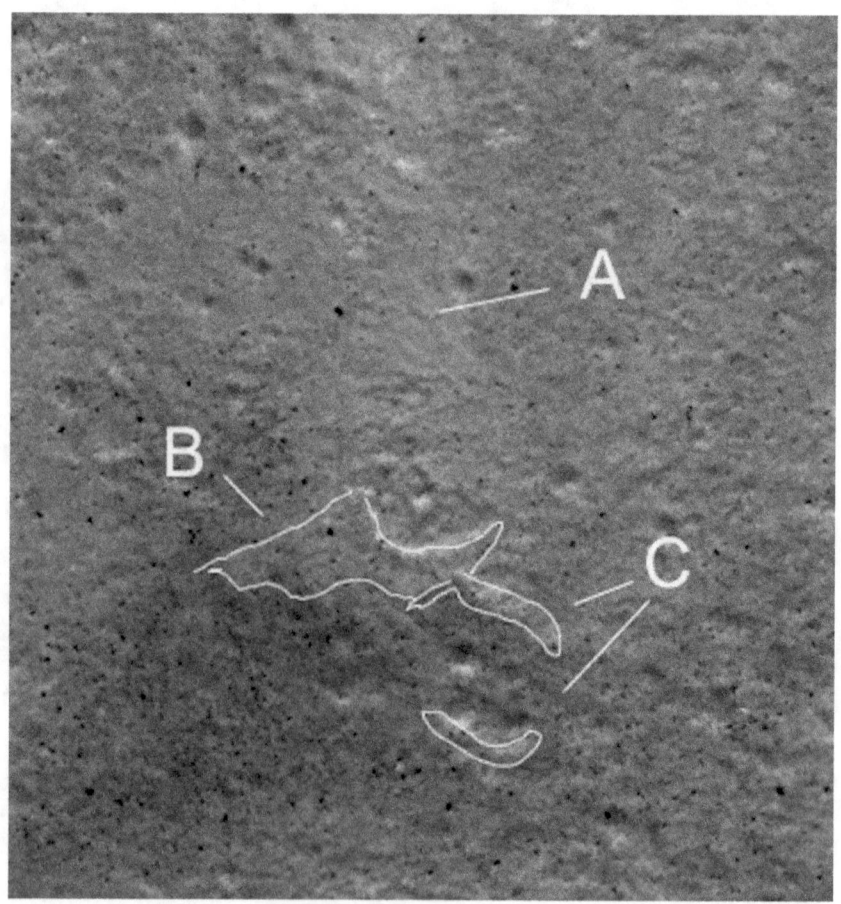

Figure 17

A shows a lighter area where the bridge of the nose is. **B** shows a nose tip shape like the eroded one on Face One and as we shall see later like the nose tip on Face Three to the right. **C** shows the edges of the right nostril which may define its shape with shadows. The left nostril is in darkness and has no details visible.

Figure 18

This shows the mouth and jawline of the Crowned Face. While there is less erosion the lack of shadows makes the features hard to see.

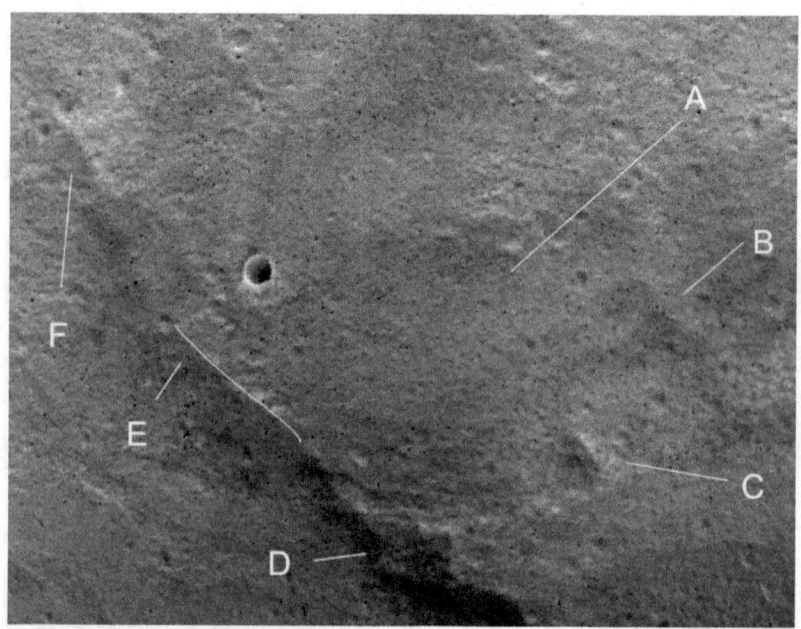

Figure 19

A shows where the mouth would be, without shadows there are few details. It may be they did not have teeth like ours; there are no indications of them in any of the Faces. B is a vertical ridge seen in the Mars Orbiter Camera photos; it is harder to see with less shadow. C shows the other vertical ridge, perhaps the edge of Face Three. D shows an artificial looking rectangular shape. E shows the left jawline. F shows an upside down U shape which looks more artificial in the Mars Orbiter Camera photos; pixilation and compression may have enhanced its shape.

Face Three

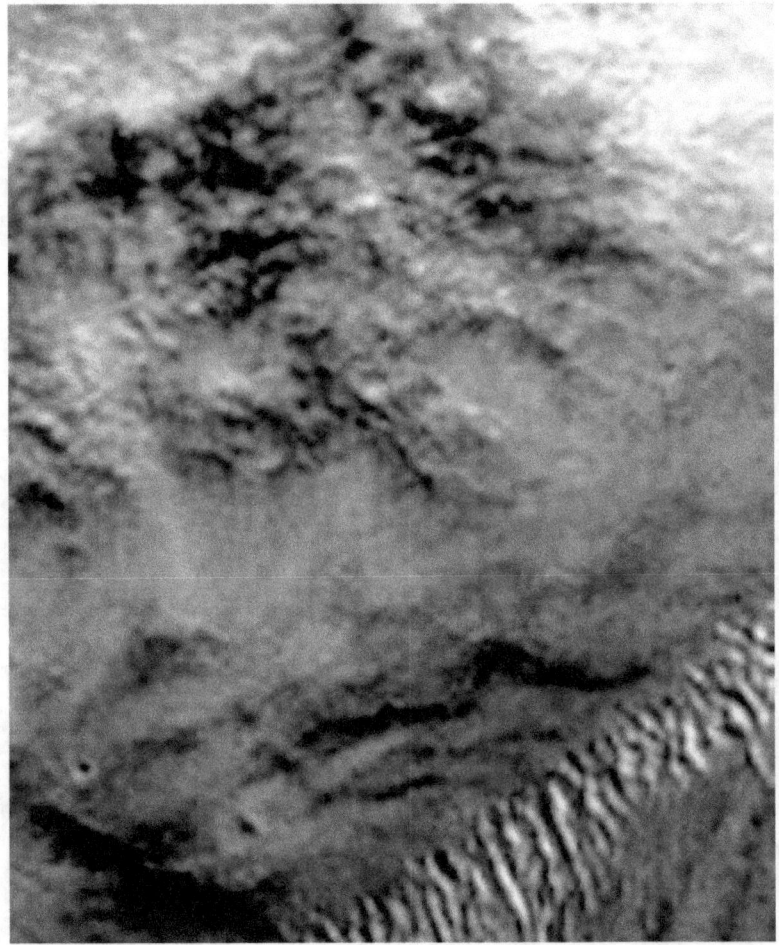

Figure 20

This is Face Three.

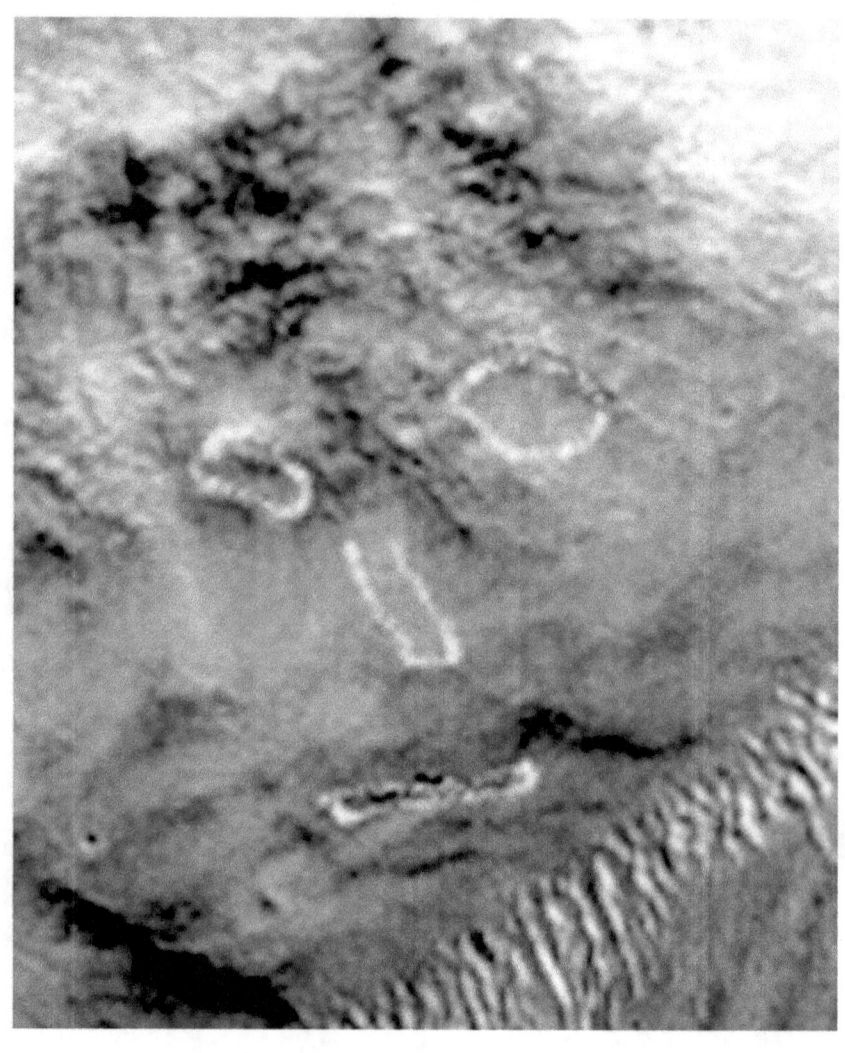

Figure 21

Face Three outlined. The left eye of this Face is the same as the right eye of the Crowned Face so it need not be shown again.

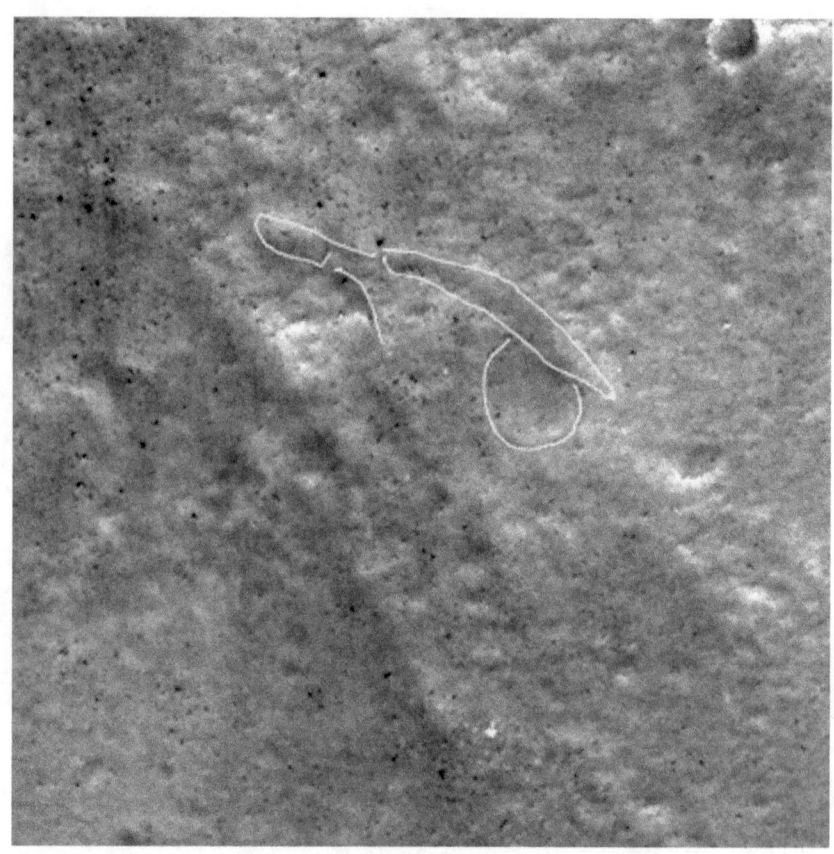

Figure 23

There is a rounded iris here and an upper eyelid with a tear gland shape to the left. Much of the eye may have been damaged then the area to the right eroded away leaving a cavity.

Figure 24

The lack of shadows again gives little detail. This is not necessarily against the idea of artificiality as many more details have already been confirmed with the HiRise image such as in the eyes of Faces One, Two and Three.

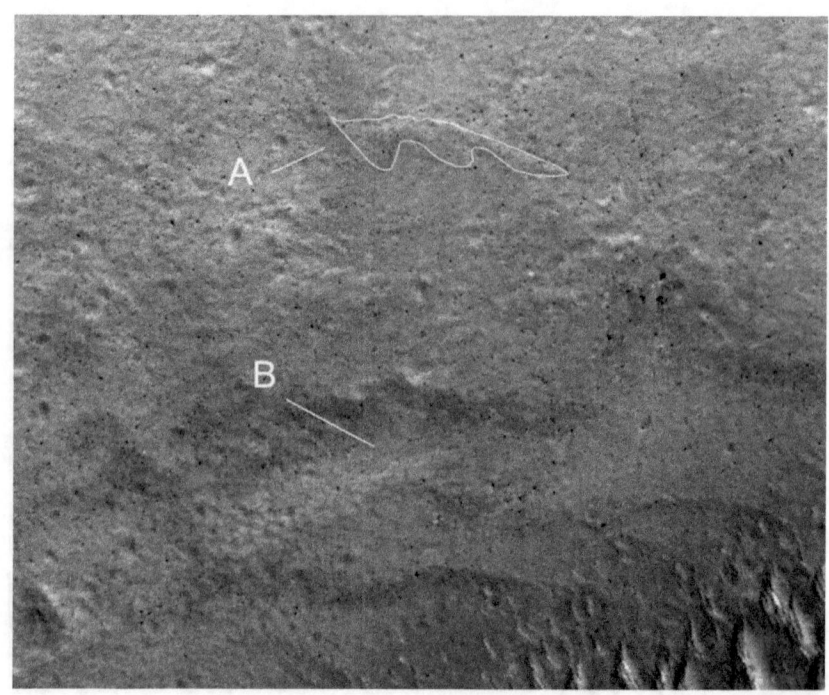

Figure 25

*A shows the shape of the nose tip which is similar to that of Faces one and Two, it has two notches in it to give shadows for the nostrils. **B** shows the mouth cavity, again with no sign of teeth.*

Below the main area of interest in photo M0203051 is seen.

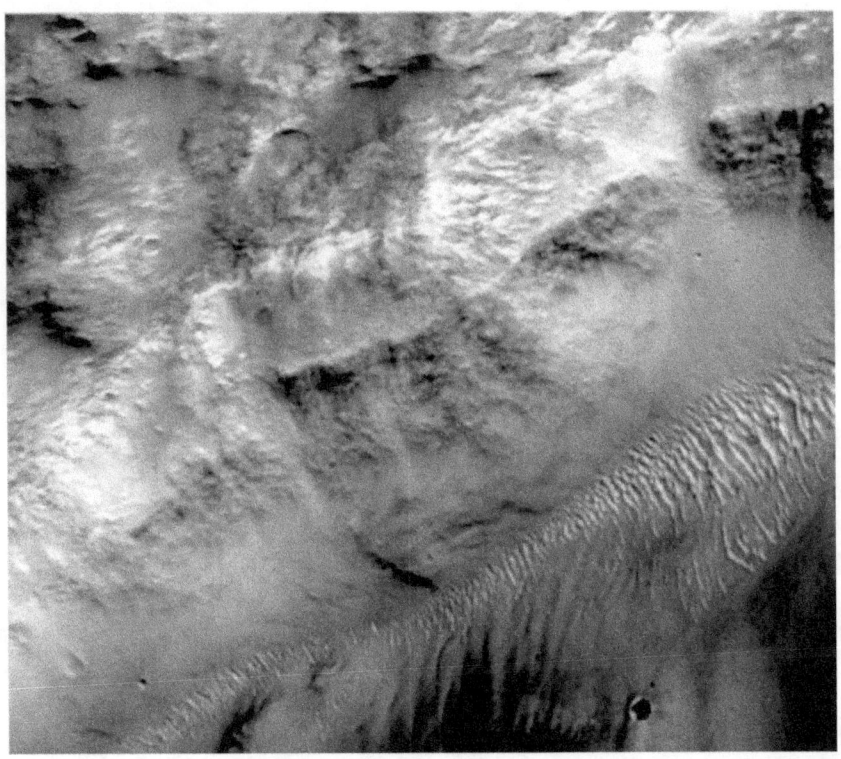

Figure 26

This gives a wide angle view of the King's Valley.

Face Four

Below shows a Crown like shape to the left of the King Face. Provisionally I call this Face Four assuming it is the top of another face.

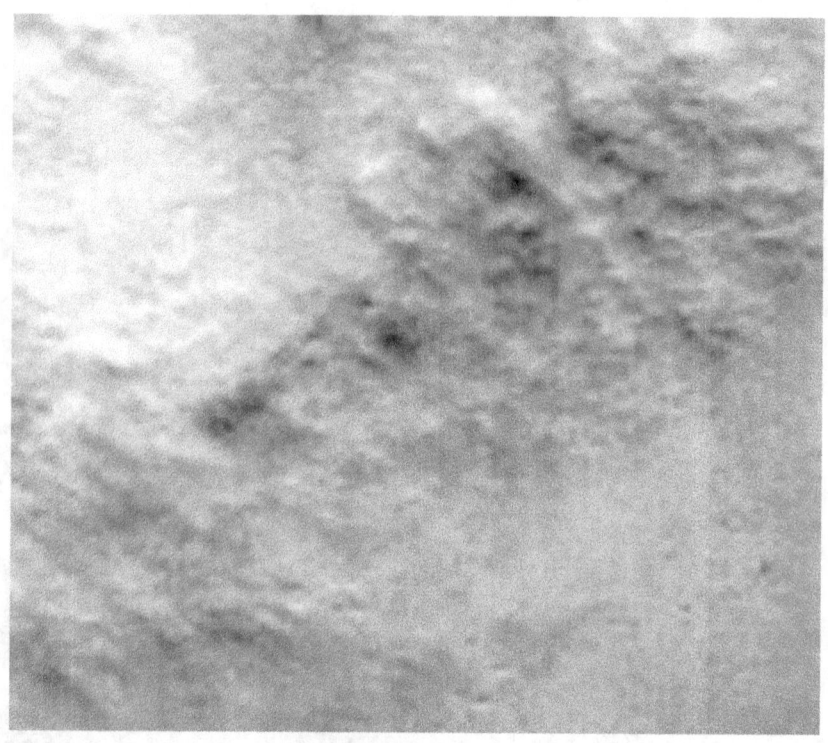

Figure 27

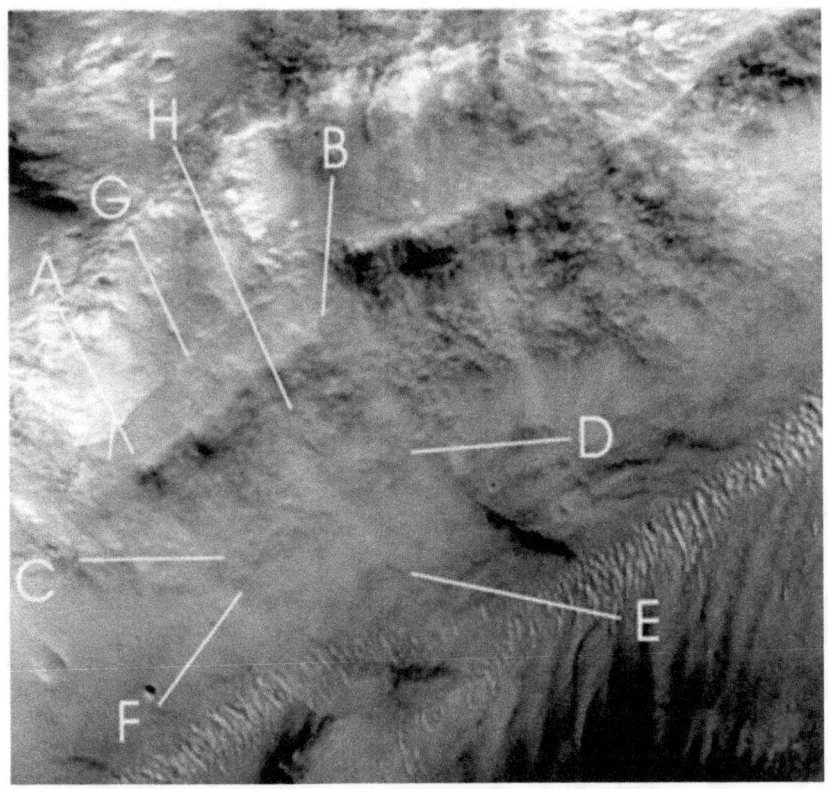

Figure 28

Here the Crowned Face is overlaid onto the top of Face Four to show its crown width is the same; there may be more features of a face buried here.

The corner **A** is aligned with the two formations. The right prong of **A** points to this lineup and the left prong points at a line which matches on the two faces. **B** shows the other edge of the crown and how it matches closely. This is from rotating the image not resizing it to fit, so this width is the same on both Faces. **C** shows an area on the Crowned Face that also serves as part of Face One. **D** shows the right hand edge of both faces aligns closely. **E** shows a ridge on both faces aligns. When overlaid this appears to be coming out of the nostril. **F** is like an ear area that matches in both faces. **G** shows a central crown like feature which sits symmetrically in the center of the King Face crown. **H** shows another area which matches both faces.

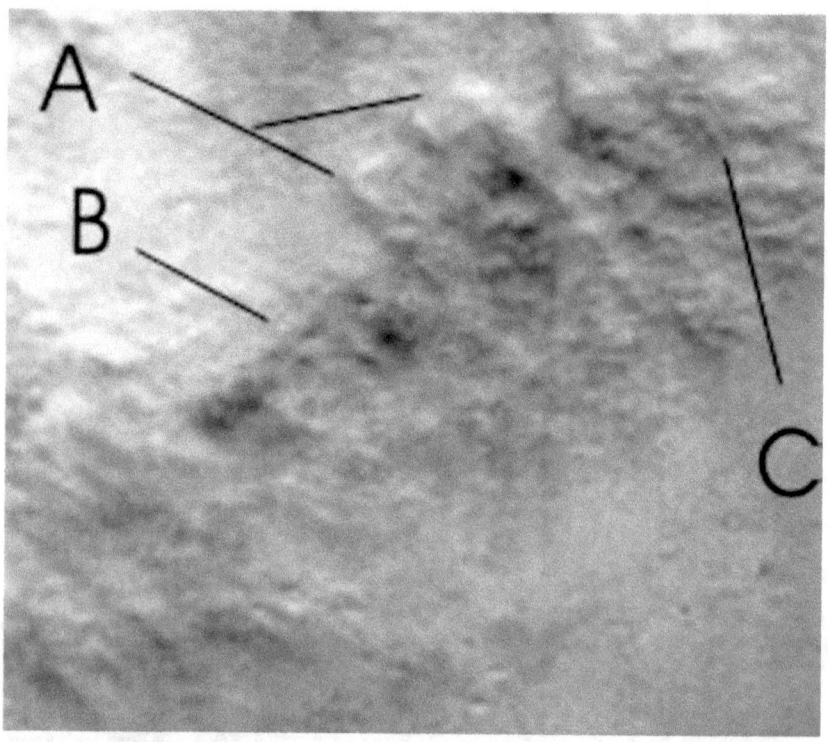

Figure 29

In the above photo **A** shows the top of two apparent ridges that come from the top of the crown and point in at each other. **B** shows the edge of a crown line which continues on to the right. **C** shows a shape which may be part of the crown, like an animal or snake head. The ground below this is very smooth and it may be buried under sand. It is possible then, with no way to tell, that there may be a face connected to the crown buried here.

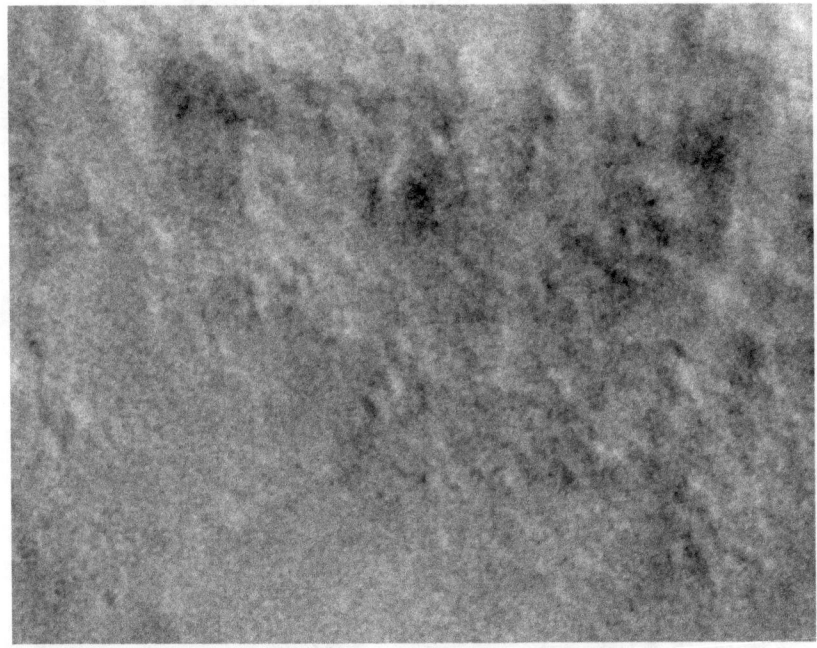

Figure 30

In photo S1402499 this area was reimaged shown above, this may make it a Dual Face with Face One to its right. The crown shape is more separate here with most of the possible Face again buried in sand.

Face Five

To the right of Face Three there is another possible face looking to the right, called Face Five.

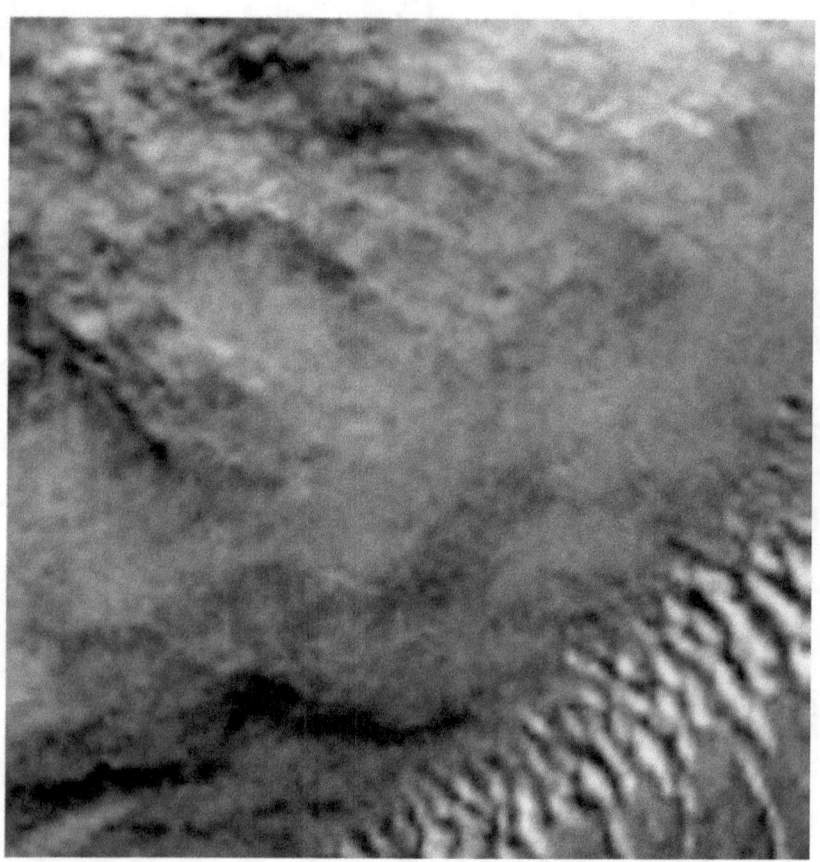

Figure 31

I saw this Face before I noticed the Profile Crowned Face mentioned later, it may mean that this Face doesn't exist separately but the area needs to be examined carefully to be sure.

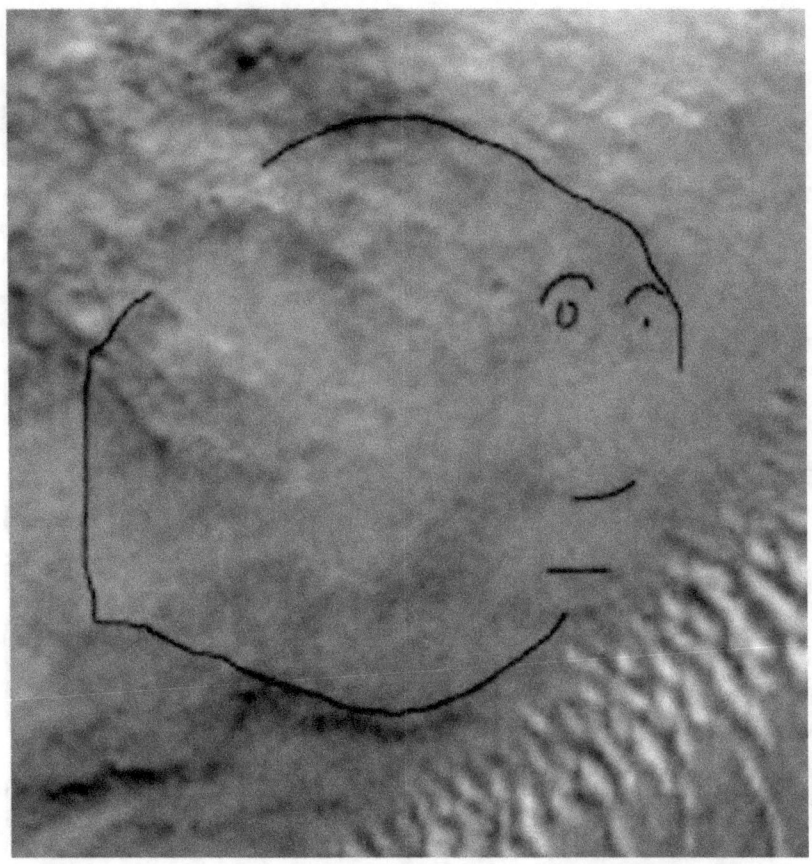

Figure 32

The Face is more distorted and cartoonish; this may mean it is just a random formation. This may be partially buried around the nose and mouth, so it is difficult to tell how plausible it is without a robotic or manned expedition to look under the sand.

Face Six

The profile Crowned Face is shown below, this is Face Six.

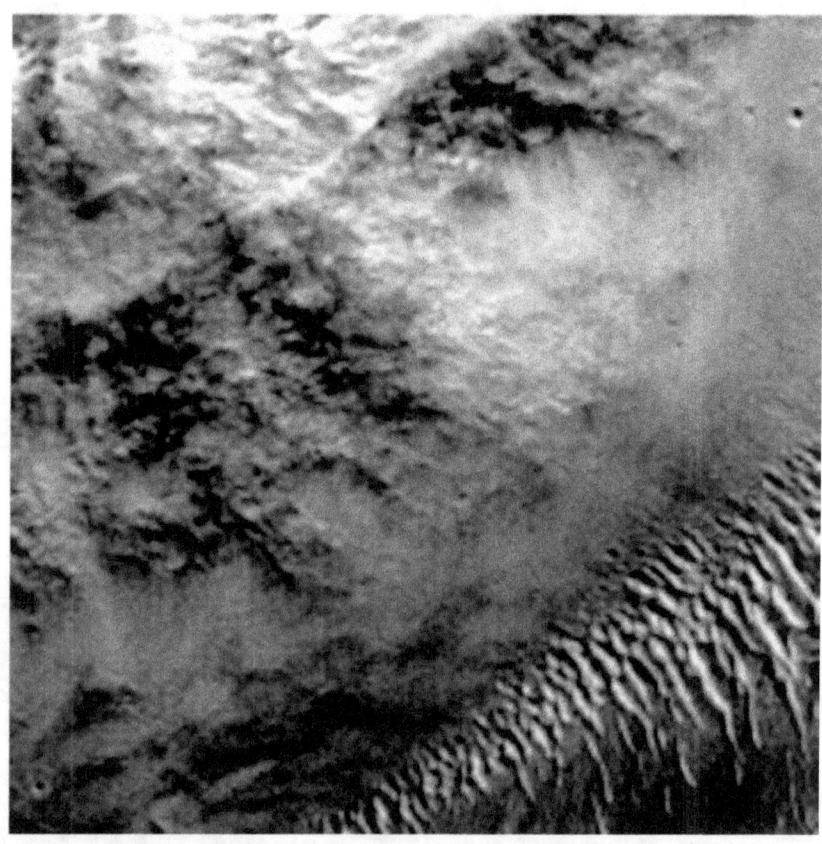

Figure 33

I missed seeing this face for a long time; though once it is seen many people have no trouble seeing it every time.

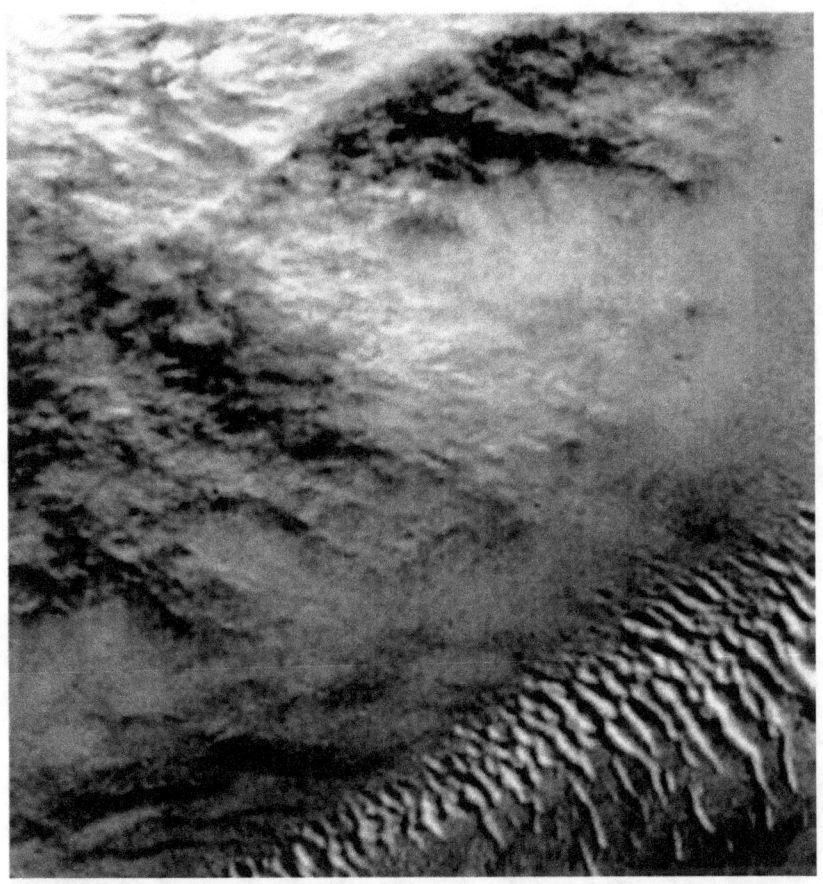

Figure 34

This is the same area on photo M0303483, the Profile Face is looking out directly to the right. The two images have slightly different shadows which allow some of the features to be seen more clearly. Each image has been resampled to 144dpi, double the original and adjusted for brightness, contrast and intensity. This is a profile image similar to the Nefertiti Face. There are several other profile faces near Nefertiti to be analyzed in an upcoming book.

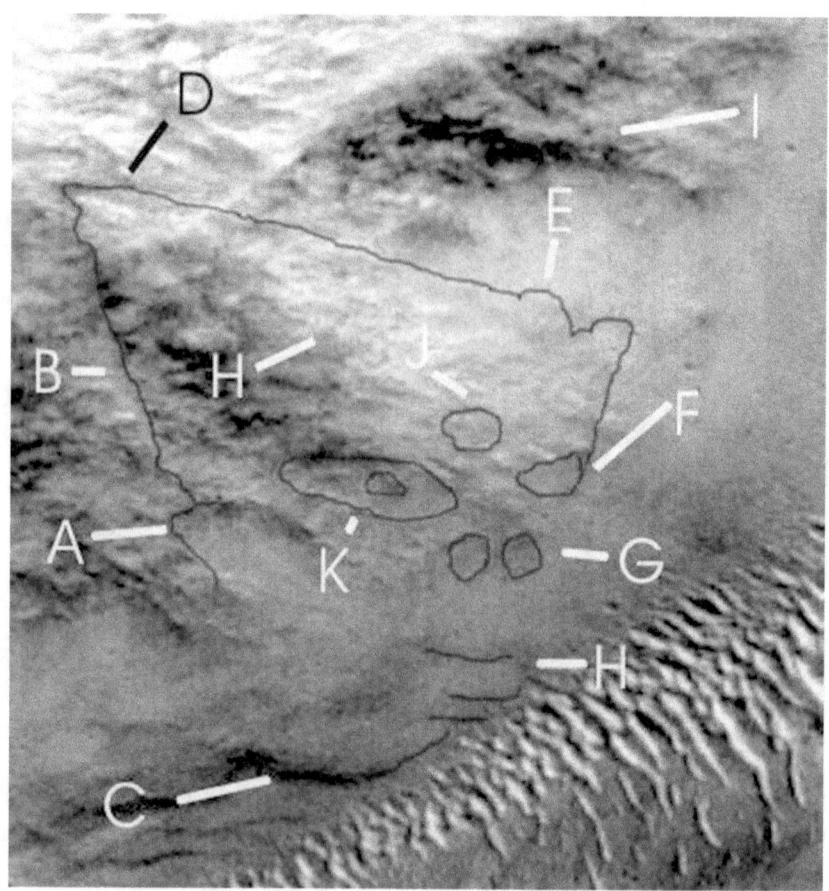

Figure 35

A shows this cavity could be an ear though the lower parts may have eroded away.

B shows a rough line which could be the edge of a crown. This would be a similar crown to the middle Crowned Face. If so it is a recurring theme unlikely to occur by random chance. For example it could have been any kind of hat or a bare head.

C shows this ridge could be under the chin of the face. Also to the left there may be a neck of this face, doubling as the cheeks of the right hand Crowned Face.

D shows how the crown edge goes over the ridge forming a triangle shape. Normally the crown shape if natural would stop at this ledge rather than continue over it.

E shows a semicircular shape on the crown edge. The crown edge to the right may also be different as shown below. **F** shows a cavity in the right position for the right eye to be in proportion. There may be another eye just to the right of this, as if there was another dual face just around the corner, with an extended crown shape going upwards. It also allows a prediction of a clear eye and eyeball shape here if the area is re imaged from a different angle.

G shows two shapes in the right position for nostrils and a ridge of more solid rock just above them. They seem elliptical in shape consistent with the profile angle. It appears as if the rock eroded here exposing these holes, so perhaps a nose similar to on the Crowned Faces was here. If so then the nostril cavities may have gone up inside the noses on the other Crowned Faces as a living creature. The area below this and around to the left ear may have eroded away, or the face may be wearing an armor like headdress that extends to here. The nostrils were also eyes of the previous Face which now appears much more dubious. The remains of the nose may be buried under the sand below this. A manned or robotic expedition could look for pieces like this as further proof of artificiality.

H is in the right position for the mouth, the upper line might also be where the tip of the nose would be, and this would make it the same kind of Face as the three Crowned Faces. In that case the lower lines would be the mouth. The two circles at **G** might be follow features like cave entrances and the nose shape might have been similar to Face Two. If you look at the eyes of the Crowned Face (i.e. Face Two) there are two small oval depressions, so these may be a facial feature.

I shows a ledge which may be a crown on this face hidden behind the Profile Crowned Face.

J shows a circular shape symmetrically placed on the forehead.

K shows a well-developed eye similar in shape to the left eye of the Cydonia Face, the KK Face as well as the Crowned Face. The pupil seems to stand out in the right position for looking ahead, has a darker iris and a pupil.

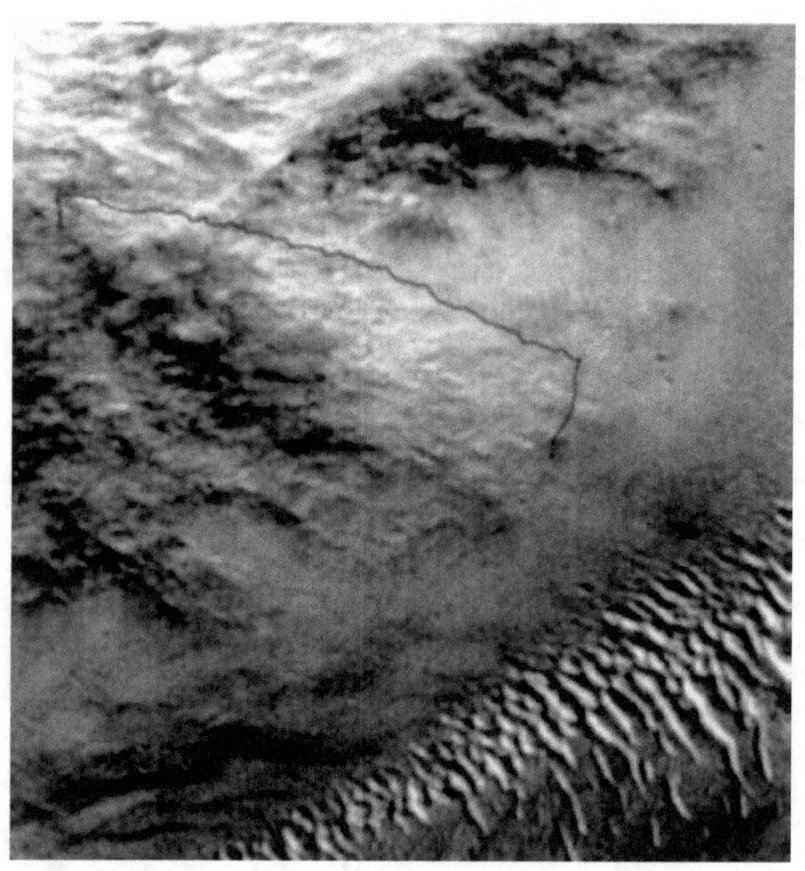

Figure 36

This shows another possible shape of the crown.

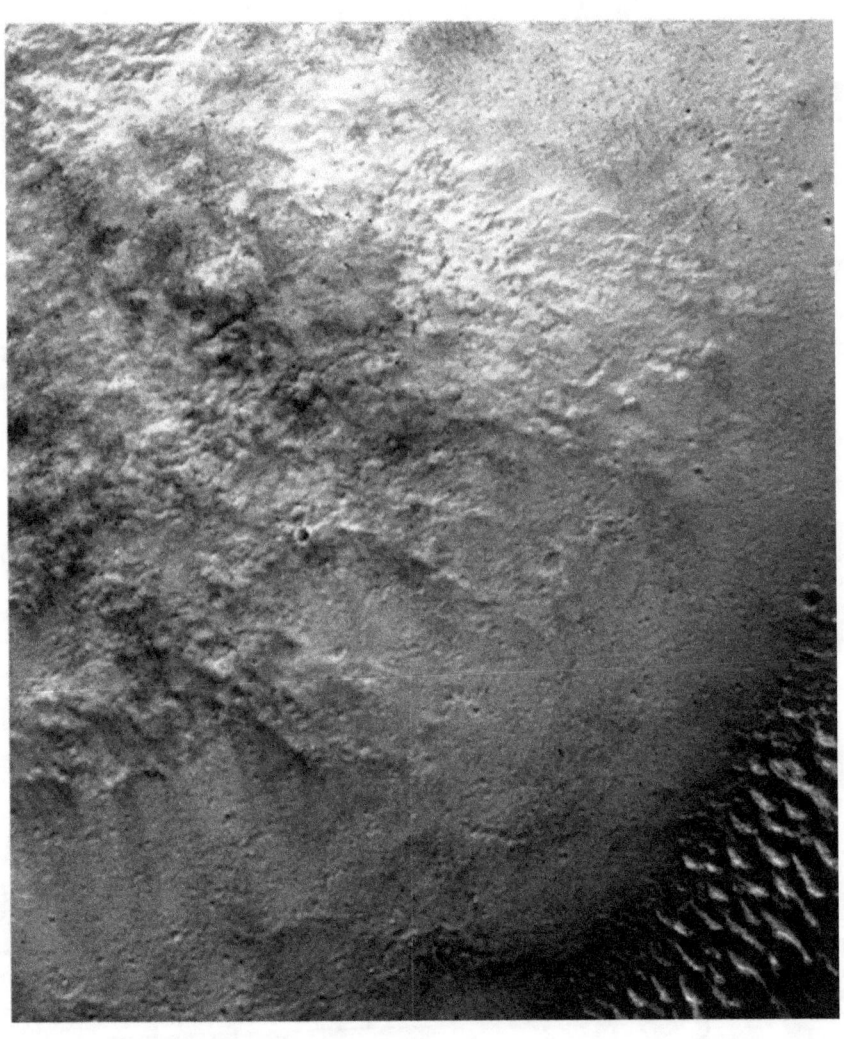

Figure 37

When the King's valley was reimaged with HiRise it also included the Profile Face.

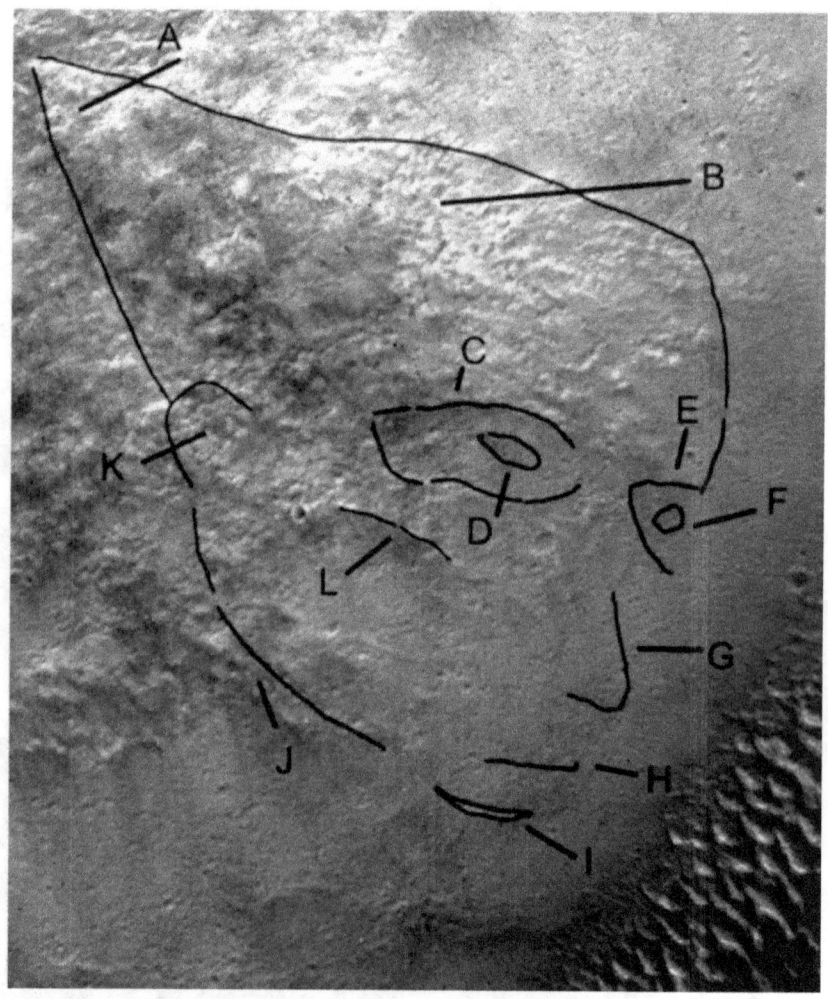

Figure 38

A shows the apex of the crown which extends past the ridge, this is unlikely to be formed naturally. **B** shows this area bulges outwards because it catches the light, this is consistent with the idea of this face being more three dimensional and turned to the side. **C** shows an eye shape with **D** showing an iris. Like with Face Three just to the left of here the iris is not circular, which is different from a human eye. **E** shows a faint eye shape and **F** another iris. **G** shows a faint nose shape. **H** is a ridge consistent with a mouth shape and **I** is another ridge consistent with a chin. **J** is a jawline shape. **K** is an ear shape. **L** is a cheekbone shape.

Figure 39

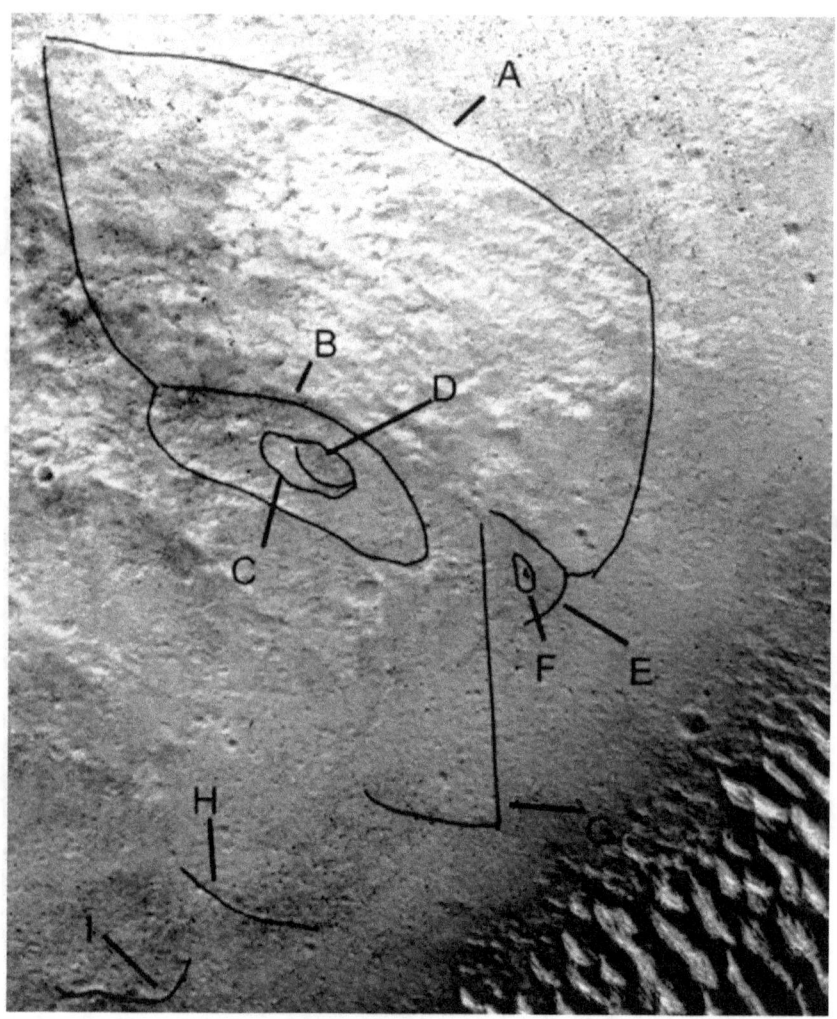

Figure 40

Above is the HiRise reimaging of the Profile Crowned Face. **A** is the Crown, **B** is the eye cavity. **C** is the eye and **D** the iris. **E** and **F** show the right eye. **G** is the nose, **I** is the mouth and **J** is the chin.

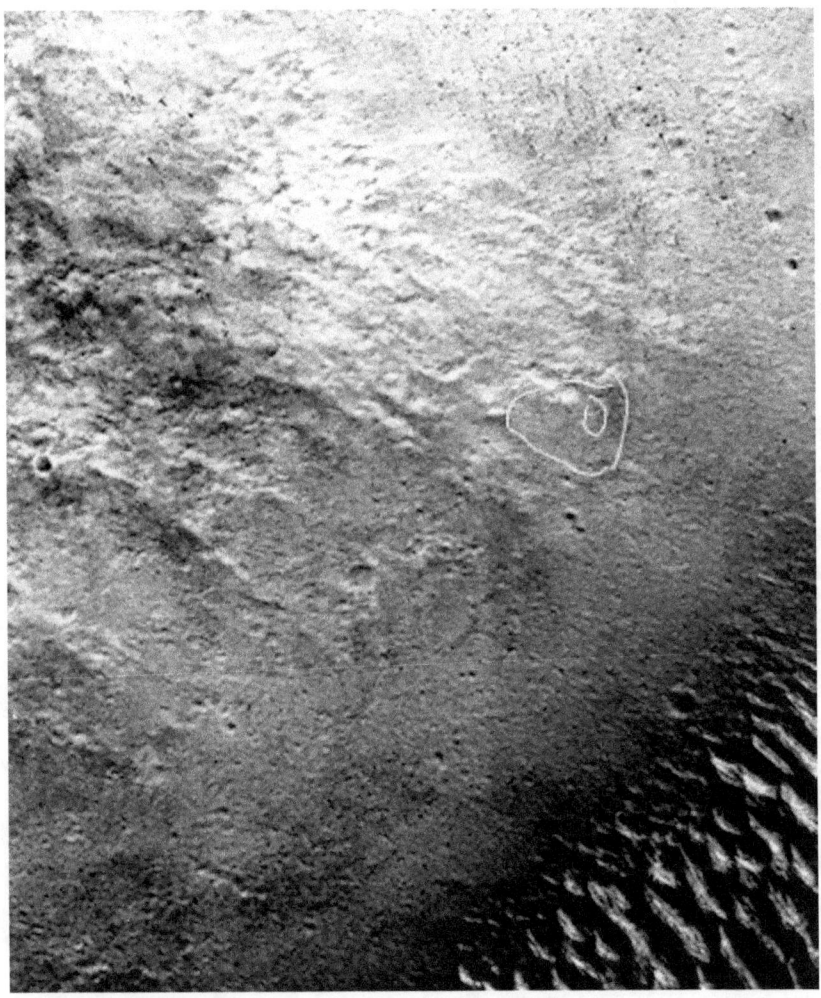

Figure 41

The right eye might also be higher up as shown above, like it appeared in the Mars Orbital Camera images shown earlier. Alternatively there may be two eyes to change the facial expression with different times of the day and seasons as with the Crowned Face. There may be another Face around the slope to the right or partially buried at **A** in Figure 42. The image is too vague to tell, but if the area is re imaged then it is a falsifiable prediction. It doesn't seem to appear in the HiRise images but this may be from a lack of shadows. It may be intended to appear as another dual face sharing the Profile Crowned Face's right eye.

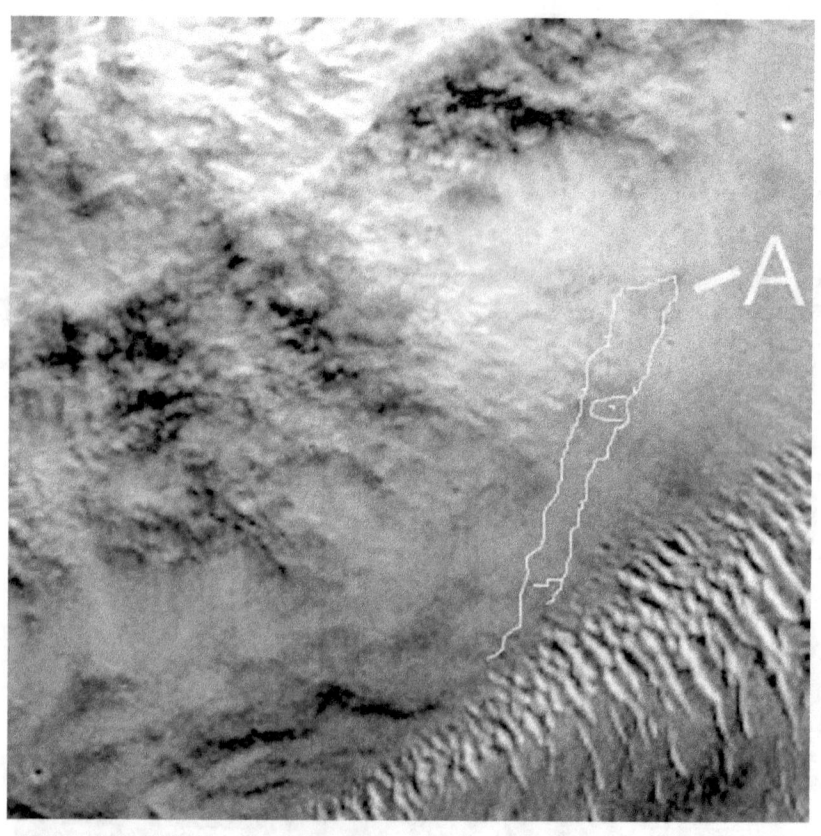

Figure 42

*The Profile Crowned Face may be intended to be another dual face, shown at **A** but there is too little detail to be sure.*

Face Seven

Just above the Profile Crowned Face there is another possible face looking down, I call this Face Seven.

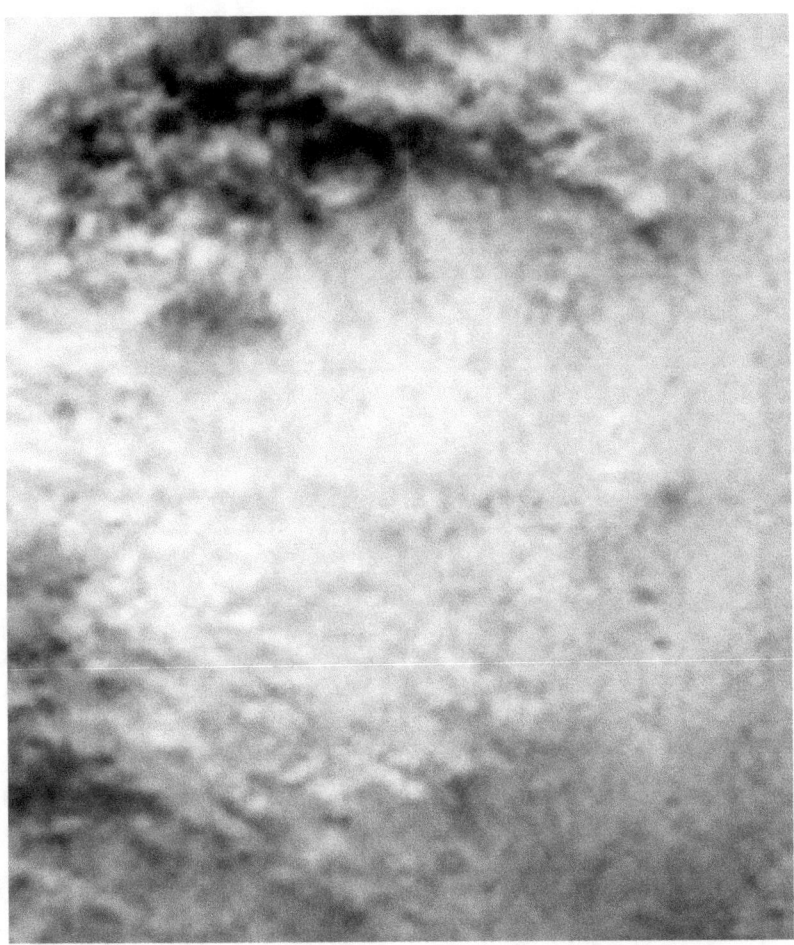

Figure 43

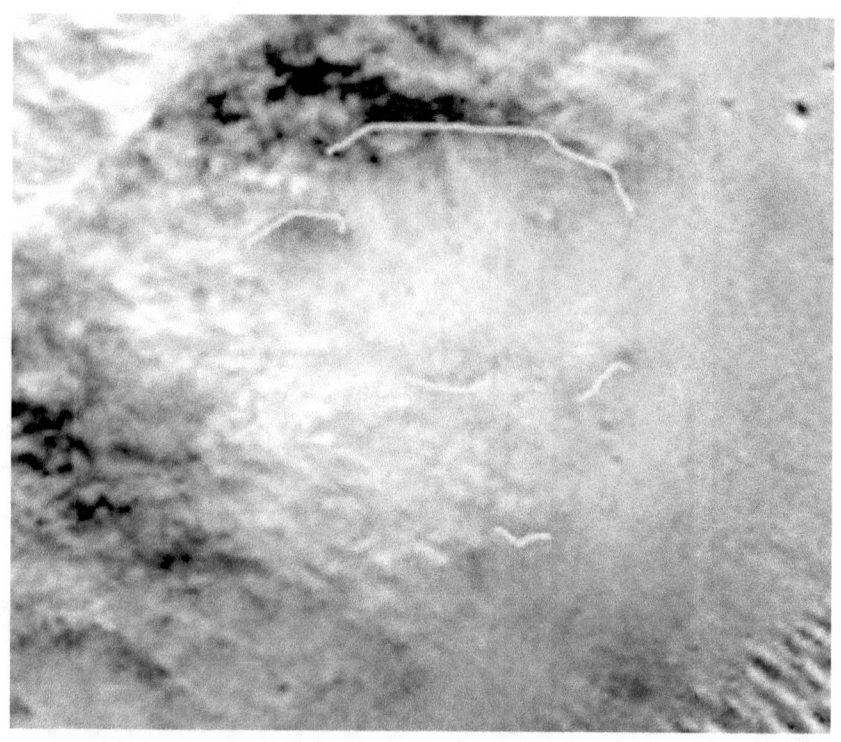

Figure 44

This shows an outline of Face Seven.

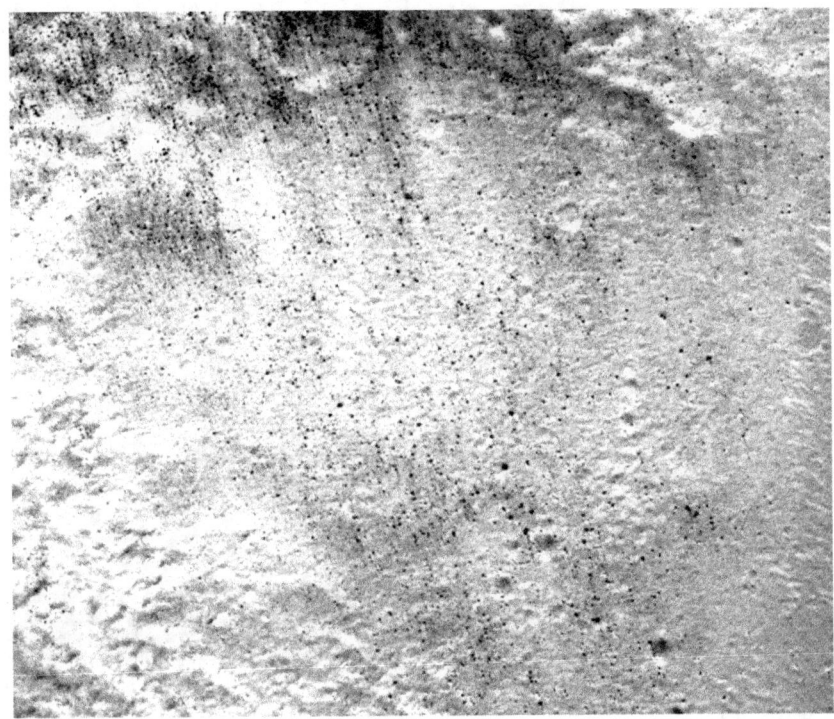

Figure 45

A HiRise remaging of Face Seven, the eyes are a little clearer but it may be too buried in sand to see more. This could also be pareidolia where the a few suggestive details and the mouth hidden from view make it more Face like than it should be.

There is another face like shape on the forehead of the Profile Face this may be a random formation or an ornamental part of the crown so I have not given it a face number.

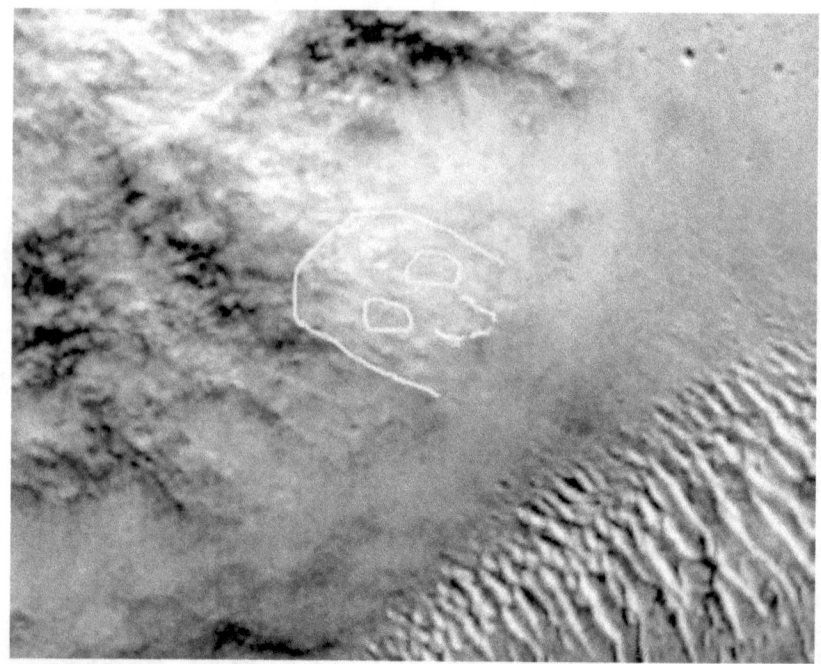

Figure 46

Face Eight

Next there is a face with a crown and torso.

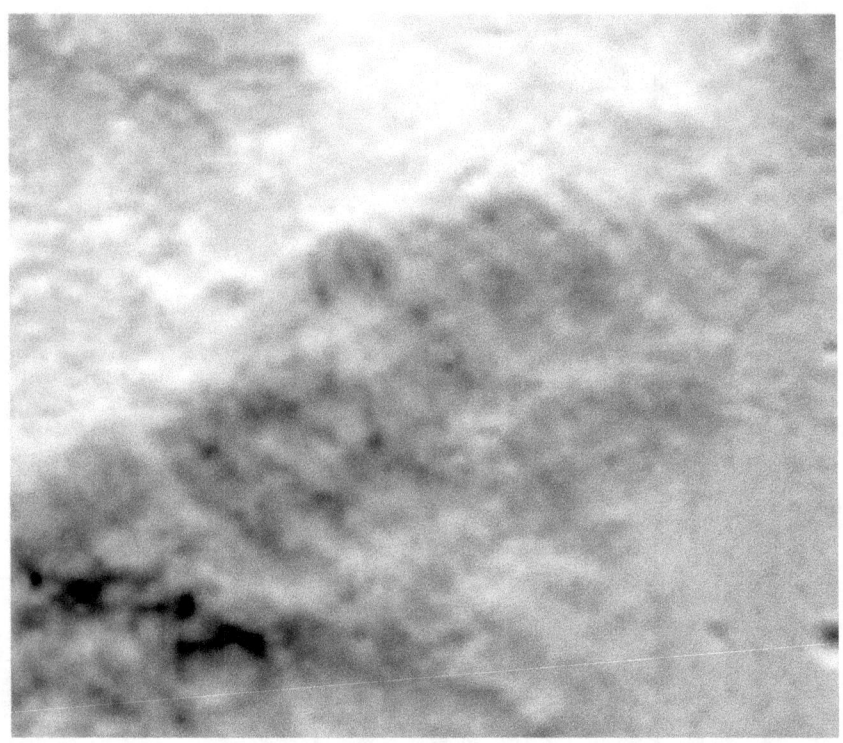

Figure 47

Below this Face is rotated and outlined. I call this Face Eight.

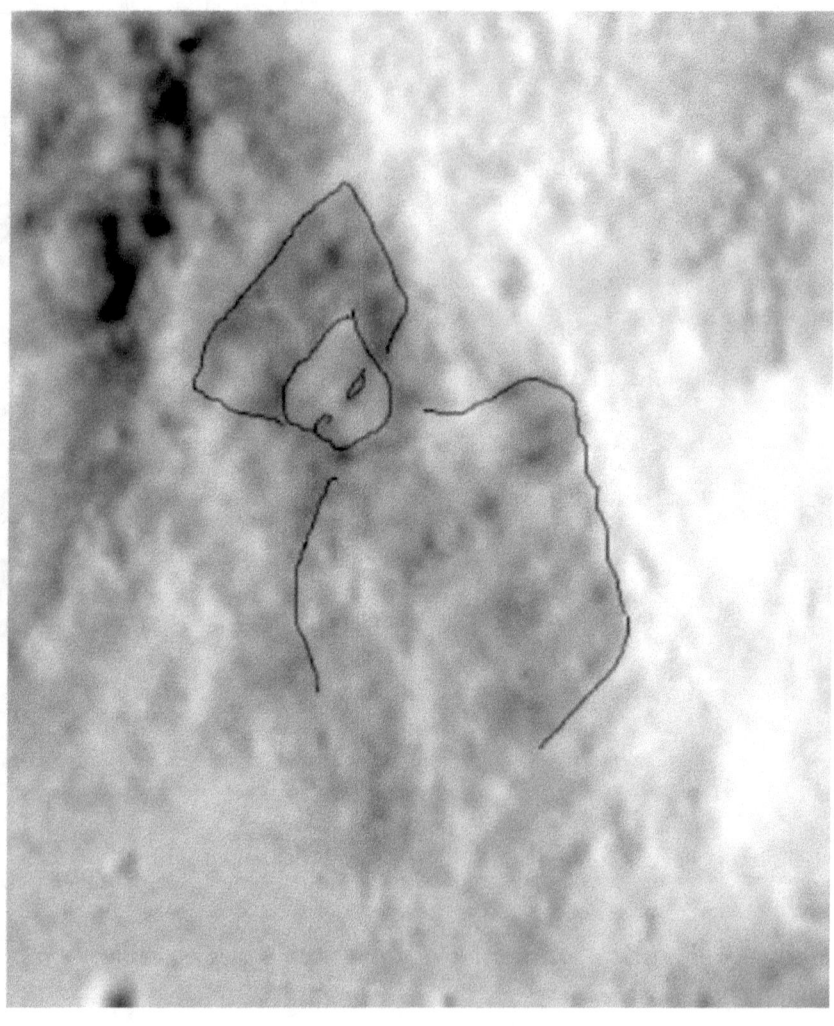

Figure 48

This is Face Eight outlined; it has a body attached to a Crowned Face.

It may also be a random formation, it appears less artificial in the HiRise reimaging of the area but there may be rocks and grooves which normally cast shadows. If the HiRise image is taken from the wrong sun angle this may not stand out, like with the Profile Face.

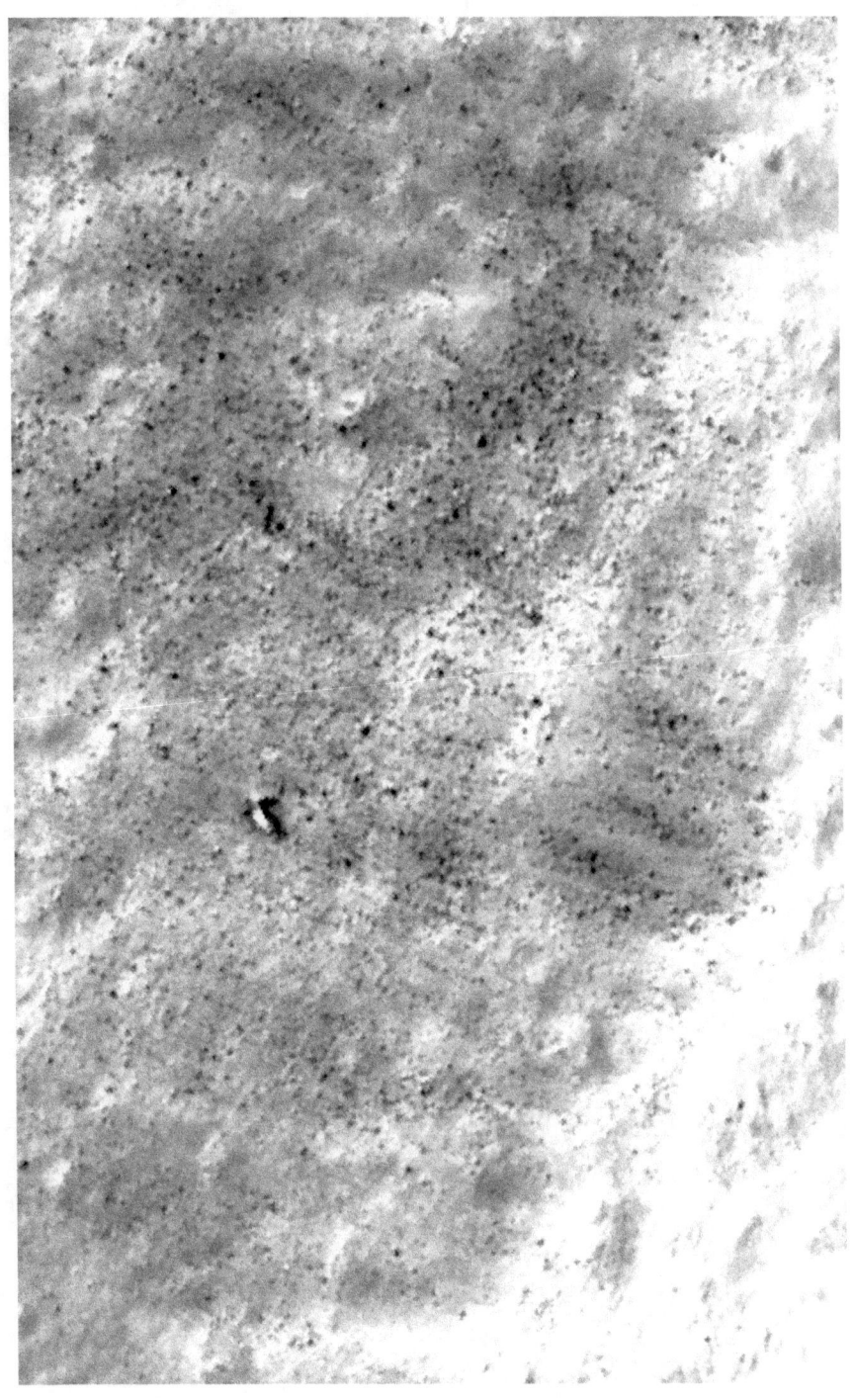

Figure 49

It appears much fainter perhaps from the sun angle, but the head and crown shape are clearer. There doesn't seem to be any facial features, which should be apparent at this resolution unless they have eroded away or are covered with dust. A manned expedition to the site may be necessary to tell if this is artificial. The dark spots are boulders.

Face Nine

Next there is a face shape on the left and part of one on the right, called Faces Nine and Ten respectively.

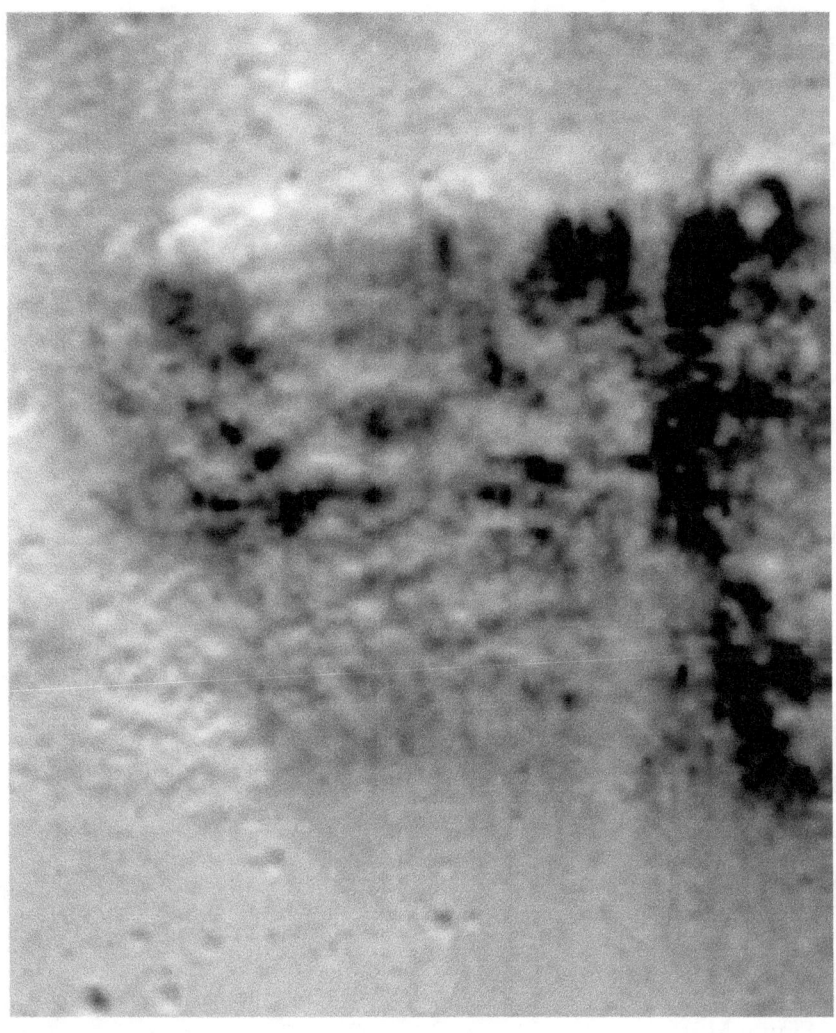

Figure 50

Face Nine is more clearly seen in Photo M303483 later. Next is the outline of both faces. This area has been reimaged with HiRise and the results are ambiguous, they may be natural formations. However given their closeness to the Crowned Face any face details should be closely examined.

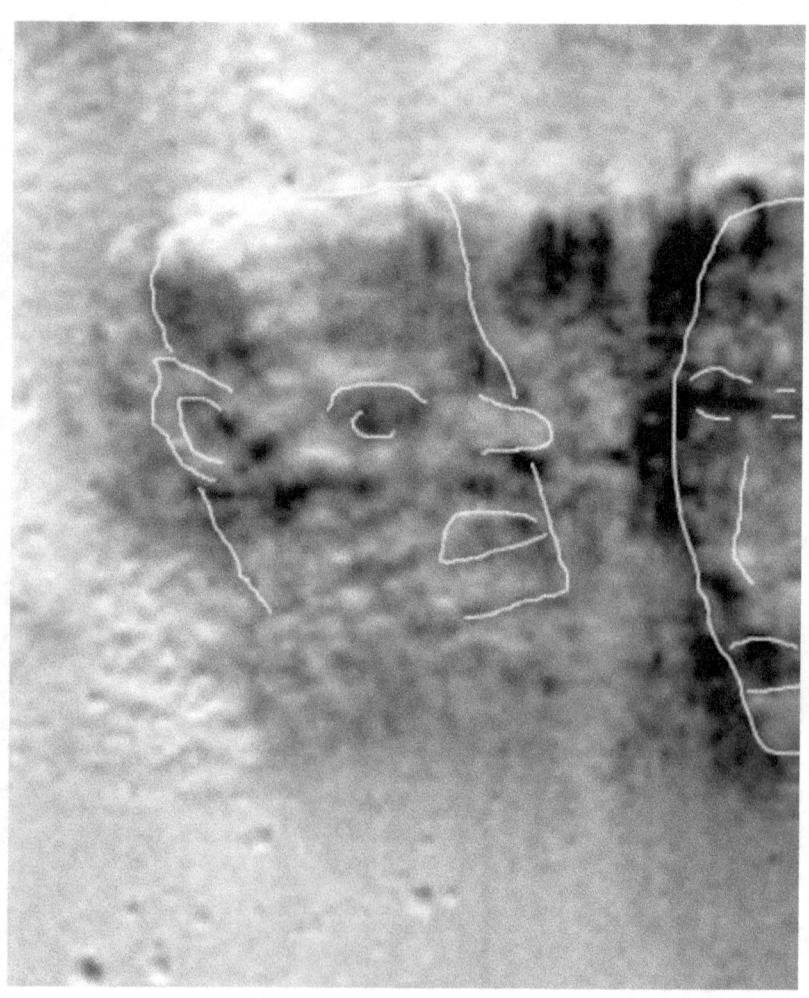

Figure 51

Faces Nine and Ten outlined.

There is little real detail, enlarging it shows how few pixels there are:

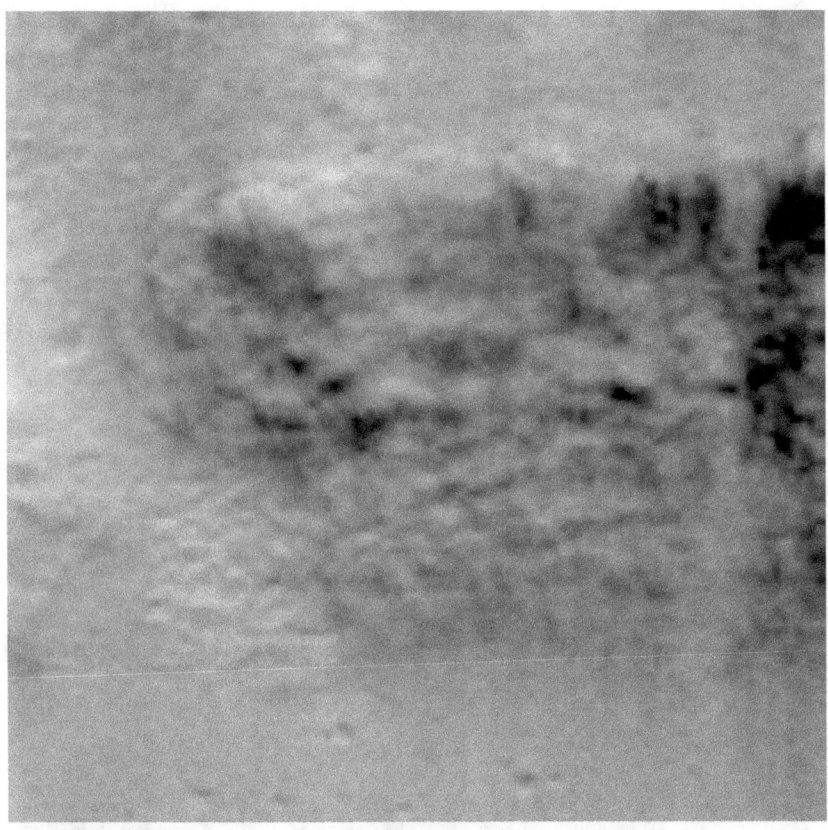

Figure 52

If natural then this might give more credence to the notion that cliffs like these tend to look more face like than other geological areas of Mars, this is a hard argument to sustain though because faces on Mars are generally very rare even by the most fanciful analysis but in this valley they are quite plentiful. The shape is mainly useful to testing the hypothesis that faces that are artificial should be more face like under higher resolution, and conversely natural formations should appear more random the more detail is added by higher resolution. This is an important point because analyzing so many formations on Mars under low resolution is difficult so a methodology needs to be worked out to avoid false positives.

Erosion is known to occur in certain ways and with this knowledge we can also differentiate artificial processes from natural ones. Erosion tends to make shapes more random and rounded while a sculptor might be expected to make noses, eyes, etc. with more details. A nose for example would be expected to jut out from a face while erosion would tend to remove protruding rocks, the nose in the image above if it stood out this far then it might be more likely to be artificial as it should have eroded to be much smaller by now. The eye appears to have an iris shape in it but this could just be from shadows in a hollow or from pixilation, it is also less likely to be from a crater impact because craters are usually more rounded. There are many random shapes from erosion in this image so it might be that the face if artificial has eroded away to almost nothing. Perhaps water or sand running down the channel to the left of the formation eroded away most of its features while the Crowned Faces were more protected.

Another possible face can also be seen in this photo, but this also has the problem of being exaggerated in shape and not like the Crowned Faces.

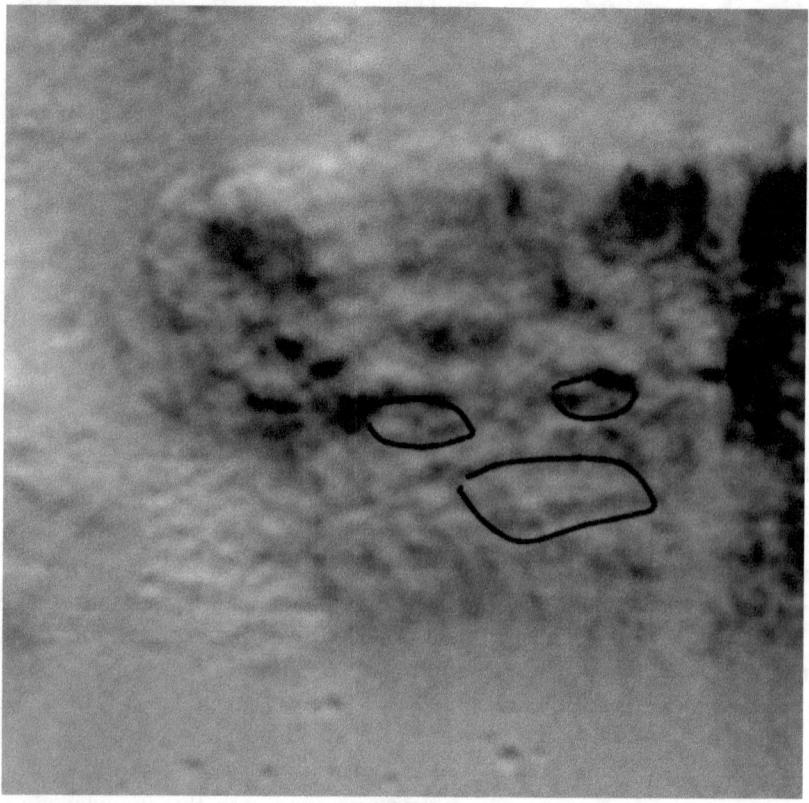

Figure 53

There is also a possible dual face here which would be a similar theme to the Crowned Faces and be less likely to occur by chance.

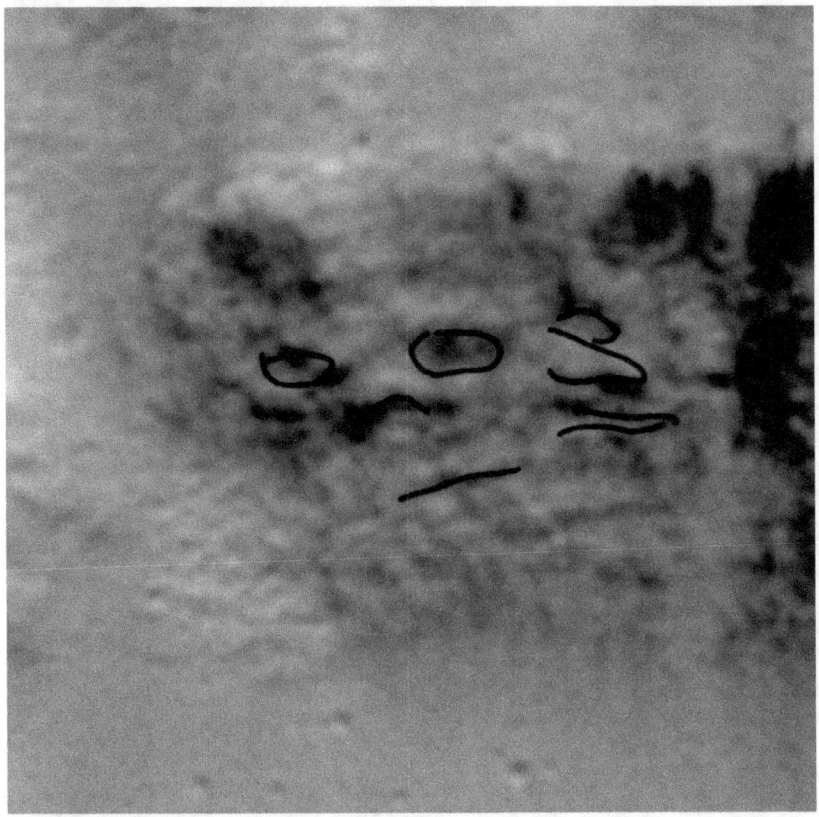

Figure 54

Here there are three eyes, two noses and two mouths, though this might also indicate that dual faces also occur randomly. The area is then an interesting one for reimaging with HiRise to not only answer these questions but to see how these kinds of formations resolve under higher resolution. Below is the face from the HiRise image, the forehead and crown are clearer and there seems to be an iris in the eye. The nose shape is still different from the Crowned Faces as it is more bulbous. It is difficult to say if this improves its likelihood of being artificial or not.

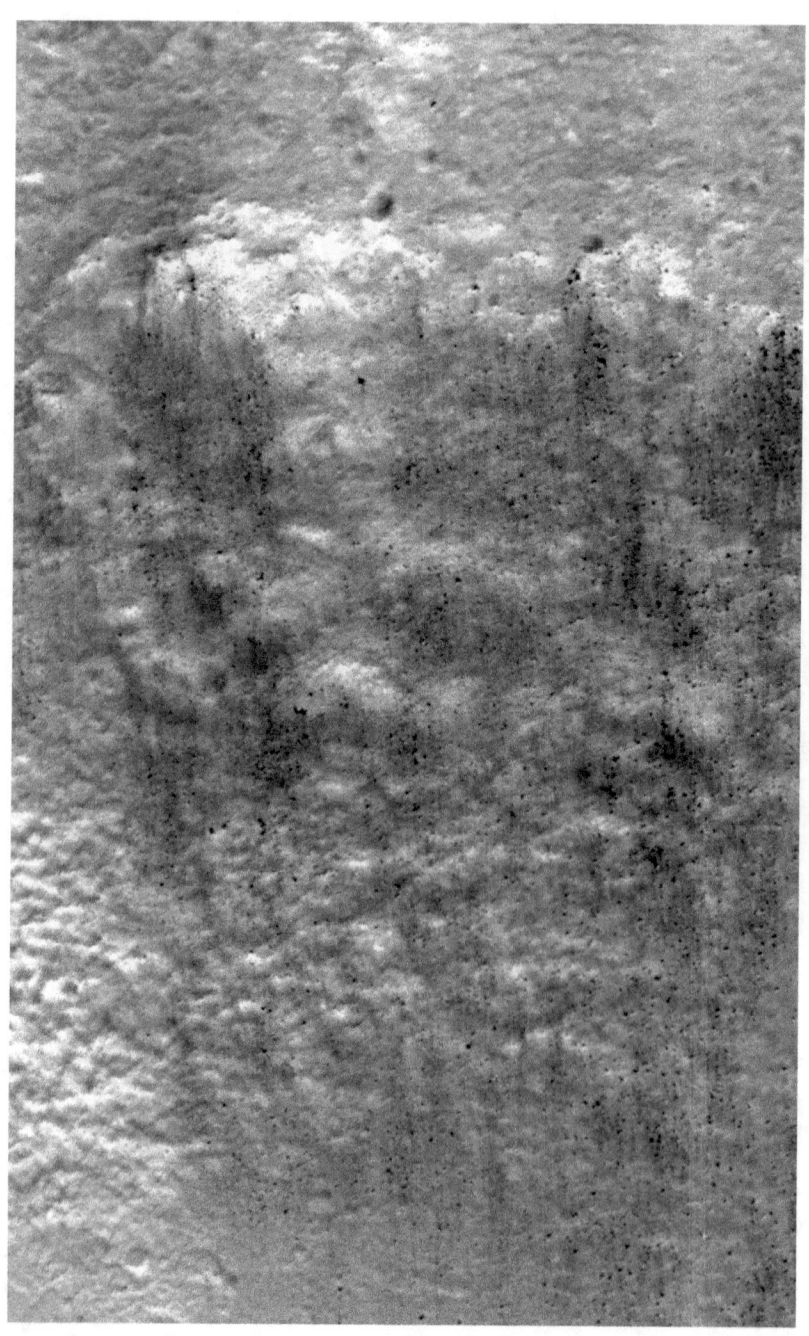

Figure 55

Face Nine also appears Face like upside down which may mean so many random features on these rocks are allowing the brain to find more Faces.

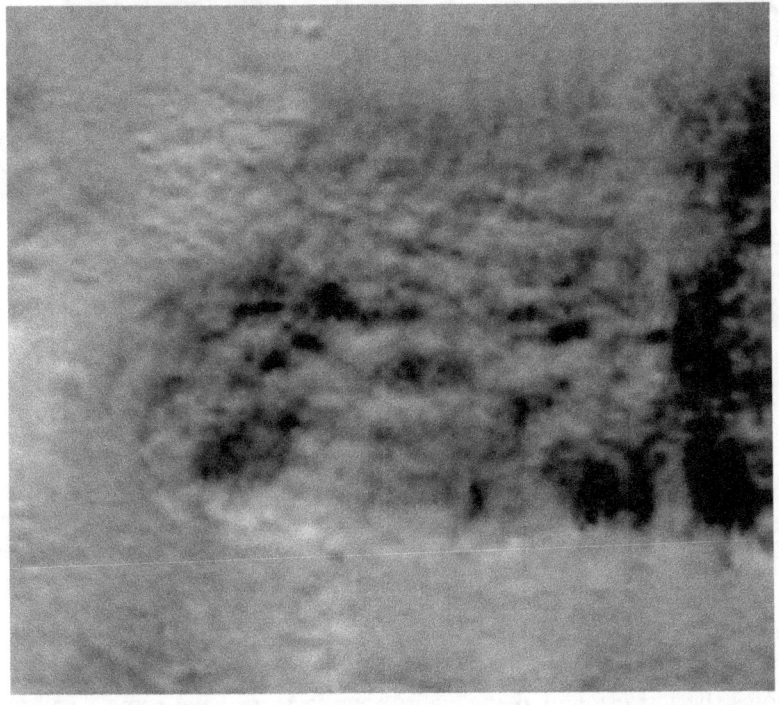

Figure 56

Face Nine inverted.

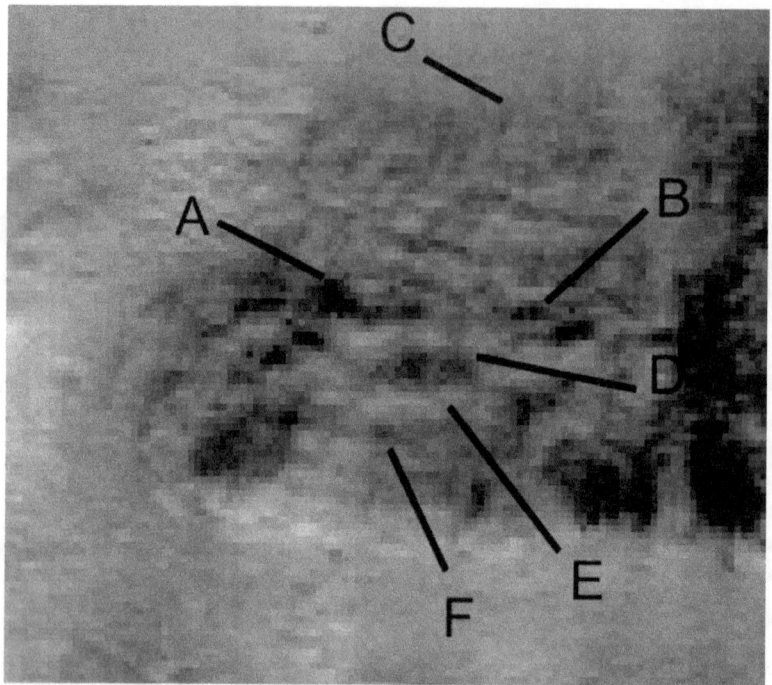

Figure 57

A shows the left eye and **B** the right; both seem to have some detail in them. **C** shows the top of the crown, there is no real nose at **D**, **E** shows a dark area for a mouth and **F** is where a chin should be. Again when the face is blown up 400% the details become very pixelated, but it seems quite different in style from the Crowned Faces. The whole formation may be either random or too eroded to see except from the ground. The image below is from HiRise and as can be seen the details are not becoming more face like, the nose and mouth appear less distinct and the eyes have little detail. This is what we would expect from a formation that randomly looks face like but this impression begins to go away as more detail is added.

Figure 58

The next face here is more elongated which is also suspect because it gives potentially another alien shaped face.

Face Ten

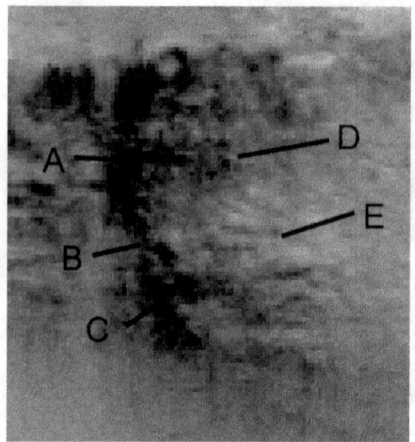

Figure 59

This is Face Ten.

A and **D** show eyes which are also highly pixilated, the nose is only a mound, and the mouth very indistinct. **E** shows Face Eleven (seen next) which joins into it. There is another face on this cliff which may be partially eroded, shown below. **C** merges into Face Eleven.

In the HiRise image below of Face Ten there is little sign of a face so it's likely this was an optical illusion. However the cliff is a good position for a face so it may have eroded away. A manned expedition should be able to examine this area to see if there is anything here.

Figure 60

Face Eleven

Further to the right along this cliff face there are another two possible faces, Face Eleven and Face Twelve. These have been reimaged with HiRise and contain many more artificial looking features.

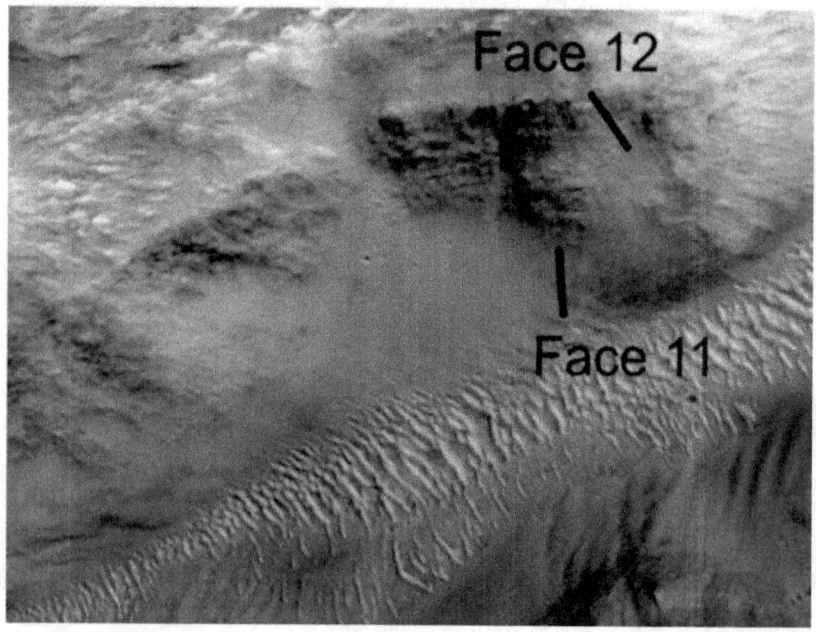

Figure 61

Face Eleven is shown below. It has the disadvantage of appearing different from the other faces and somewhat cartoonish, perhaps like Ernie from Sesame Street. This is the problem with looking for faces; the mind tends to recognize them not only in living things but also from cartoons. There appears to be a beard, the nose is larger and the ears much more prominent so it may be an example of a chance alignment of rocks appearing face like. However it is quite compelling and its being so near to other better defined faces makes it more interesting.

Below the face is blown up 400% to show how pixelated the features are:

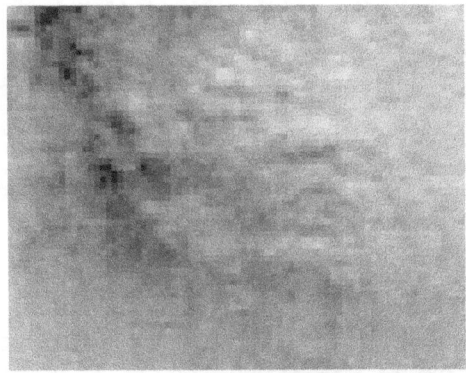

Figure 62

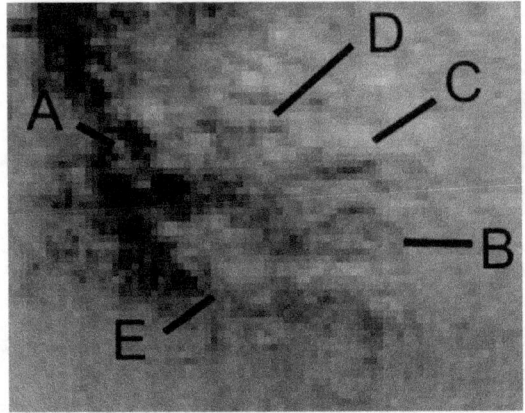

Figure 63

The ear is shown at **A** and is rounded while the other ears on the Crowned Faces seem more pointed. The eye at **D** is highly pixelated from its small size, it can be seen here that the slit of the eye is a straight line from pixels rather than an actual feature. **B** shows the mouth which is again a different shape from the other faces; **C** shows the nose which is quite large. **E** shows a jawline also highly pixelated.

Below is this face reimaged by HiRise. The lack of shadows makes many of the features so indistinct that it is again hard to judge artificiality.

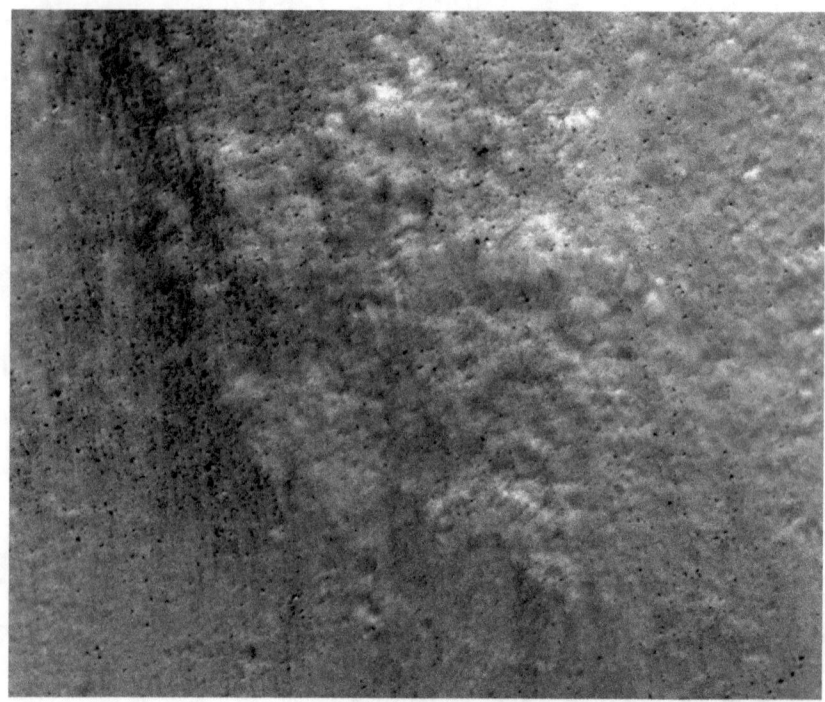

Figure 64

The shape around the face is fairly symmetrical with a crown like peak at the top, with vague appearances of eyes, a nose and a mouth. While it appears less face like with higher resolution this may be because of the lack of shadows, the symmetrical shape around it though is unusual and is like a crown at the top.

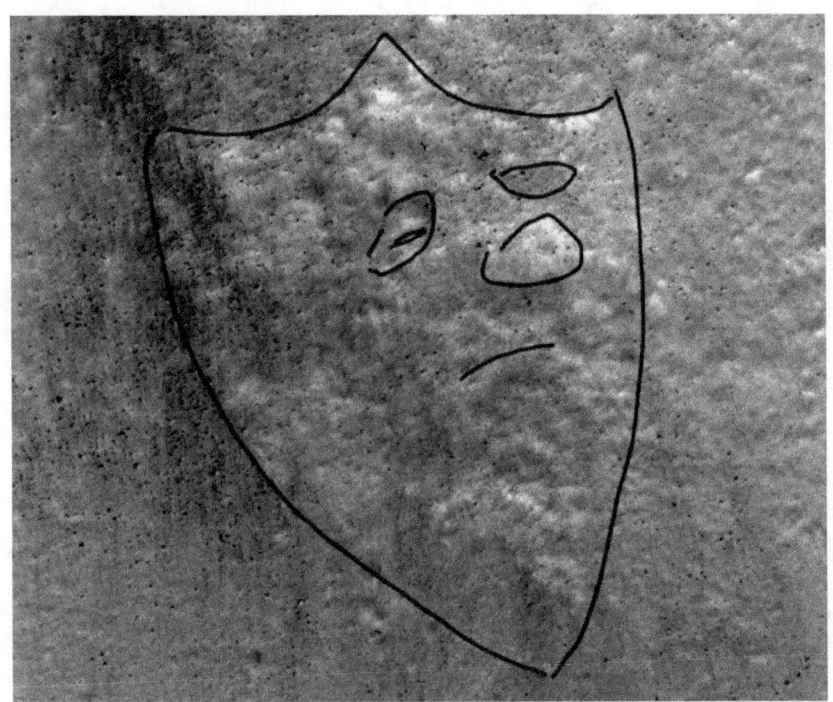

Figure 65

Face Twelve

Face Twelve is seen below.

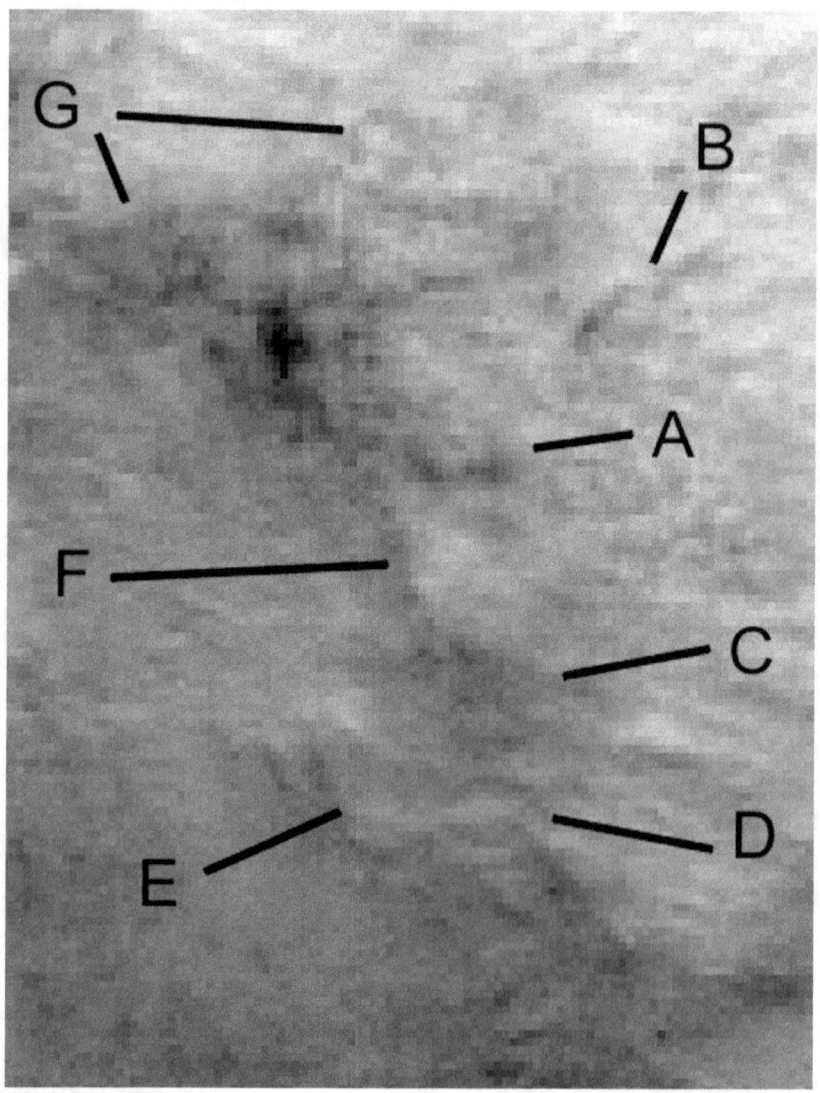

Figure 66

A shows an eye which has some structure though magnified 400%. B shows a possible ear which is pointed more like the other Crowned Faces. C shows a shade suggestive of a mouth, D and E show a chin outline. F shows a nose about the same type as on the Crowned Faces. G shows a possible top to a crown with an oval shape suggesting the crown may be intended to be rounded in shape like with Nefertiti as we will see in an upcoming book. This may have been another Crowned Face, the image below shows it reimaged with HiRise.

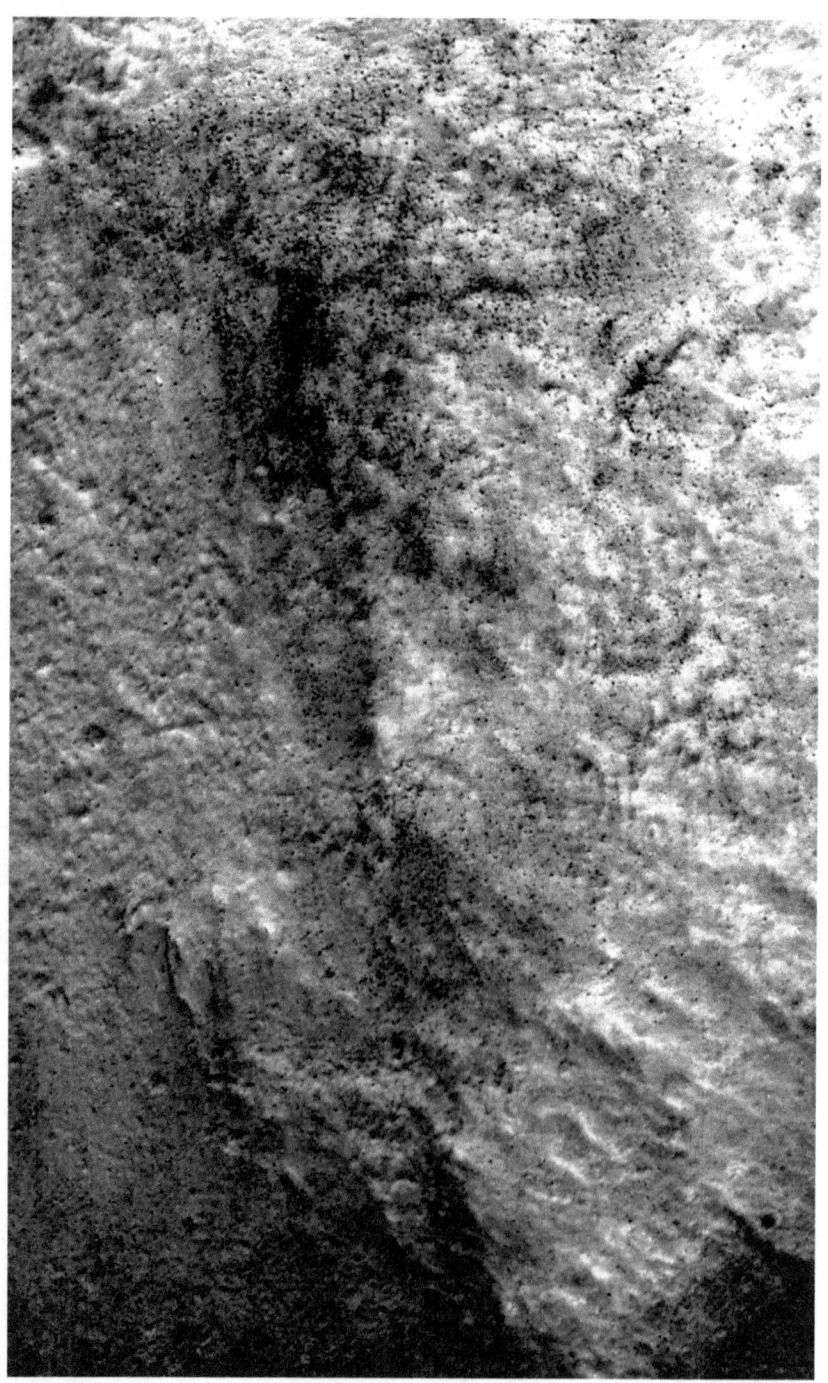

Figure 67

As seen below the features have become more face like under higher resolution, a hat shape has appeared with an oval top implying it is rounded like a crown. It is also similar to the crown on the Nefertiti face as will be shown in the next book. The ears are pointy and somewhat similar to the Crowned Faces, the nose is long and slender like on them as well. There are two eyes whereas only one was visible in the MOC image, so there is a successful prediction of an eye in the right place. Both eyes seem to have a dark spot in them like an iris. The similarities to the Crowned Faces make this unlikelier to occur by chance. It also has an angry expression like Face Three though of course their facial expressions probably were meant to convey different emotions and body language.

Figure 68

234

Face Thirteen

The King's Valley was partially reimaged in photo E1100360, there are more face like features on this side. There are two possible faces below called Face Thirteen and Face Fourteen.

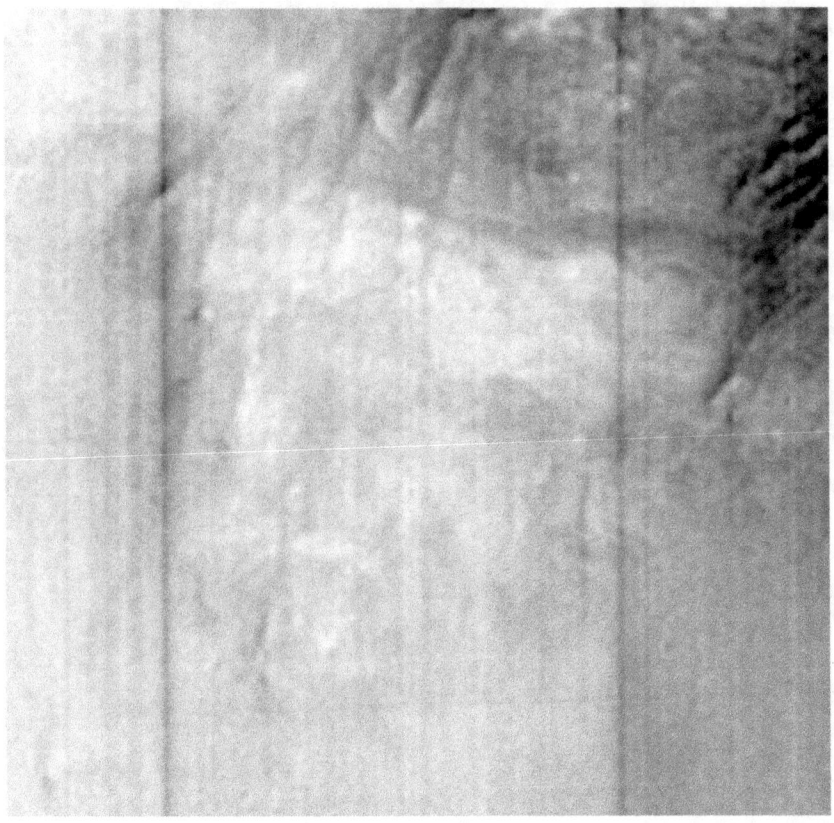

Figure 69

This is the whole formation; it appears to be another dual face.

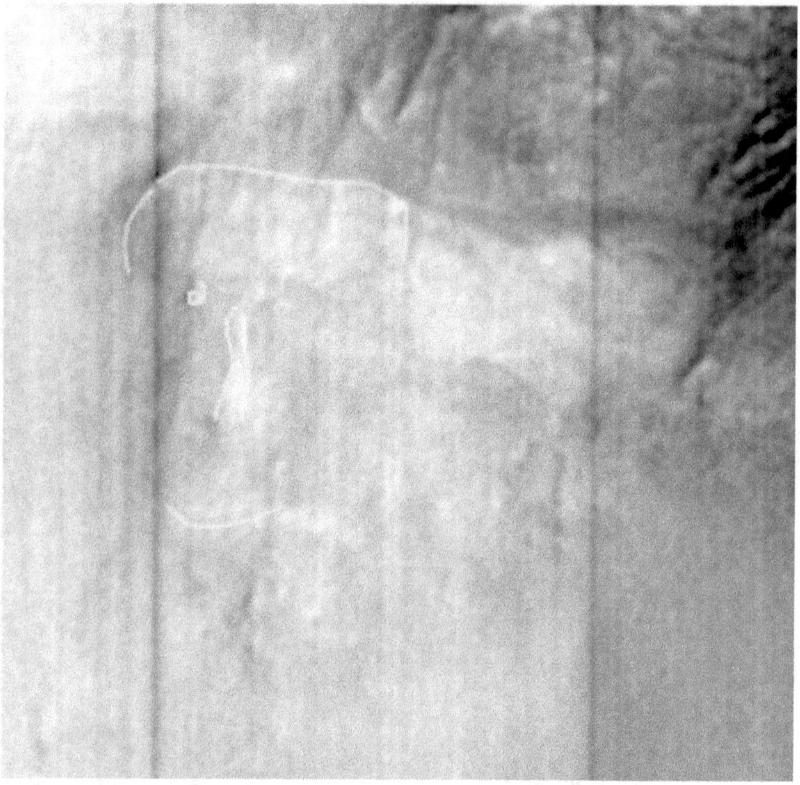

Figure 70

This is Face Thirteen; it is highly eroded but may have a crown, two eyes and a nose.

Face Fourteen

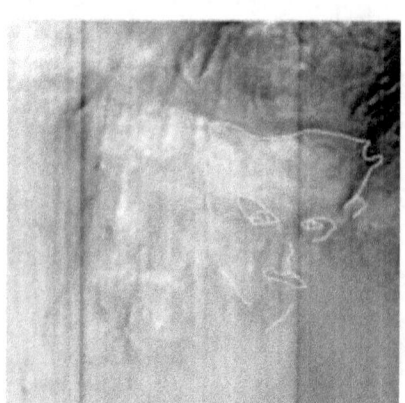

Figure 71

This is Face Fourteen, it has a crown like head and two eye cavities, both faces have noses that are curved to one side like the Crowned Face.

Face Fourteen was also in photo M0203051 seen below.

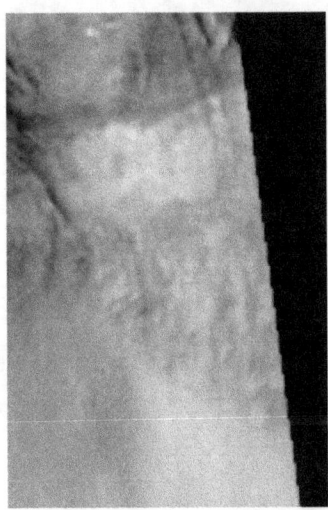

Figure 72

This Face has some facial features; it is however smaller than the main Crown Face so some of these features may be distorted from pixilation.

Face Fifteen

The two faces can also be seen as a single face called Face Fifteen seen below.

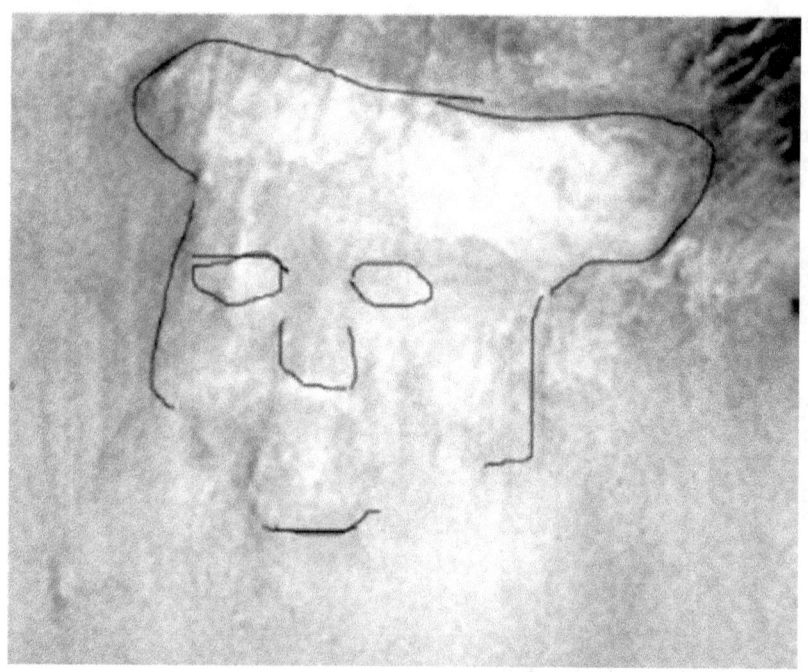

Figure 73

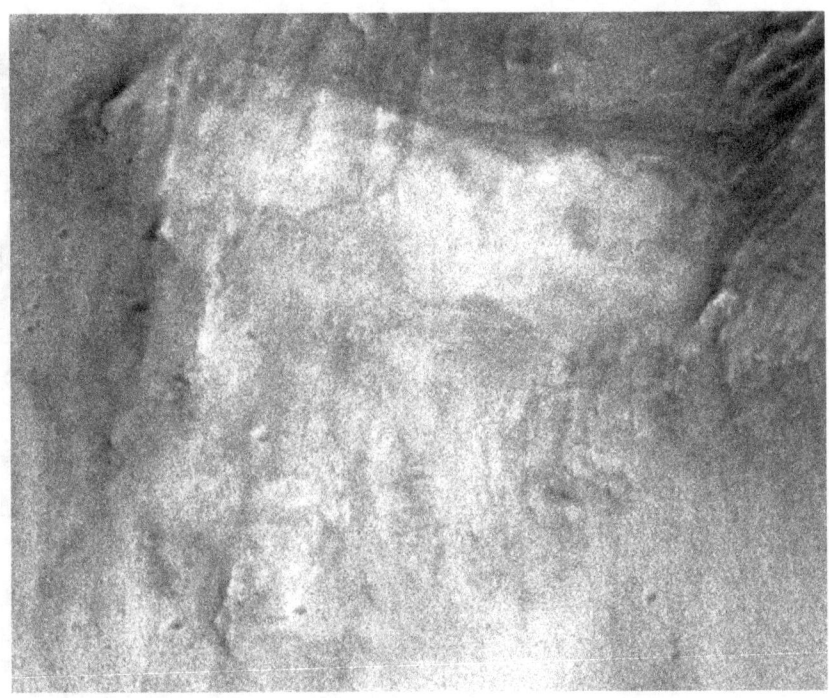

Figure 74

This area was reimaged in photo S1402499.

Face Sixteen and Seventeen

Opposite the Crowned Face there are two blank areas which may have been faces, were going to be faces, or they may be natural formations. Provisionally they are called Face Sixteen and Face Seventeen, shown below. An expedition to the area may be able to see eroded features on them.

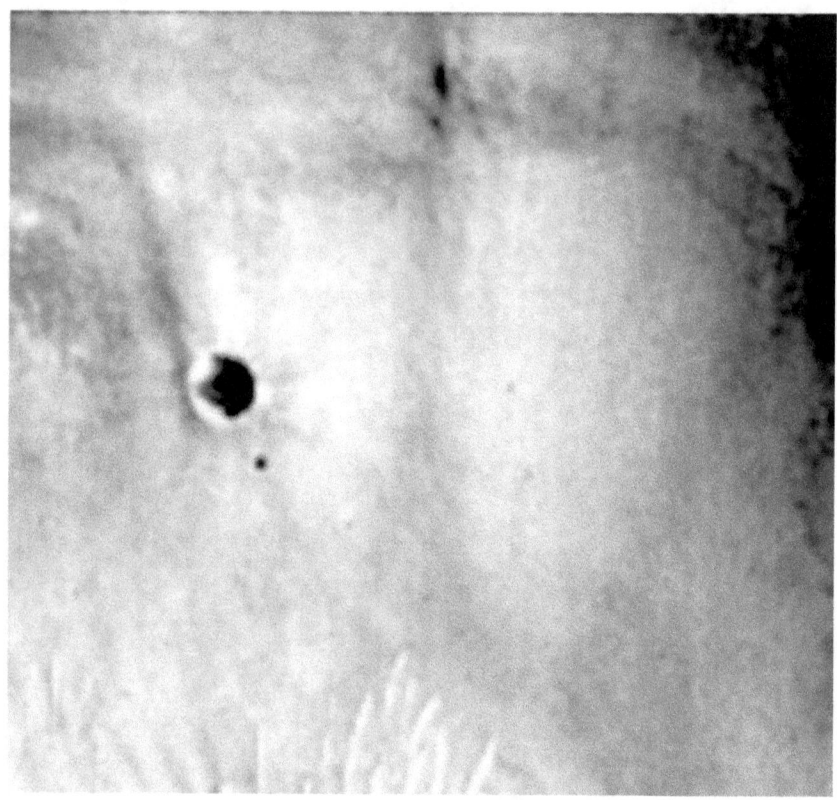

Figure 75

Below is the King's Valley to the left of the main Crowned Face, there are some interesting features here. The Crowned Face edge is in the upper right corner. There may be some other faces buried here under the sand.

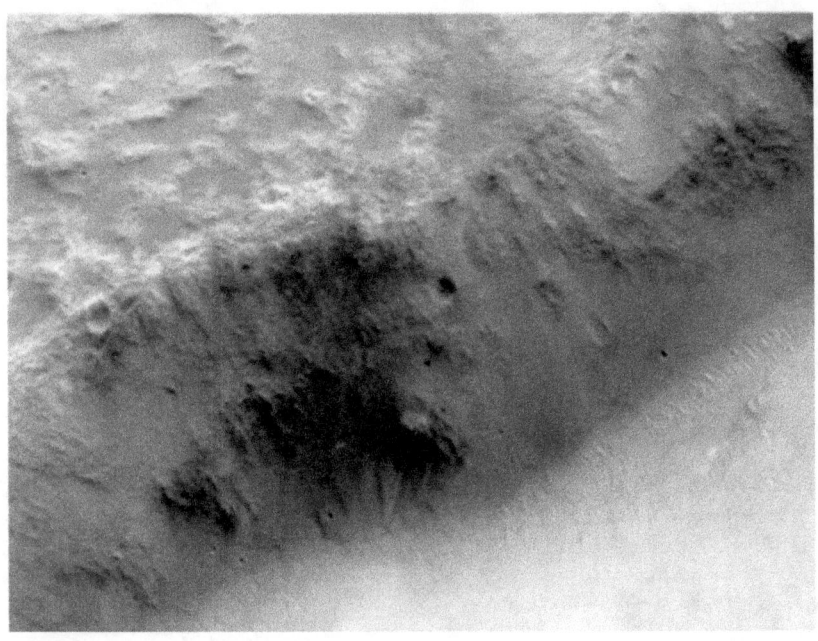

Figure 76

The formation outlined below may be another crown and face buried here.

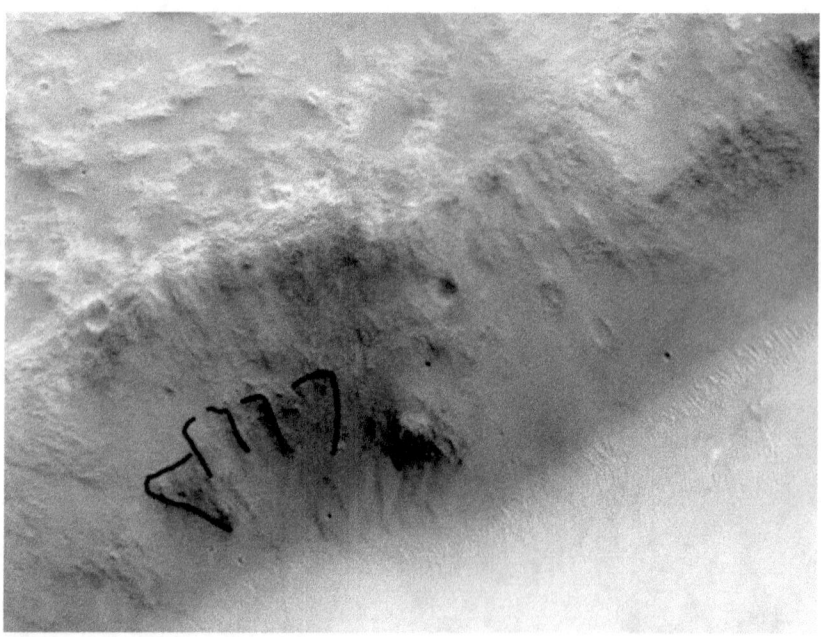

Figure 77

Below are face like shapes on the opposite side of the King's Valley, taken by HiRise. They may be highly eroded faces or could imply Face like shapes are being created here by geological processes. They might also be uncompleted faces, for example as placeholders for rulers or public figures to be memorialized later.

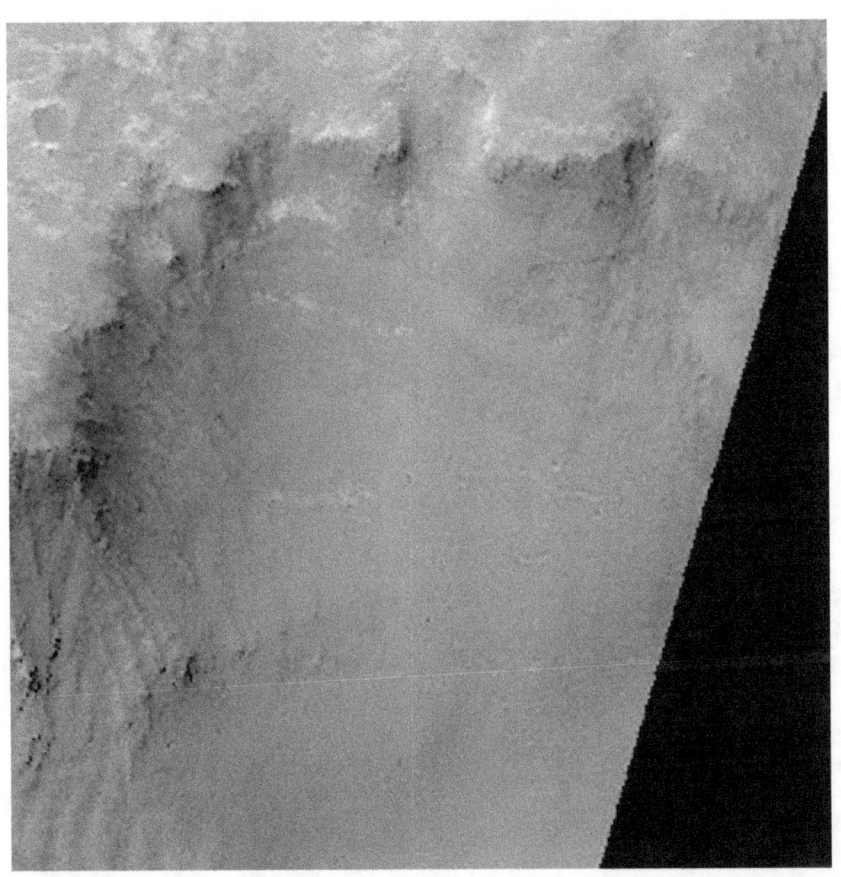

Figure 78

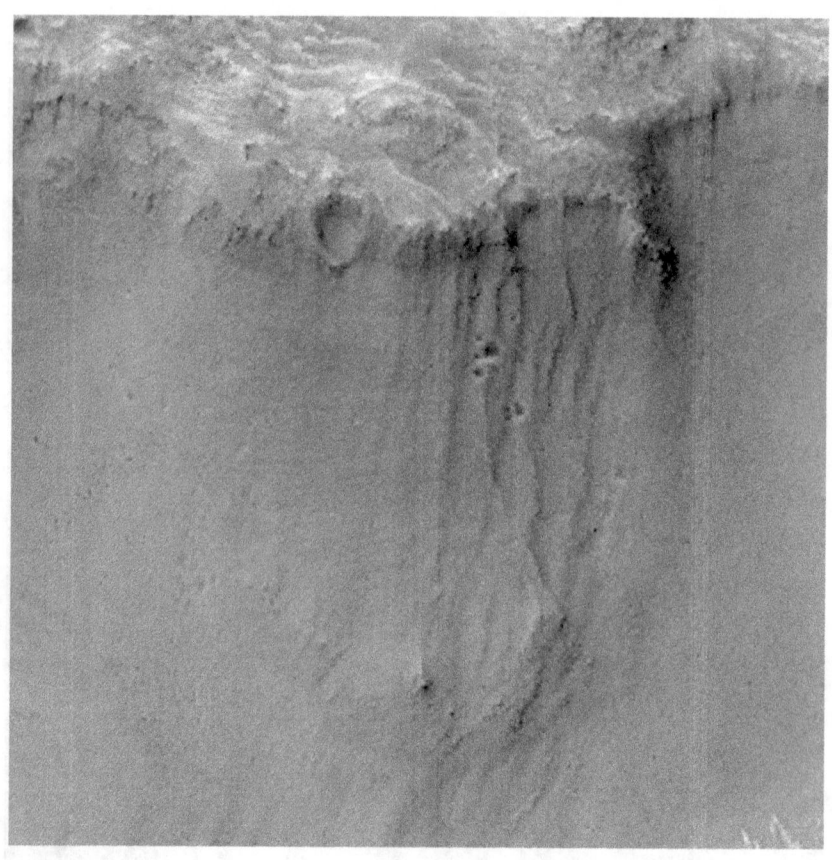

Figure 79

Face Eighteen

This is the original MOC image of another possible face, called Face Eighteen.

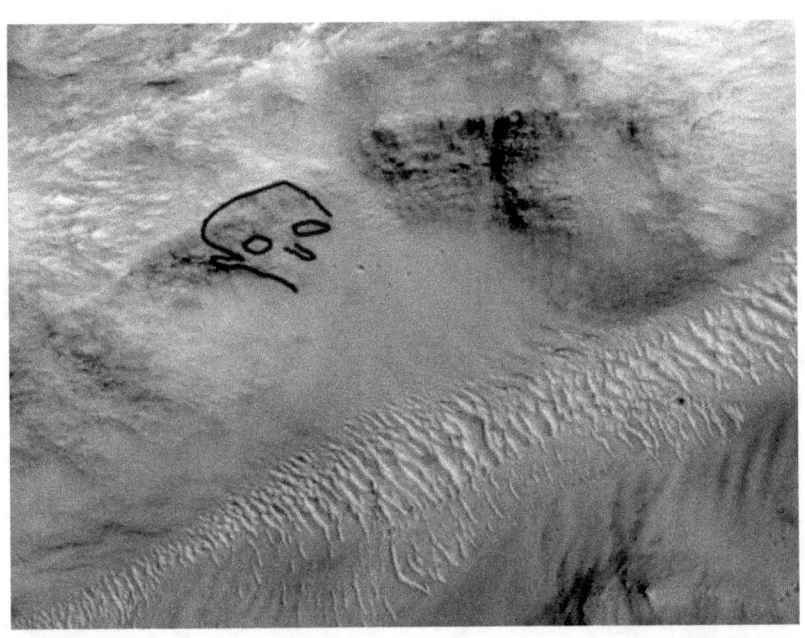

Figure 80

This is from the original photos of the King's Valley photo M0203051.

Figure 81

This area was reimaged by HiRise with the Crowned Face.

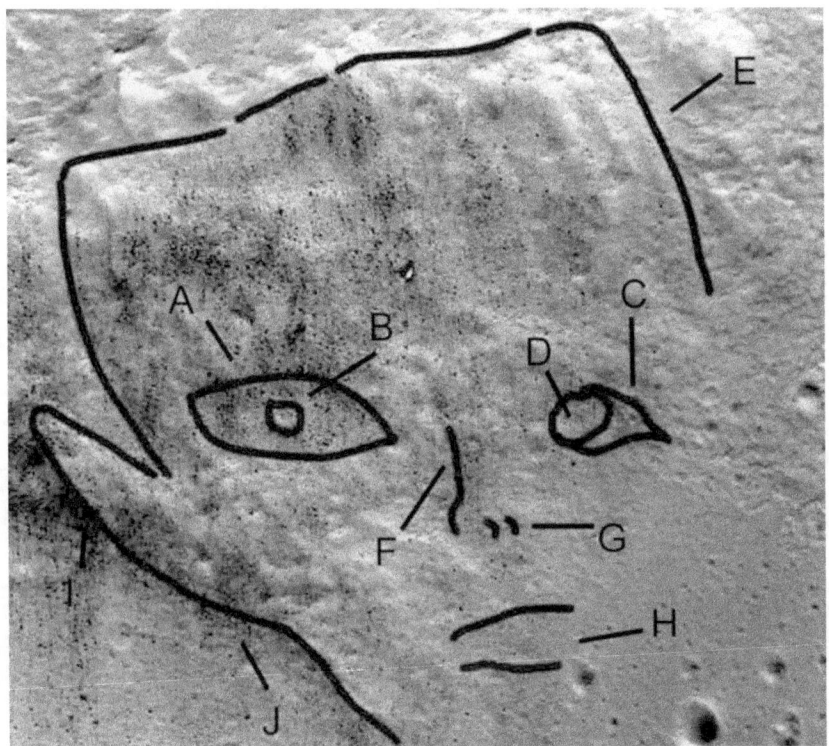

Figure 82

This area is looking more face like compared to the original MOC image above. **A** is an eye shape with an iris shape at **B**.

C is another eye shape with an iris at **D**. The right eye is slightly below where it would be on a human like with Faces One, Two, and Three. **E** is a crown shape similar to the main Crowned Face. **F** is a line like the ridge of a nose with **G** having two marks like nostrils. **H** looks like a faint mouth. **I** looks like a jawline and **J** like an ear but much larger than on the Crowned Face. Overall there are many similarities to the Crowned Faces so this is increasingly unlikely to occur by random geological processes.

Face Nineteen

This may be another dual face, shown below there is Face Nineteen to its left.

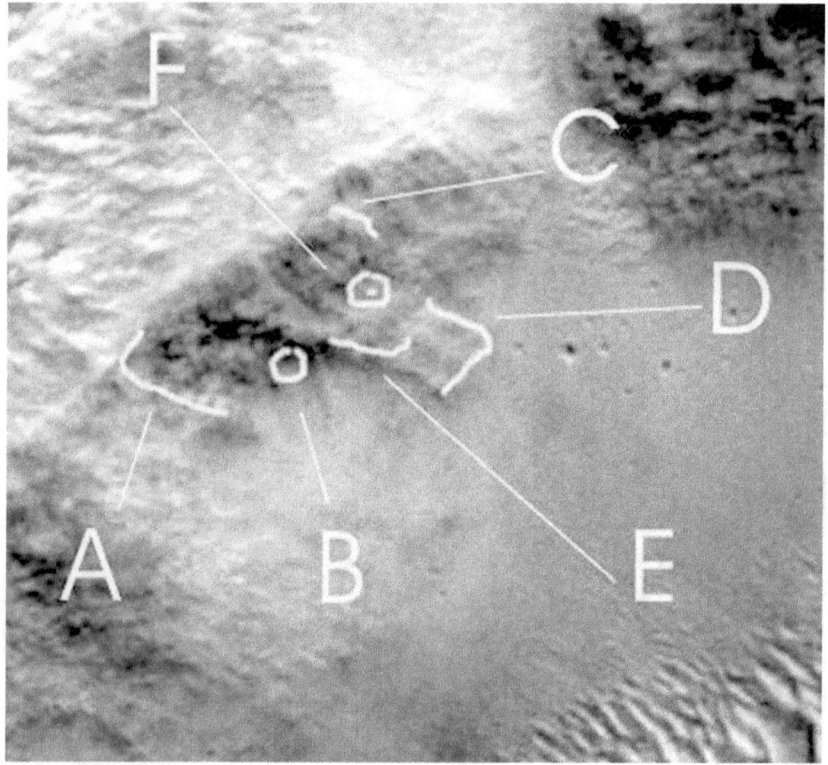

Figure 83

A represents a ridge which defines the left side of the Face. B and C are eye shaped and dissimilar to crater shapes. D represents another ridge which could throw a shadow and define the right side of the face. E is a nose shape and F would be the right eye being part of a dual face with the face imaged above. This would be a repeat of the motif of dual faces which seems to occur in the King's Valley as well as with the Cydonia and Meridiani faces. Again there seems to be little evidence of a nose and mouth and little of a chin but this area also seems buried in soil. If for example the rest of the face did not appear buried then this absence of features would argue it was not a face, and that face shapes might be occurring by chance. There is another a priori prediction that a human or robotic expedition would find more facial features buried here. Arguably there is another a priori prediction in positions like this, that missing features of a face should be buried or otherwise damaged to account for their being missing. It is unlikely a builder would fail to finish a face unless a cataclysm happened such as a major meteor impact.

Here is this possible face reimaged by HiRise.

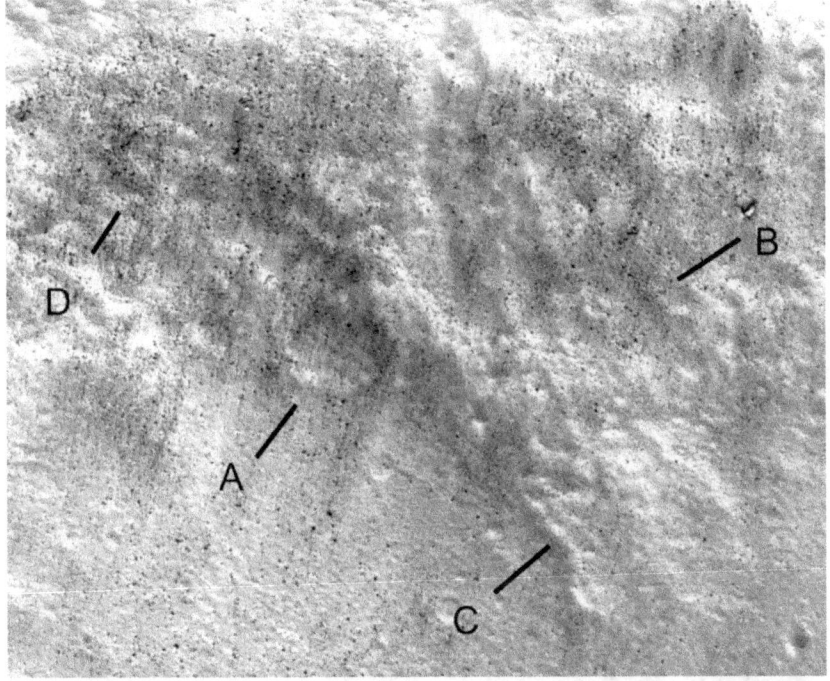

Figure 84

This is the HiRise image of Face Nineteen.

A shows no more eye detail here, it looks more like a crater except it is in the exact position to be the left eye. The left ear of the face above appears here between the eyes, this is similar to the motifs with the Crowned Faces where parts of one face blend into another. **B** is the left eye of Face Eighteen to the right; it is consistent here with the right eye of this face. **C** is consistent with a nose line; this would give a long and slender nose like the Crowned Faces. The whole area to the left of this line might have broken off; the area buried below this in the valley should be searched by a manned expedition. It appears similar to the broken off nose of Face One. **D** may be the edge of a crown shape. It is difficult to say whether this is another possibly artificial Face or an optical illusion caused by a crater but a Dual Face is consistent with the motifs seen in this valley.

Face Twenty

Face Twenty below is more dubious because of its exaggerated features, but reimaging may resolve this. There are some other possible faces in the King's Valley; more images from HiRise may make them plausible enough to be given a designation.

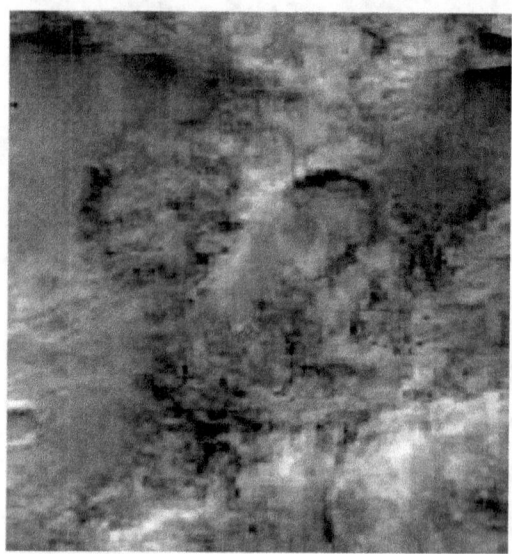

Figure 85

I didn't find this; it was pointed out by a female researcher on a forum many years ago. I will update their name if I find it out later. All the formations I refer to are found by me unless otherwise mentioned. It is highly pixilated in the image because it is small.

Face Twenty One

The Face below is new from a recent HiRise image. I call it the Mug Face because it appears that a crater blew off the top of the head, it gives an impression like novelty coffee mugs with a face on them. This is Face Twenty One.

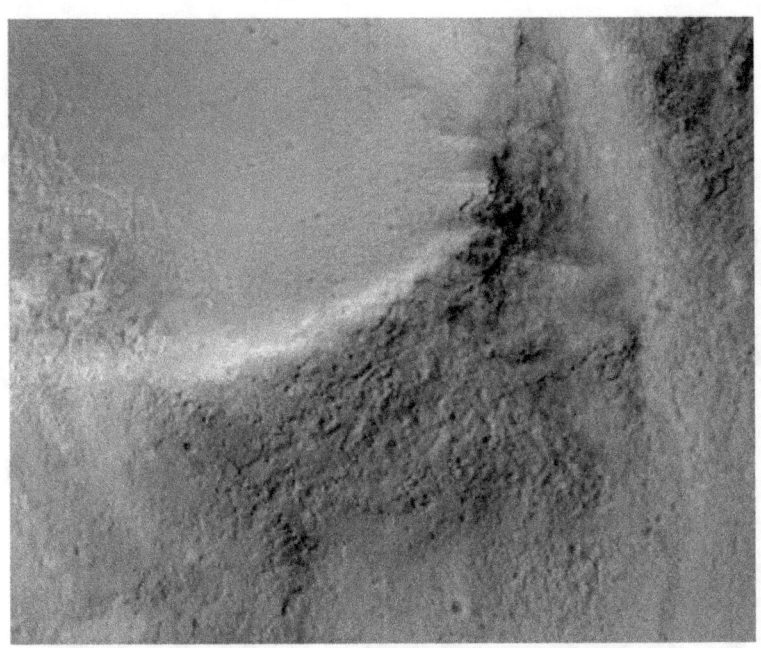

Figure 86

The crater may have been formed later blowing off the top of the face or it may have been incorporated to give the impression of a crown. This is Face Twenty One.

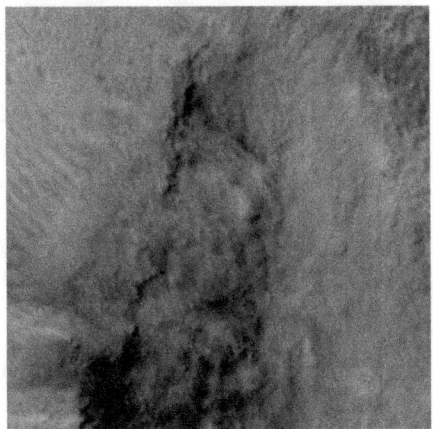

Figure 87

Above may be another face looking right, not included in the numbering here, also perhaps damaged by the crater impact.

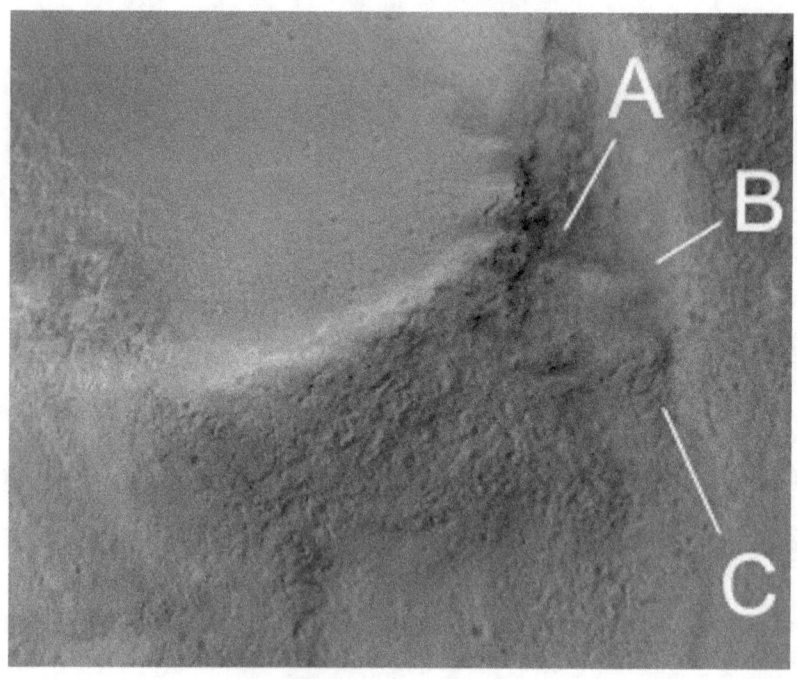

Figure 88

A shows the eye, ***B*** *the nose and* ***C*** *a less clear mouth.*

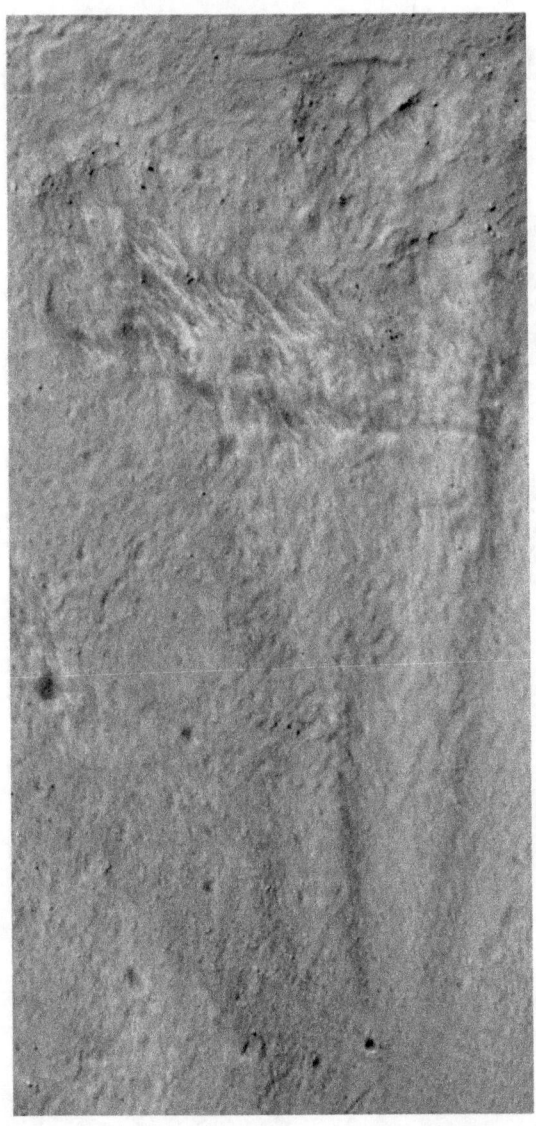

Face Twenty Two

Figure 89

This is from the HiRise image in the King's Valley; I call this the Sunglassed Face because of a resemblance to wearing sunglasses. Provisional names like these are just to make them easier to remember; no doubt someone will come up with a better name later.

253

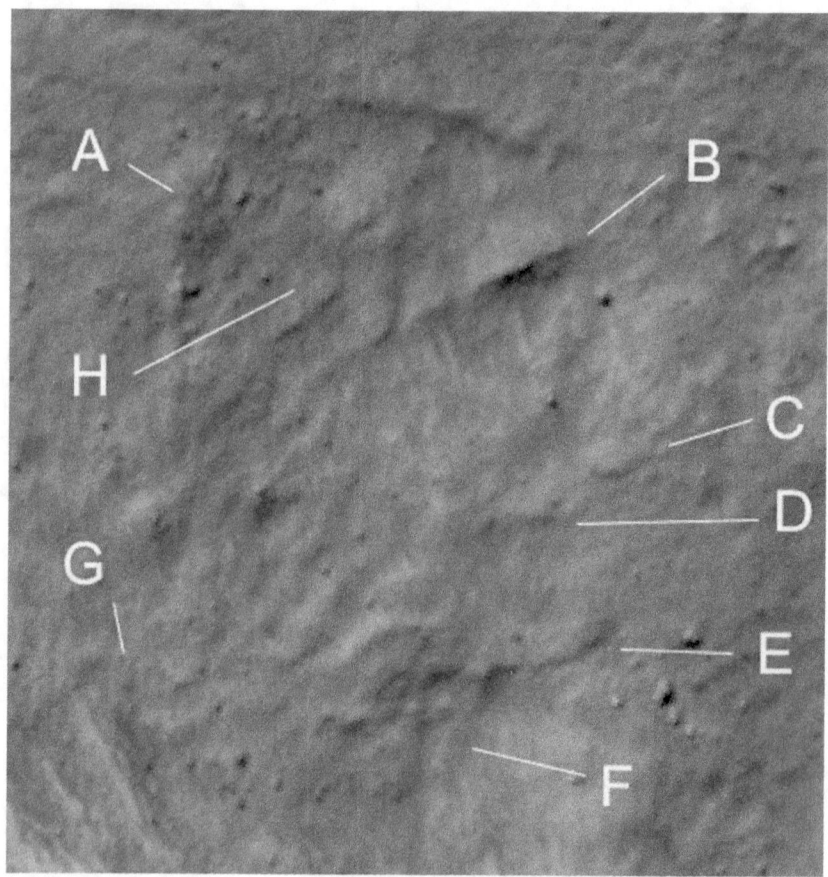

Figure 90

The shape of the head at **A** is rounded rather than wearing a crown; it may also be a more rounded helmet. **B** shows either the edge of a helmet or an eyebrow, the eyes are less clear in this Face but this area also looks like it suffered some erosion. **C** shows a rounded nose shape consistent with the Crowned Faces. **D** is a mouth in approximately the right position but there is no lower lip, this area also appears highly eroded so this gives an excuse for it being missing. **E** gives a rounded chin though the Crowned Faces have a more pointed chin. This can be a natural variation in a species; this does not look like the same humanoid as the Crowned Face. **F** shows a neck width similar to our own, there is an Adam's apple shape but different to in a human. **G** shows a possible shoulder or this could just be sand dunes. **H** shows possible waviness like hair or the side of a helmet.

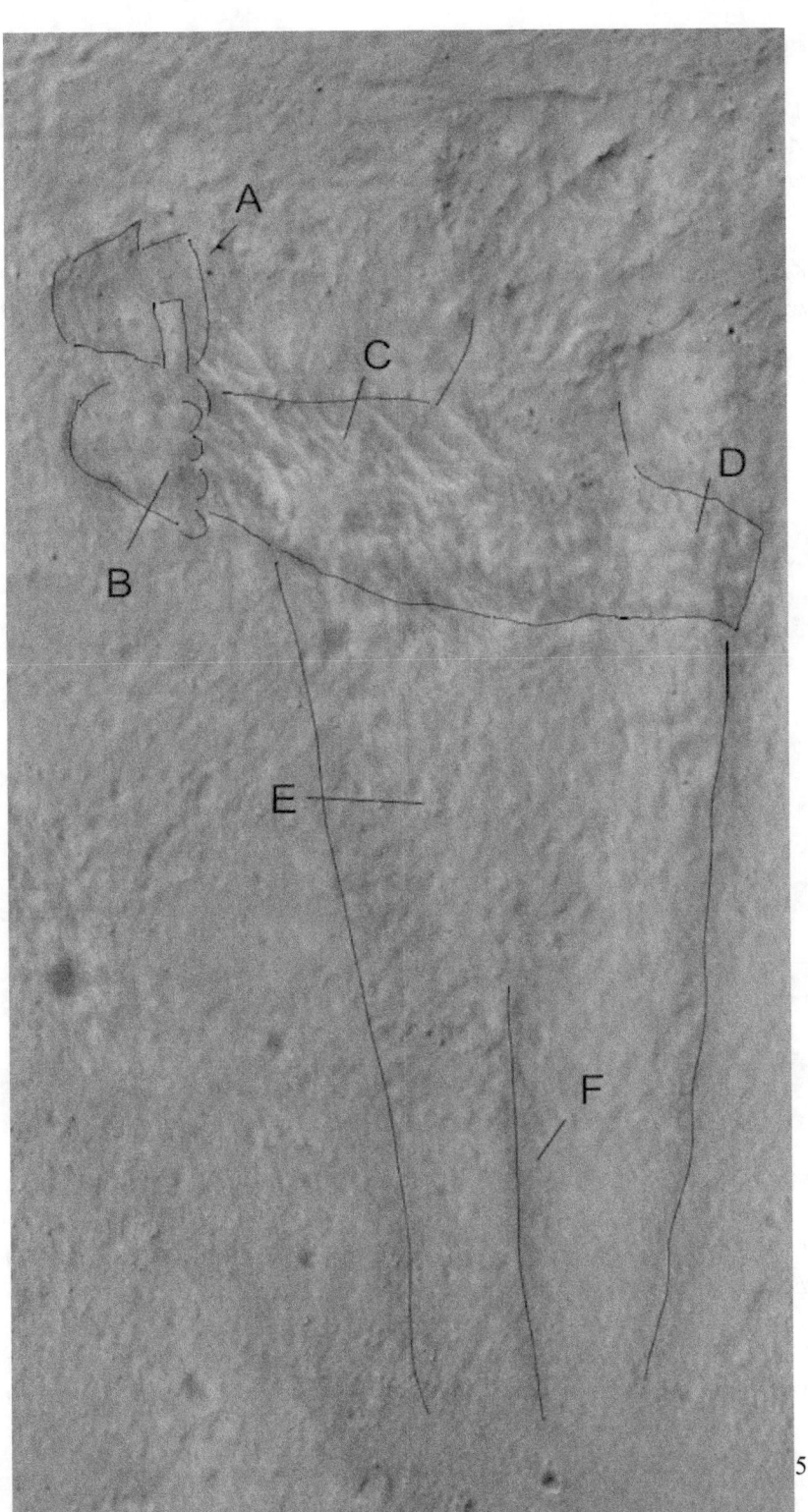

Figure 91

There appears to be nearly a full body attached to this face, the first clear face with one. A seems to be something that is being held, it looks like the old style flashes cameras used but may be light source like a candle with a parabolic mirror behind it. B seems to show a hand with 4-6 knuckles holding the device. C shows an arm, there may be a short shirt or armor being worn, D shows the right shoulder. E and F show the possible torso, this is a natural part of the cliff but the coincidence of it being like a 2 legged body was probably intentional. That assumes of course that they had two legs.

Face Twenty Three

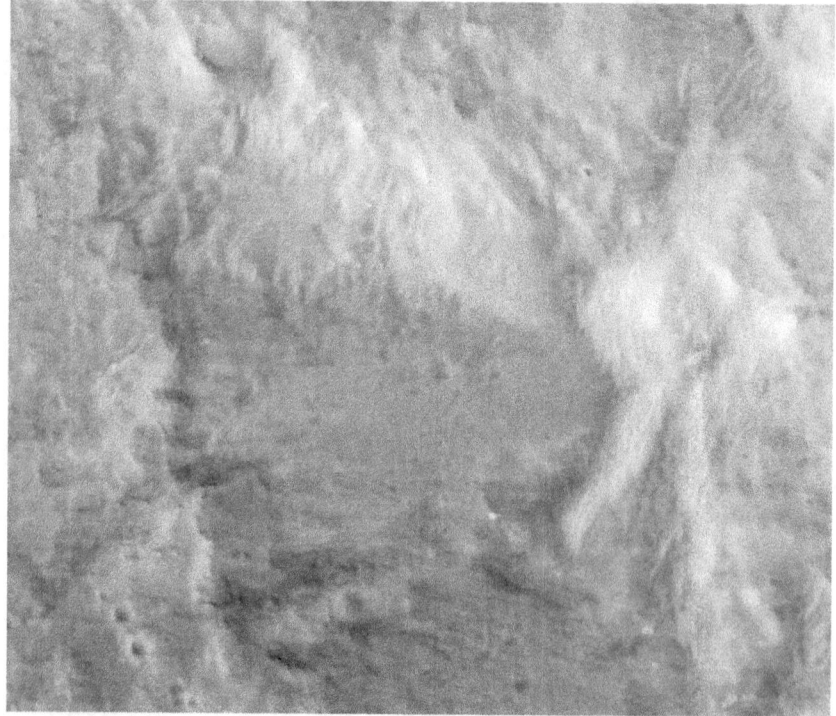

Figure 92

This is called the Dante Face, because its crown or helmet is similar to Dante's hat in the Statue of Dante. This is Face Twenty Three.

The HiRise image fulfills an a priori prediction of more faces appearing as well as more features of the larger faces. However it does not seem to produce more faces outside of this valley so it does not seem to be a fault of the way the camera takes pictures.

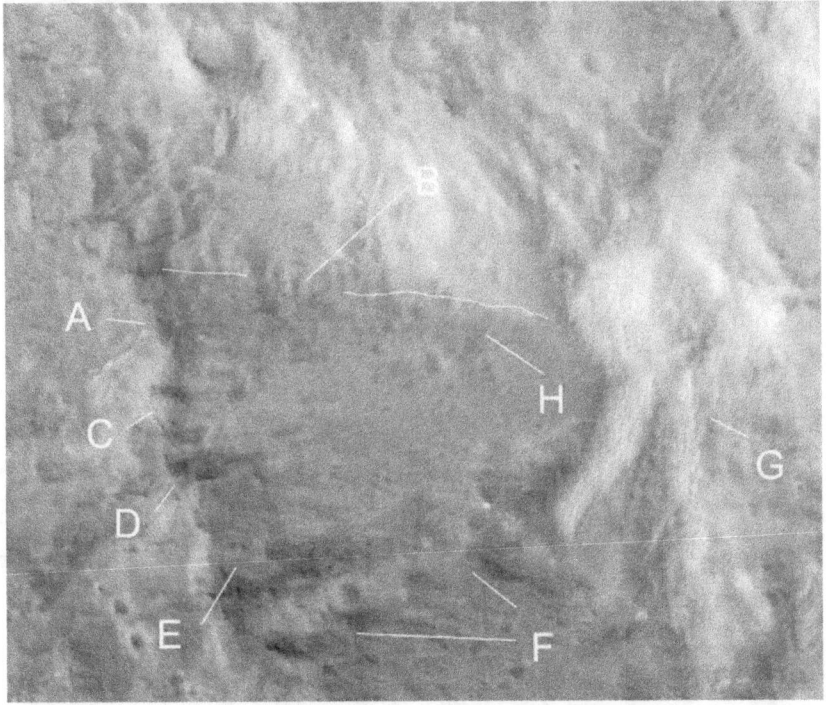

Figure 93

A shows an eye with some detail as does **B**. **C** shows a nose with a rounded piece missing out of it, this may have been caused by a meteor crater and mars the facial shape. **D** shows a mouth with a clear lip but there is no lower lip. **E** shows a rounded jaw. **F** shows a possible top of a jacket or armor under the jaw. **G** shows a ribbon like shape on the edge of the helmet but this may be sand dunes on the side of the crown or helmet. **H** shows a line where the helmet or crown fits onto the Face.

Face Twenty Four

The next formation is in a fork in the river just near the King's Valley, called Face Twenty Four. Long ago a traveller up the valley whether by boat or walking might have seen this large face and then have come to the main King's Valley with twenty three other faces so far. Below it appears face like even in the standard Google Mars background.

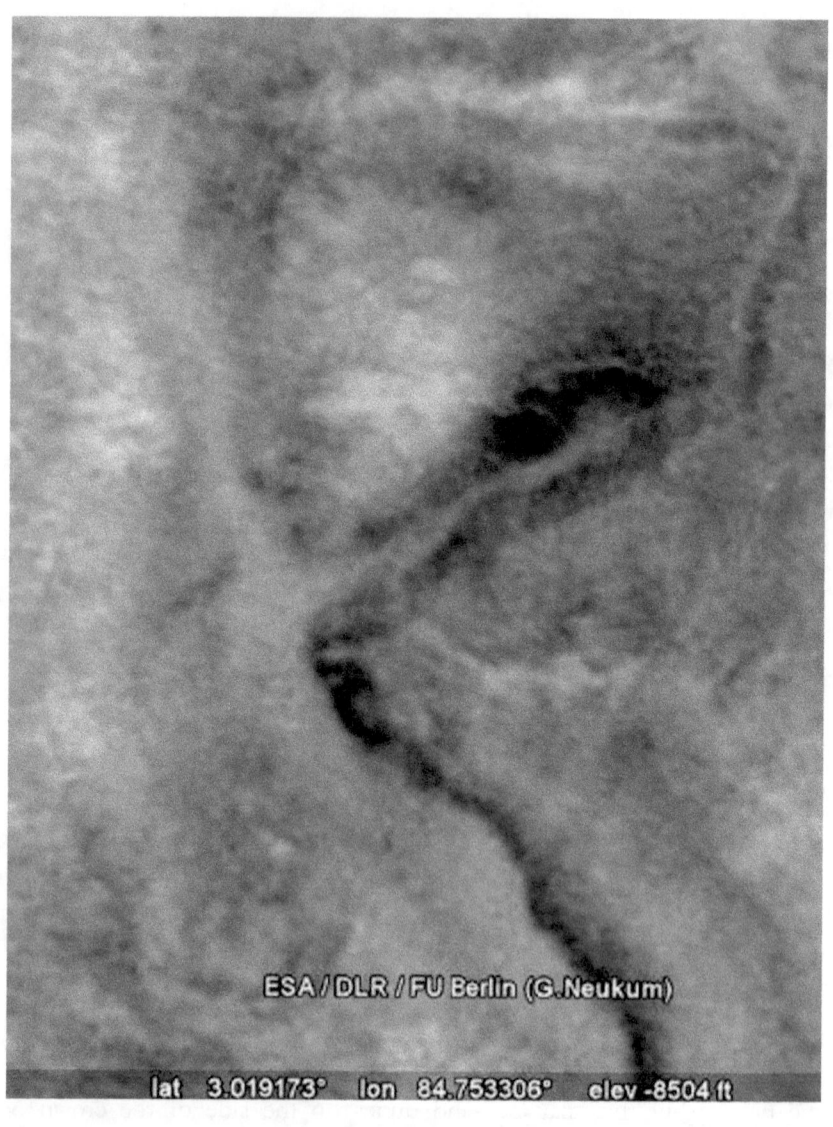

ESA / DLR / FU Berlin (G.Neukum)

lat 3.019173°   lon 84.753306°   elev -8504 ft

Figure 94

The Fork Face stands out with much lighter colored soil than its surroundings on the right. It has a distinctive crown shape which is similar to the faces in the King's Valley. The more similarities there are to the other faces the less likely it is a random formation.

258

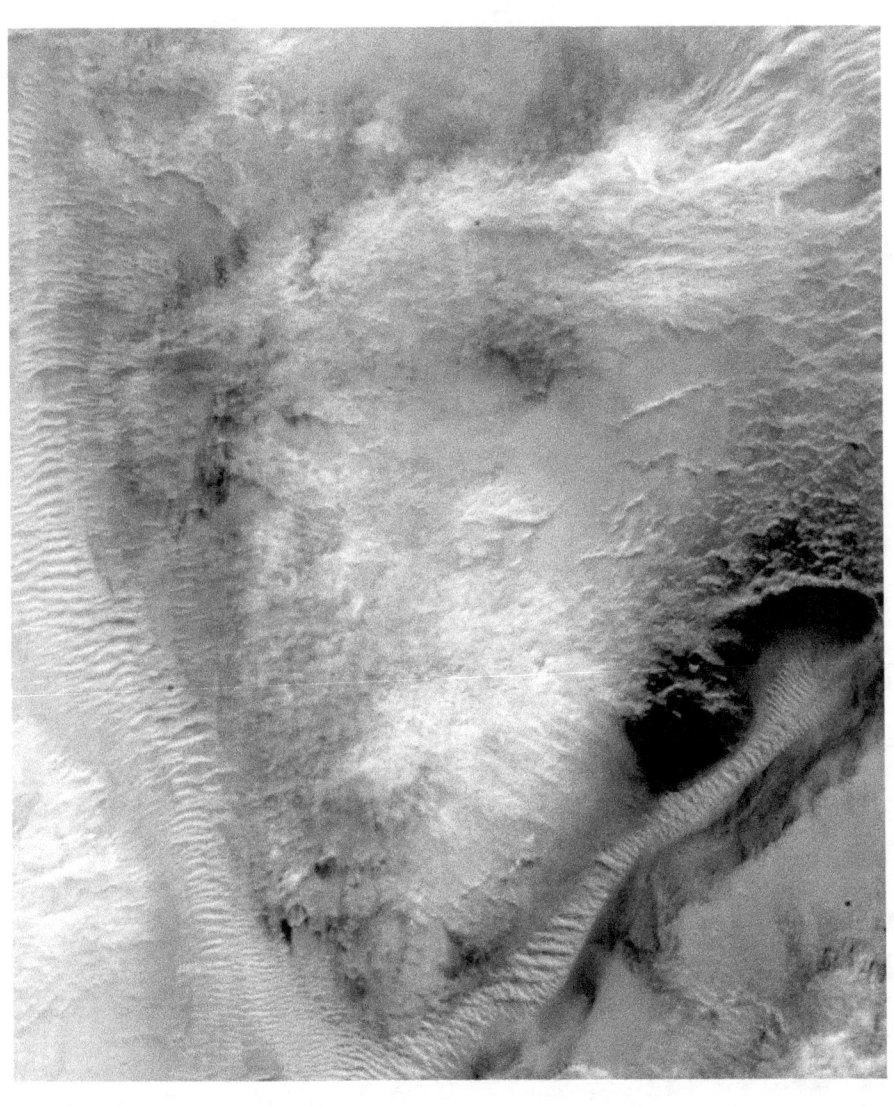

Figure 95

In Photo R1901247 this formation was reimaged. If artificial it is highly eroded and the sand dunes obscure some of the features.

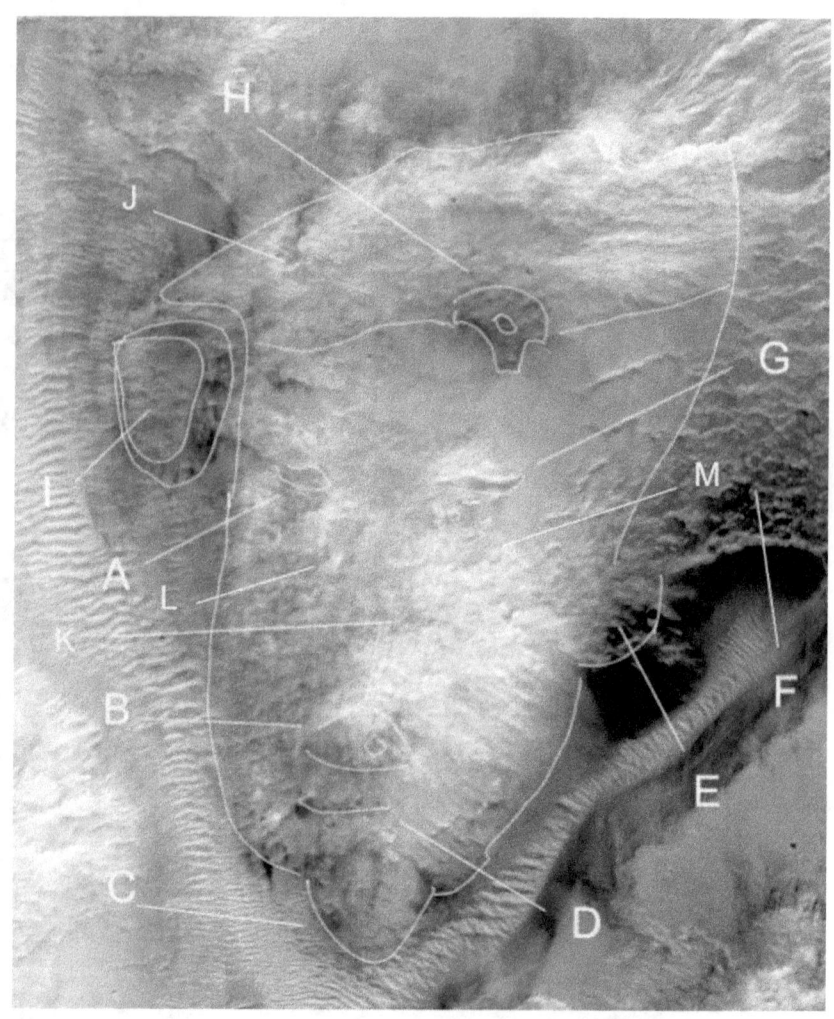

Figure 96

A *shows a left eye though there is little detail.* **B** *would probably be the tip of the nose, it seems much higher than the other parts of the possible Face but it seems a little too far down the Face.* **C** *has a chin that is elongated like the Crowned Face, it also has a chin like Face Fifteen below.* **D** *would be the mouth but there is little detail.* **E** *protrudes in the right place for an ear.* **F** *shows dunes positioned to look like hair but this should be a coincidence if the Face is hundreds of millions of years old.* **G** *shows the right eye which is fragmented in shape like on Face Three.* **H** *shows a strange dark shape with a dark spot in it, this is like a dark spot that appears on the Crown Face's crown.* **I** *is shaped like another crowned face, wider on the top and with a dark outline around most of it. It seems unlikely for this shape to be the same as the larger Face by*

260

chance. *J* shows the crown as looking more like a beret with a clear line across the forehead like in Face Fifteen shown again below. *K* shows a strange mark in the shape of an eye. *L* and *M* are an alternate position for the eyes, there is an eye shape with a pupil at *L*. Reimaging this feature with HiRise may resolve some of these questions, its proximity to the other faces makes it likely to be artificial.

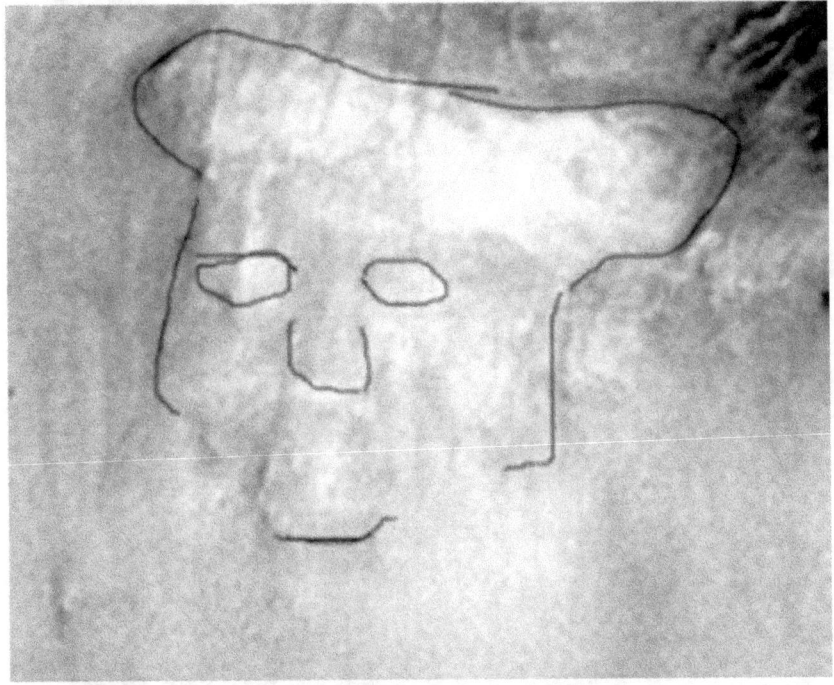

Figure 97

Face Fifteen on the valley wall opposite the Crowned Face was reimaged in photo S1402499∞. It also has a crown that is larger and sticks out on the sides like the Fork Face, and a rounded chin. It may be of a different but similar humanoid.

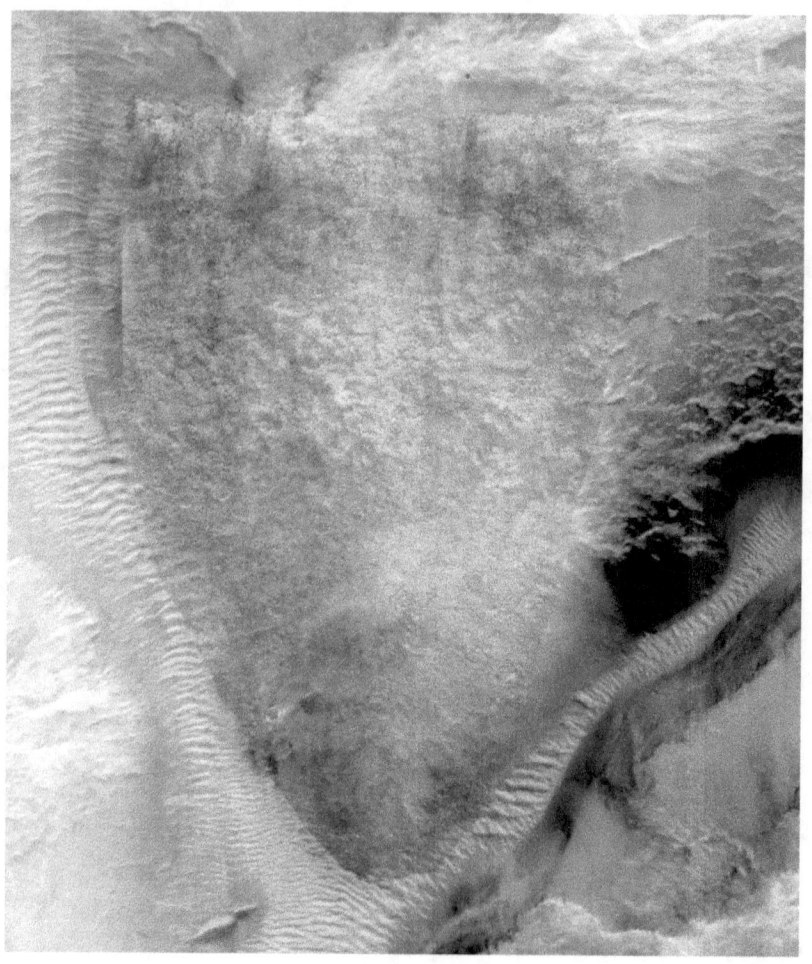

Figure 98

This is the Fork Face overlaid with the Crowned Face reimaged by HiRise to compare the proportions of the features. When the chin is fitted to the fork in the river the crown fits and the nose tip on the Crowned Face fits quite well on the highest point of the Fork Face, unlikely to occur by chance. The width of the face is also quite similar except the crown is wider on the right, but this can be because of the arbitrary way I have selected the crown edge where it blends into the crown of Face Three. The eyes end up in a lower position seen as **L** and **M** above which also appear to be eyes on the Fork Face; this could then be plausibly a face originally very similar to the Crowned Face. A fit this good is not likely to happen randomly, I didn't have to distort the images in any way to fit them like this.

Fish images

In the HiRise image approximately opposite the Crowned Face there are two fish representations, plus some unclear carvings possibly representing seaweed. The King's Valley would have had water in it and the dam upriver may have had fish.

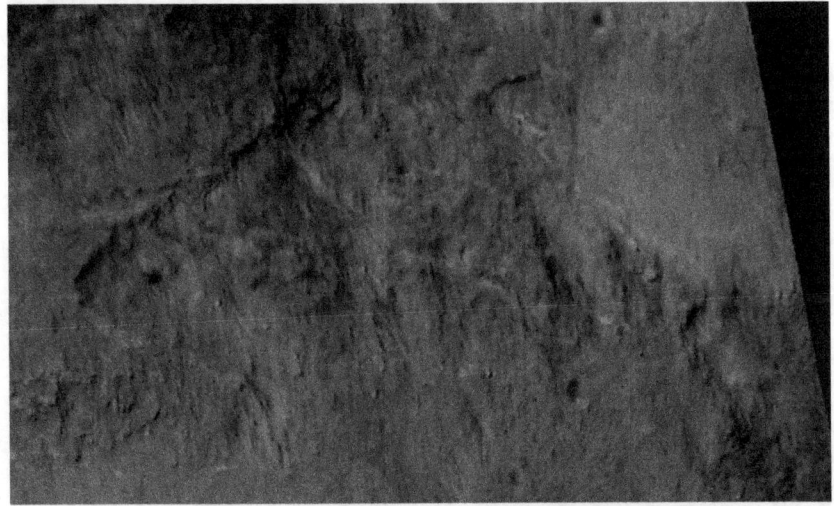

Figure 99

This may be one or two fish; perhaps the one on the right is attacking the other or they are mating. The fish on the left is Fish One, the one on the right is Fish Two.

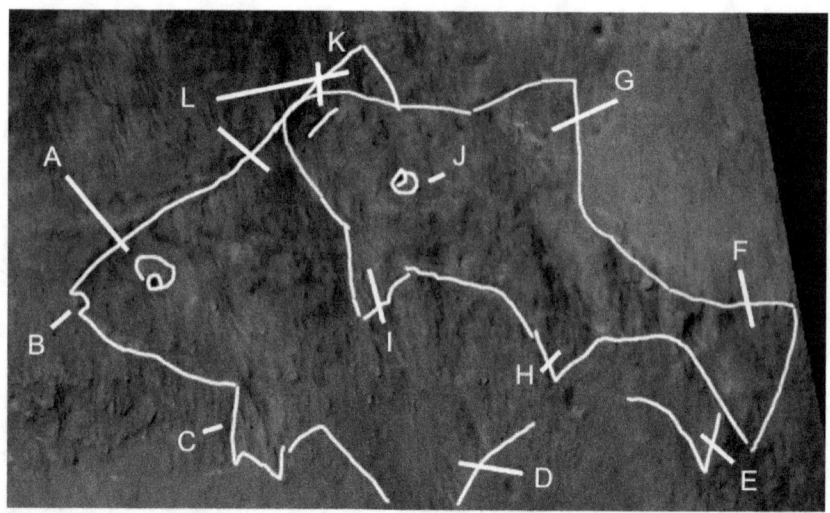

Figure 100

The picture is not very clear, below is one interpretation of the two fish shapes. **A** shows an eye with an iris like on Fish Two at **J**, **B** shows a mouth and Fish Two may be a view from above with the mouth at **K**. **C** shows an extended gill shape like on Fish Two at **I**. **D** shows a fin like on Fish Two at **H**, **E** shows a very faint tail like the one on Fish Two at **F**. **G** and **L** show similar top fins on both fishes.

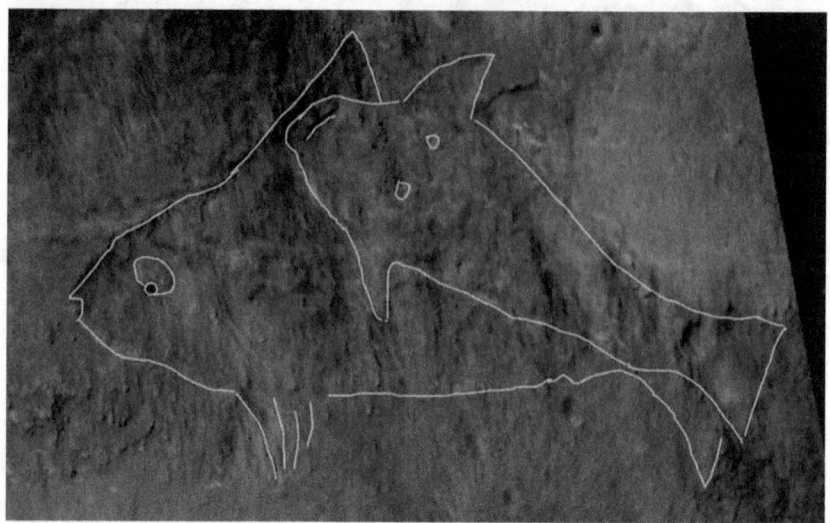

Figure 101

Above is another version, here Fish Two on the right is from an overhead view. It is unlikely of course for there to be on fish image, to have two next to each other is highly improbable because they are so alike, but there is also another possible fish nearby looking to the right called Fish Three.

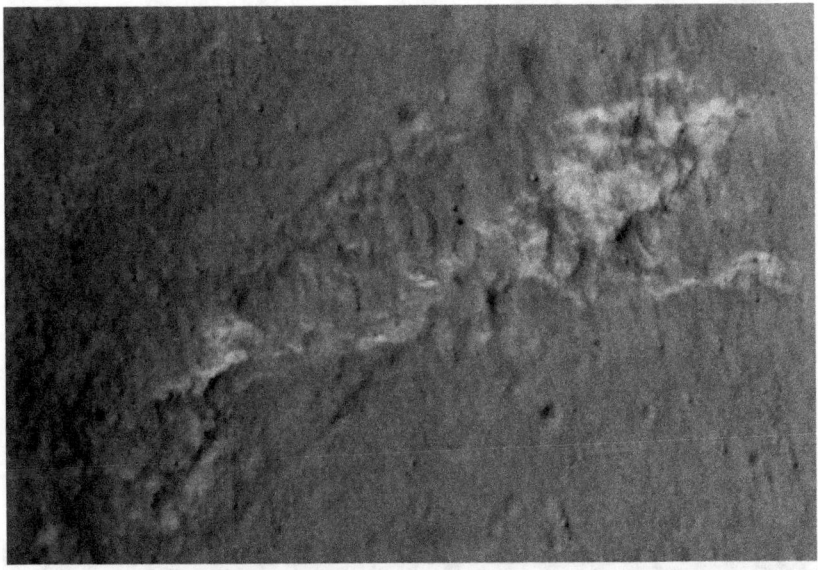

Figure 102

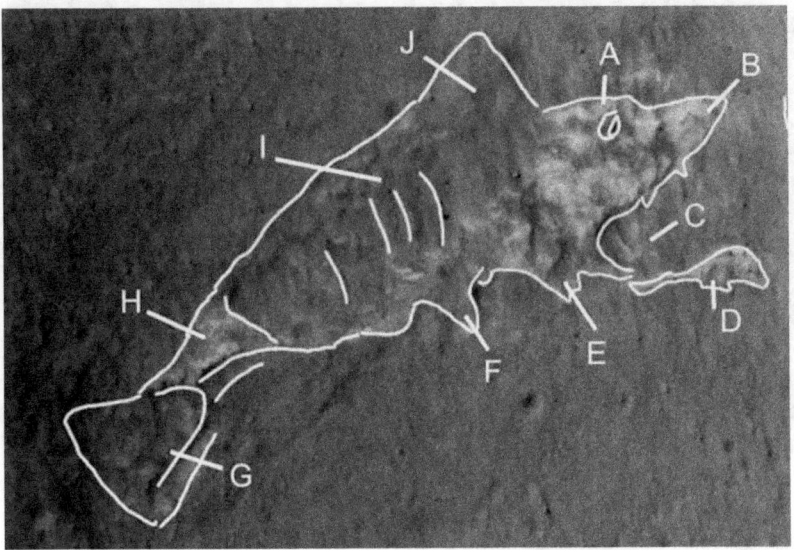

Figure 103

Fish Three may represent one of the other two having been eaten, their being close together may represent they were caught and were waiting to be eaten. **A** shows a similar eye to Fish One, **B** has a similar snout, **C** shows an open mouth, perhaps the jaw at **D** has been ripped off while being eaten. **E** shows a similar enlarged gill or fin to Fish One, **F** shows another similar fin. **G** shows a similar tail and **H** perhaps part of Fish Three still intact. **I** may show the bones after the flesh has been eaten or they could be scales or markings. **J** shows a similar fin to the other two fish. There are six similar features to the other two fish at **A, B, E, F, G,** and **J.** If there was just one fish shape it may not have been worth pointing out, but the position opposite the Crowned Faces deserves more attention. For there to be two or three fish images so similar to each other is extremely unlikely to occur by chance. For example six similar features on two fish at odds of 10 to 1 each would give 10^6 or 1,000,000 to 1 against chance. If Figure 100 is accurate then there are 12 similar features on three fish which gives 10^12 to 1 or 1,000,000,000,000 to 1, a trillion to one against chance. The fish serve as a good prediction that more details will be found on reimaging this area, there may be more fish to the right of this image.

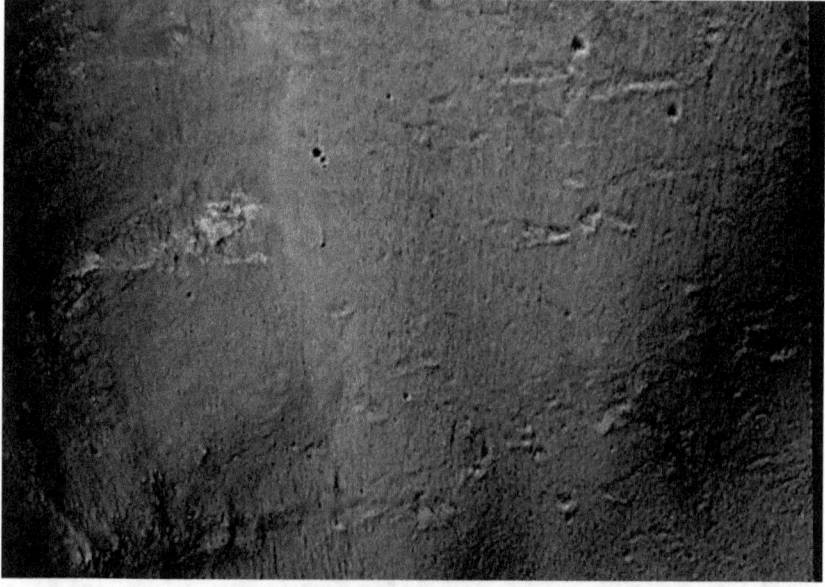

Figure 104

There may be something else here, Fish Three can be seen on the left and Fish One and Two are above this. The wavy lines on the right may represent seaweed or they could be many fish that are too eroded to see clearly. The shape directly to the right of Fish Three looks vaguely like another fish but is highly eroded.

Chapter Twelve

The King's River

One reason the King's Valley Faces are so well preserved is that the water may have been dammed very close to them up the river valley.

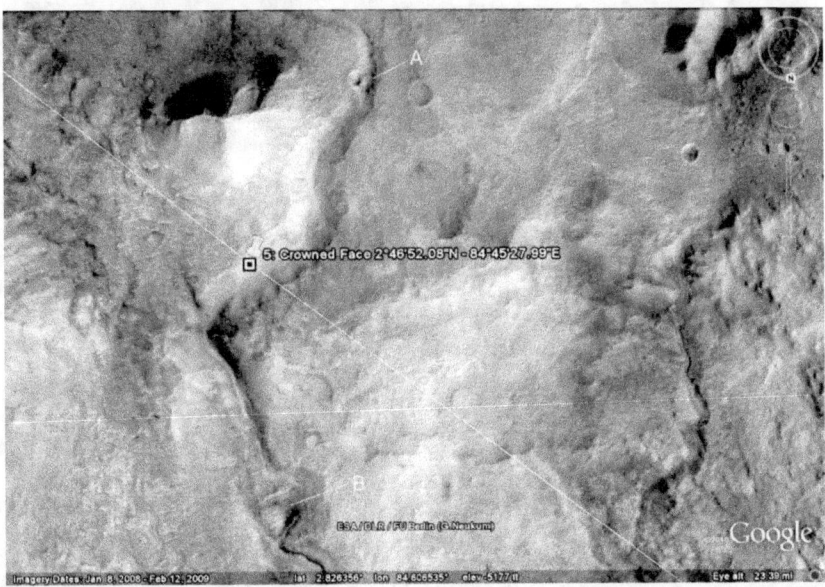

Figure 98

A shows the possible dam next to the crater, this would prevent some water from moving down to the King's Valley Face area and then down into Isidis Planitia and the ocean. B shows the Fork Face which someone would see first moving up the valley from the ocean, perhaps arriving there by boat. It might also be intended to attract attention from space so someone would look closer at the King's Valley. The line is the old equator that the King's Valley used to be on.

Figure 99

The river seems to be blocked next to the King's Valley with all the faces.

270

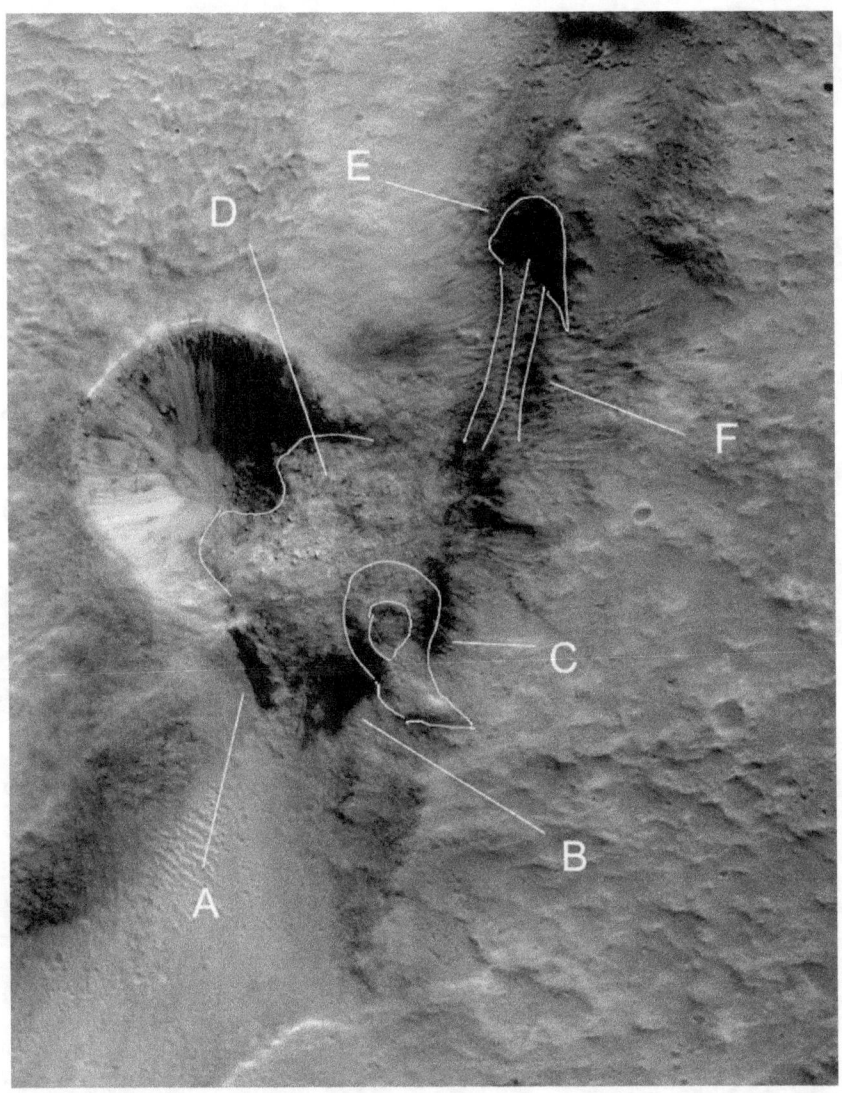

Figure 100

A shows a straight edge like the wall of a dam. Perhaps water was accumulated in the deeper area around the letter **A**, this seems more likely as on the other side of the dam there is no sign of water backing up. It might also be a spillway for water filling up in the crater. The water may have diverted to another river when the dam was built, photos later may show this change in its flow. **B** shows another straight edge where the dam abuts the wall of the valley. **C** is an odd shape perhaps artificial. **D** shows a large mass of material that is far bigger than ejecta from the crater. Where could this material have come from, there is no hollow nearby it could have come from naturally as a crater would spray ejecta in all directions much further than this. It fills in part of the crater showing it could not be ejecta; it must have after the crater was formed. **E** shows a possible cave that the road or channel **F** comes out of. If this is a cave or lava tube it may be an ideal location for a base for a manned expedition. There may be underground ice in this area left from when there was a river here.

Near the king's Valley

There seem to be no indications of building shapes near the King's Valley with the possible exception of the angular wall like shapes below. These may just be random formations; they may be worth reimaging or visiting if a manned expedition comes to the King's Valley.

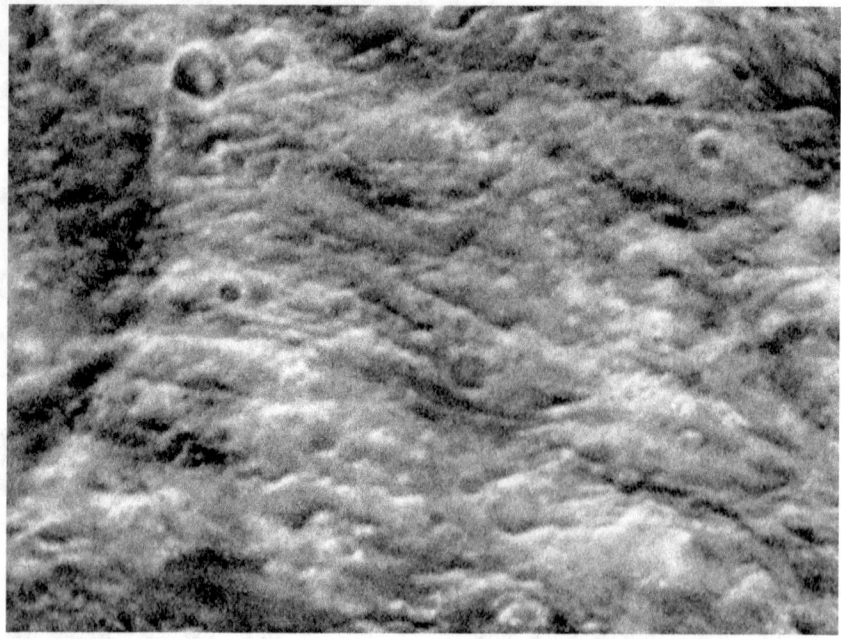

Figure 101

Photo AB110703, one of the earliest photos taken by the Mars orbital Camera which is why the resolution is much poorer.

272

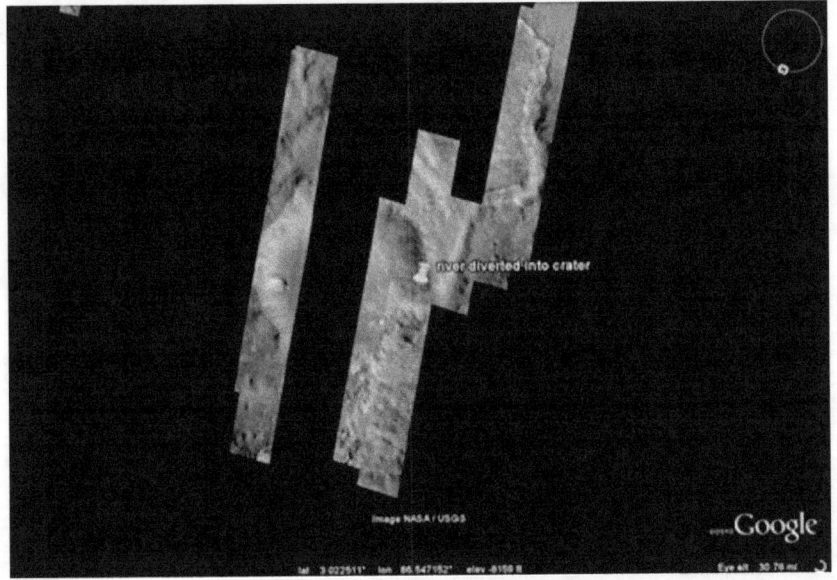

Figure 102

Close to the King's Valley there is an unusual occurrence, a river going directly into a crater which would have acted like a water reservoir. This could be checked out by a manned expedition to see if the river was diverted for this purpose. More photos later may show if this river had another natural course it was diverted from or whether it flowed through this area and then was hit here by the crater impact.

Chapter Thirteen

Comparisons of Face One and Face Two

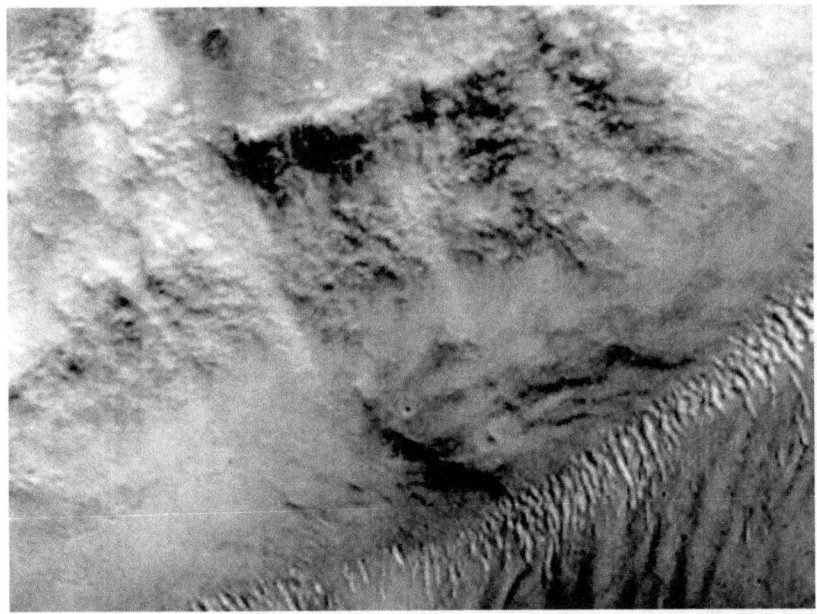

Figure 1

In this section of the King's Valley there appears to be three faces partially overlapping each other. One possible explanation is these are meant to represent three different humanoids. Another is they are of the same humanoid with different facial expressions. For example with a different sun angle each face might stand out at a different time of day or season with a different look such as happy in summer, sad in winter, or however their facial expressions might translate to emotions if they had any. The centre face is best known as the Crowned Face and is seen by most people above. There may also be a face on the left which is highly weathered.

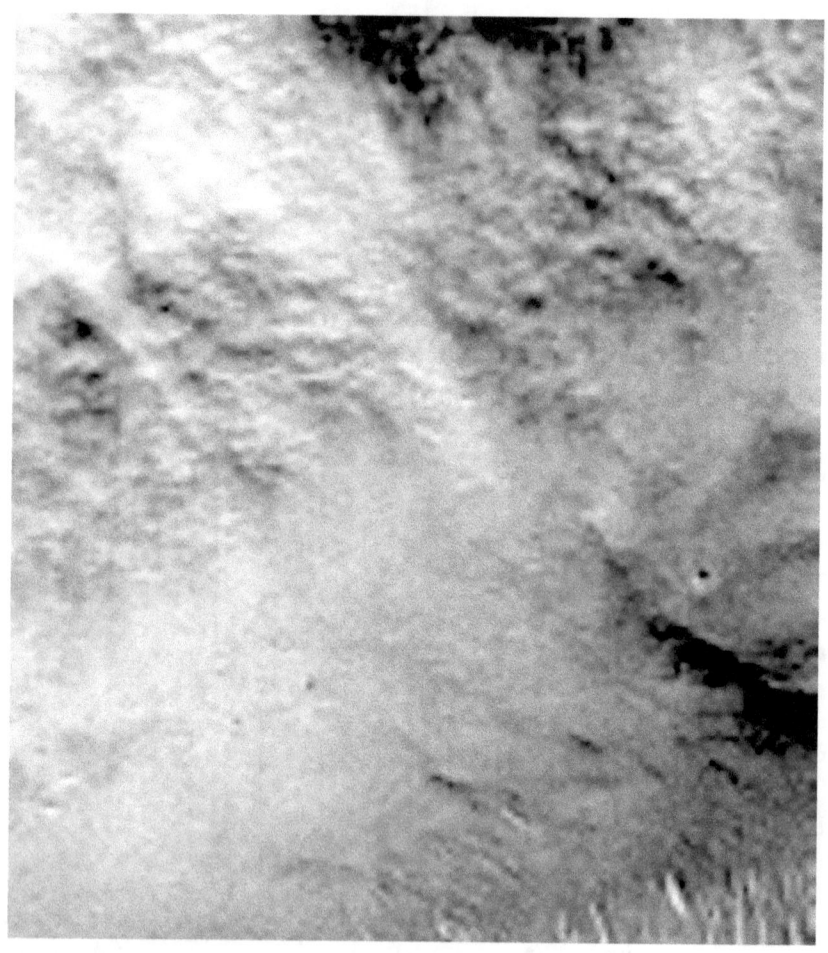

Figure 2

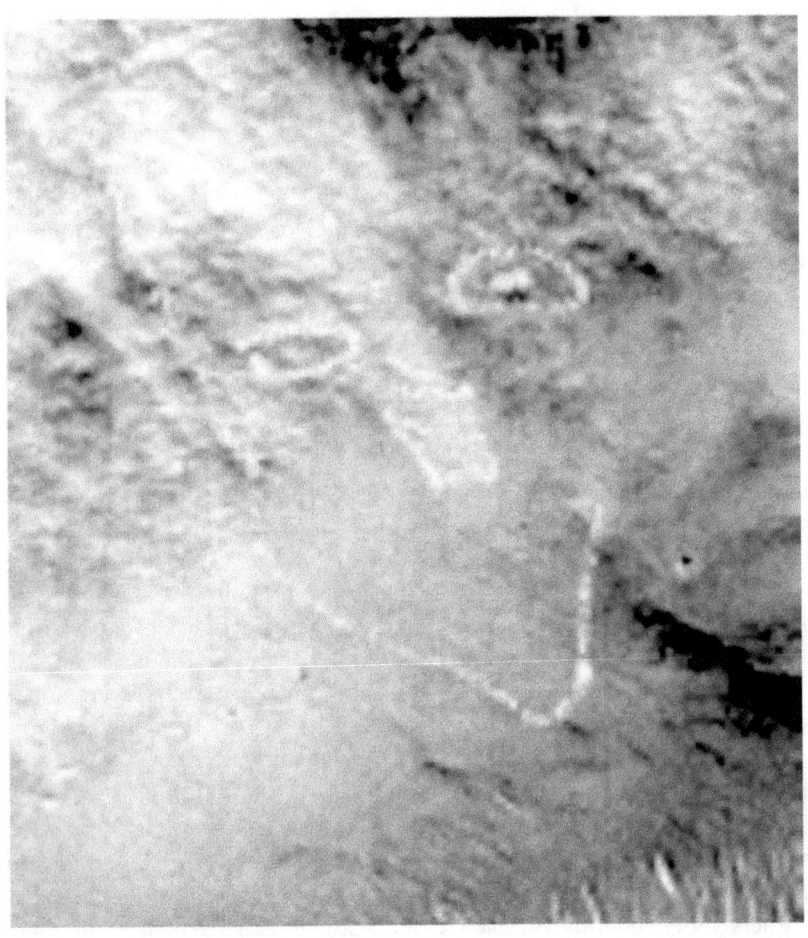

Figure 3

This is Face One; the Crowned Face is Face Two.

The third Face is on the right and also overlaps the main Crowned Face. First I will compare the central Crowned Face with the one on the left, i.e. Face Two with Face One. Then I will compare Face Two with Face Three on the right to look for similar features. I will also discuss each feature as to how it could be formed naturally, i.e. by geological processes like wind or water erosion as opposed to having been constructed artificially.

I will also attempt to estimate how improbable the similarities between the faces are. This will be estimated in the odds against these similarities happening randomly, if sufficiently improbable then it indicates that they should have formed by some common process. This can be a geological process if one could exist, for example the different faces might have similar erosion on them from the wind so some similarities can occur this way. Other similarities are harder to explain such as similarities in the eye and noses of the faces, wind and rain should erode these features randomly and not have a tendency to form faces that look like each other. If they did we would expect to see Crowned Faces all over Mars where this erosion occurred. So similarities between these faces in this small area imply a common process to form them only there, while other processes that acted all over Mars like wind, rain, sleet, snow, meteor impacts, quakes, freezing and thawing, chemical reactions such as from radiation, dust storms, moving dunes, etc. did not make Crowned Faces elsewhere. This is an important point; faces can occur in clouds because their shapes are highly random and clouds that are good for making face shapes are common around the world. However there is nothing special geologically about the King's Valley, there are valleys like this all over Mars, there is evidence of some rainfall and rivers flowing all over Mars, there is wind erosion all over Mars, and erosion by ice is also ubiquitous. So why should these supposedly random forces make 24 faces in one small part of a valley? There is no answer to this, nor should there be, the idea that rain and wind could make detailed faces here and nowhere else is absurd and is a prima facie case for artificiality.

Probabilities between independent events should be multiplied together; this means that objects that are formed independently such as by wind and rain should be unlikely to look the same. We see this in clouds for example where they look different from each other because the wind and rain acts independently on each cloud. Clouds can have faces in them occasionally but each face is different from the others because of these random variations. Imagine for example if one cloud suddenly developed 24 faces in it most with crowns while the other clouds did not, this is like the argument of random chance making all these faces in the King's Valley. Mountains look roughly similar in shape which is from the common processes of erosion but when we look closely mountains rarely look too similar to each other because the erosion acts independently on each mountain. In the same way each area on Mars should look somewhat similar to other areas because of these common processes of erosion but not too similar because these processes are still working randomly and independently.

Similarities between faces can be shown to be improbable from natural processes because faces on Mars are rare. Even with the most fanciful interpretations it is unlikely there have been more than 20 Faces discovered by researchers outside the King's Valley and yet there are 24 inside it. Some of these faces are debatable but even with 10 faces in one small valley this is extremely unlikely to occur by chance. Random events should be able to be plotted on a normal curve but it is impossible for this situation to be. If for example there was a 1 in 10 chance of a face sized area of ground looking like a face by random chance then 10% of Mars should be covered in faces. Just in the Mars Orbiter Camera images this would give over 10,000,000 faces on Mars. We know however that only 3 photos out of approximately 100,000 Mars Orbital Camera images have these Crowned Faces, add to this the similarities between the Nefertiti Face and other near it that appear to have crowns plus the similarities between the Crowned Face and the Cydonia Face as well as the Meridiani Face and there may be 10 photos out of 100,000 so there is about a 1 in 10,000 chance of randomly selecting a photo with a crowned or similar face on it. Then there is the percentage of each photo that contains a face, this would be far less than 1% so that gives an initial improbability of 100 times 10,000 which equals 1,000,000 to 1 against someone selecting an area on the MOC images that contained one of these faces. That assumes that these faces are around the same size. So if there are two faces that have a similar feature then there are 1,000,000 places where another face with a similar feature could have been formed but didn't, so this is a very rare occurrence. With 100,000 narrow angle photos in all (half of the MOC images are wide angle with low resolution) and 100 places in each photo of face like size this gives 10,000,000 places for this similarity to occur. Two formations having a similar feature is highly likely to happen but as we will see the number of similarities that do occur are far more than this.

An approximation of the odds against chance can be worked out from this. There is about a 1 in 10,000 chance of selecting a group of 10 photos with a face on Mars in it. That is, about 10 photos contain virtually all the faces so far, so having 10,000 groups of photos each with 10 photos in it gives 100,000 photos in total. Any group except the one with the faces in it can have its 10 photos selected randomly before they are examined. Time after time the faces appear in this one group of 10 photos instead of in the other groups. So with 23 faces in these 10 photos (1 is too large to be in one photo) the odds against 23 faces over and over again appearing in these 10 images is $10,000^{23}$ which is 10,000 to the power of 23. This is 10^{69} or 1,000,000,000,000,000,000,000,000,000,000,000,000,000,000,000,000,000,000,000,000,000,000,000,000 to 1. Such a result cannot be random; there must be a nonrandom explanation for this. Even assuming half of the faces are not credible doesn't change the result significantly, this would still be

10^34. The point is not whether each face is obviously artificial or not, it is that faces are in a small set of photos instead of randomly distributed over Mars whether someone thinks the faces are credible or not.

Numbers like this are hard to analyze accurately but it is defensible to use the number 1,000,000 to 1 for each similarity, however to be far too conservative I will use the number 10. Such similarities between the faces must be more improbable than 10 to 1 because as I said earlier this would mean that 10% of Mars would be covered in faces that look like the Crowned Face. First I will calculate from this base number of 10 to 1 for each independent similarity between faces and then substitute 1,000,000 for 10 to see how that changes the odds which should lie somewhere between the two numbers. This will become clearer as I examine the similar features between faces.

One objection to this method is that individual features like a nose need not make us recognize a face elsewhere. For example an eye or nose by itself would probably not be noticed by anyone so there may be far more of these than we realize. However if 10 of these features in the right proportions and positions to each other are found then some researcher would probably have reported it. At the end of the process then I will assume that 10 features of a face are needed in an area in the right proportions and positions relative to each other to be recognized as a face candidate. For example this might give an eye detail, a nose, some crown details, etc. To adjust the odds I will then divide the results by 10,000,000,000 but this hardly changes the odds against these faces being natural.

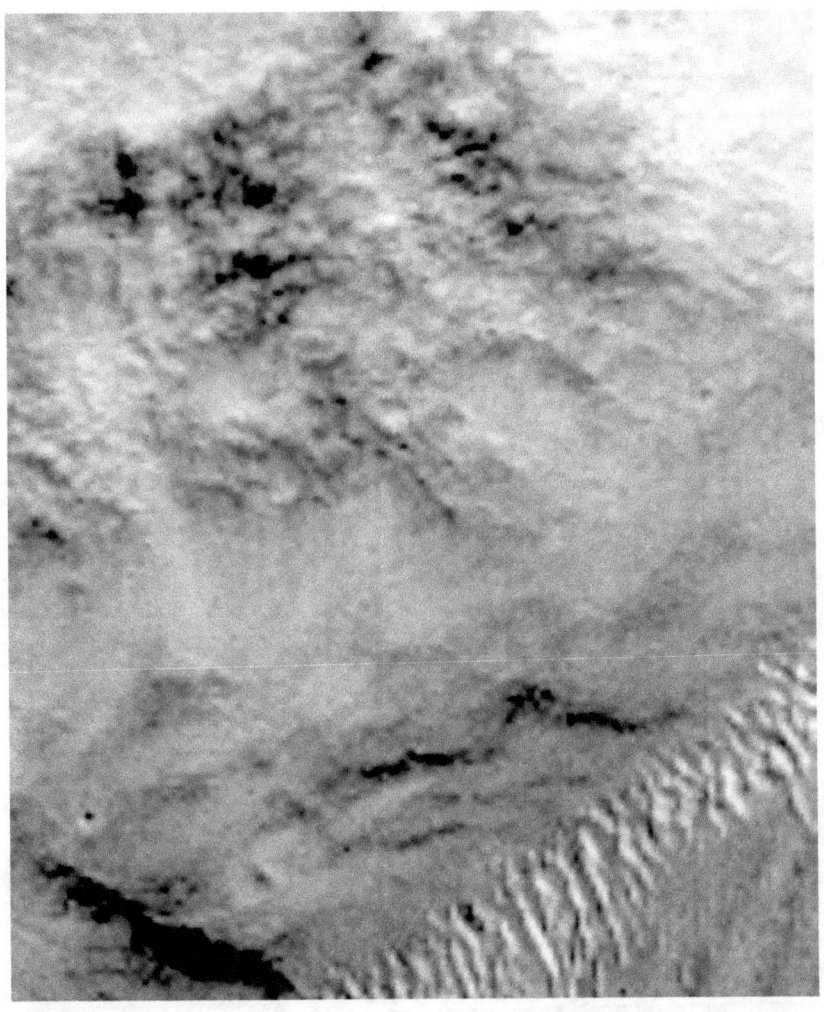

Figure 4

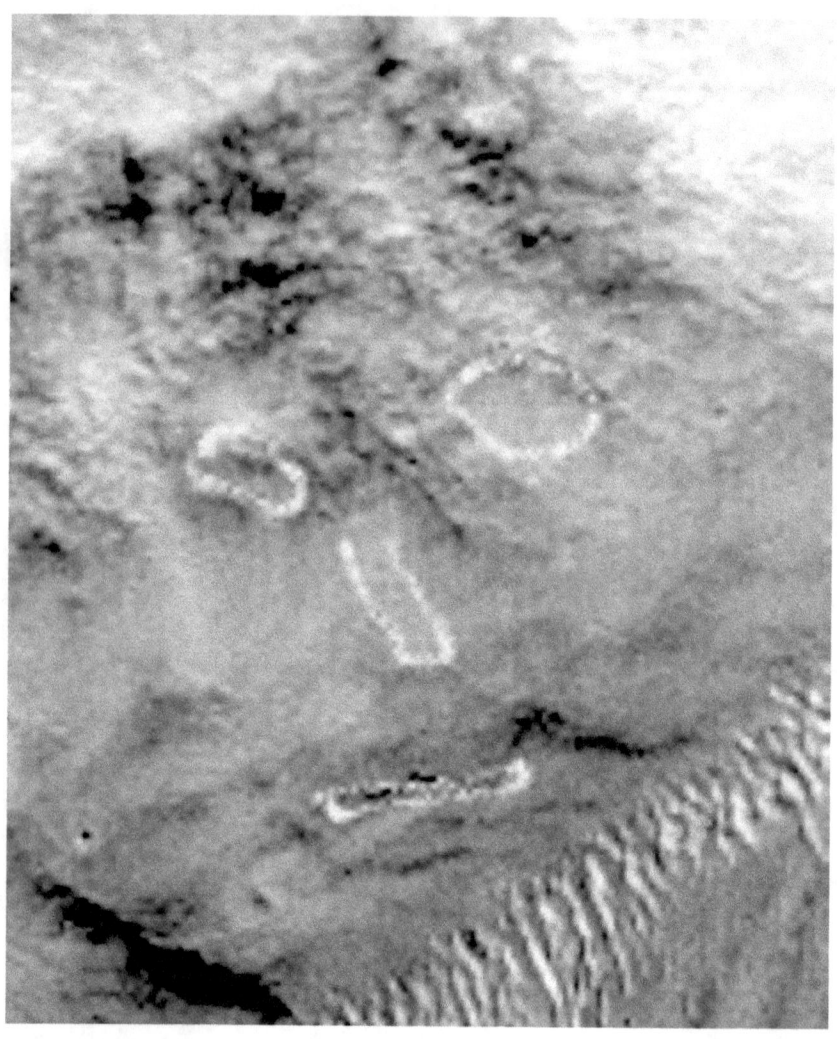

Figure 5

This is Face Three.

The right eye here is on the left hand edge of the right circled hole in the image above, this is shown in more detail elsewhere in this book. This hole may have been caused by a landslide. If these are three different Faces then there is little to be learned by comparing them unless by seeing similarities from their being in the same species. If they are of the same Face then each Face might have very similar features to the others. The more similar features the less likely this is by coincidence and the more likely they are artificial. Some of the similarities pointed out are borderline and might be argued as not exactly the same on the different Faces, but this can be resolved as the Faces are reimaged from different sun angles to have this features stand out better with shadows. Other features are extremely similar so they average out as being very similar.

The first feature to be analyzed is the top left part of the crown which overlays very closely.

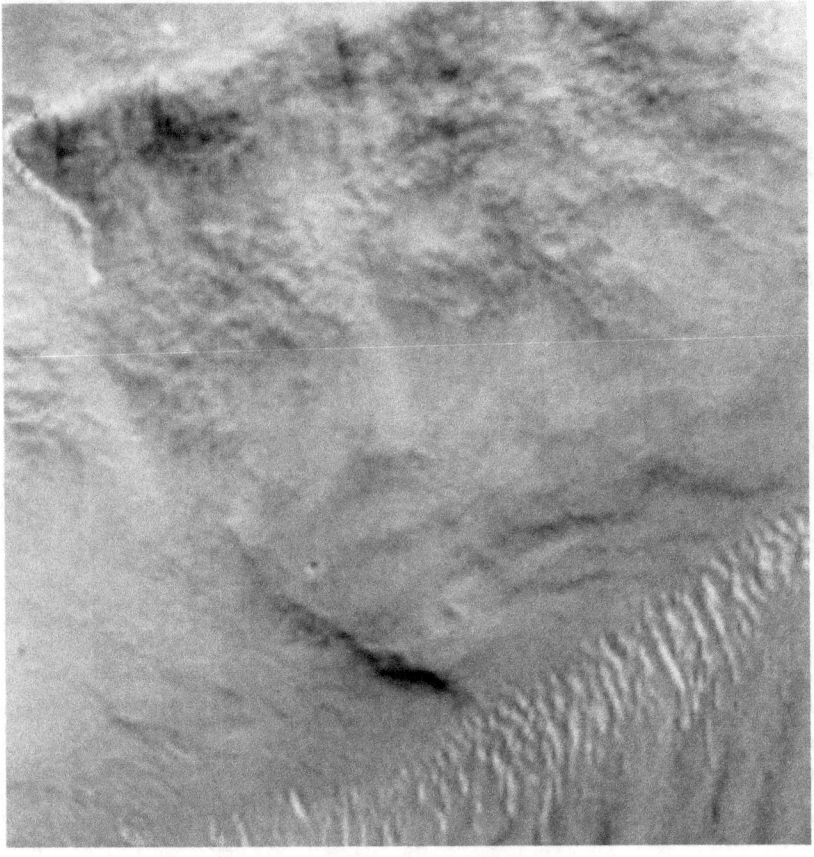

Figure 6

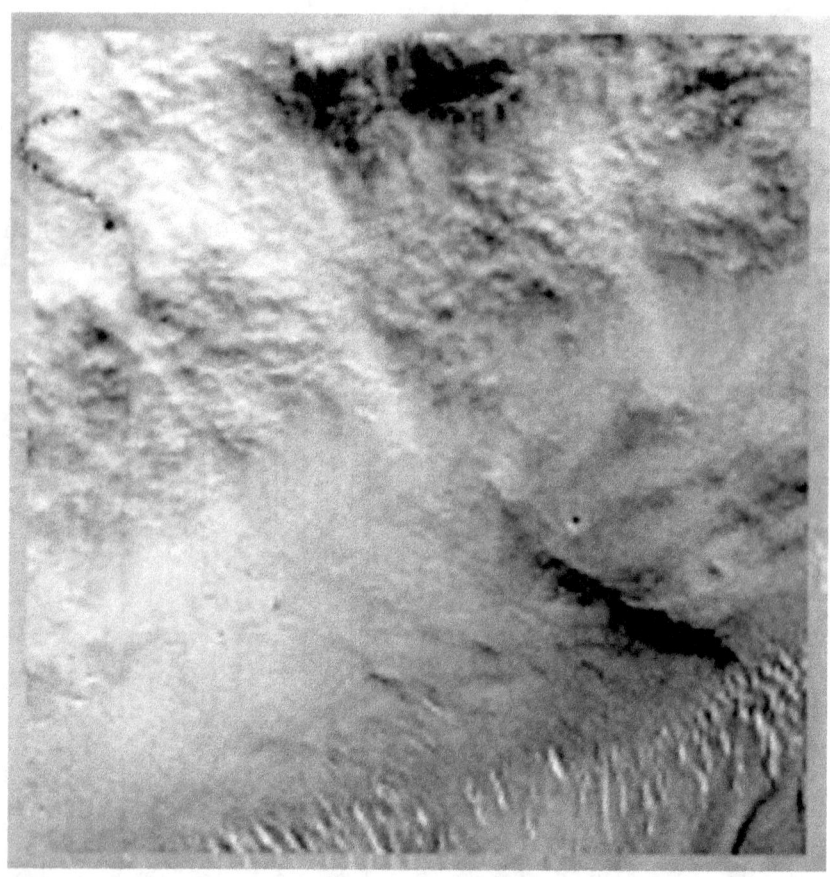

Figure 7

In Figure 6 the white line shows the edge of the crown on Face Two. In Figure 7 the black line shows a similar curve on the crown of Face One. The more similarities there are in different Faces the less likely that this can occur by chance. You can compare the outlines I make with the unaltered photo in Figure 1. It seems unlikely for the shape in Figure 6 to be made naturally because it protrudes away from the cliff and so should have eroded away. As explained earlier when similar features occur in different faces these are independent events and their probabilities should be multiplied together. The odds of this features occurring naturally in both Faces is at least 10 to 1 against chance.

The next comparison is of a small feature shown in both faces below.

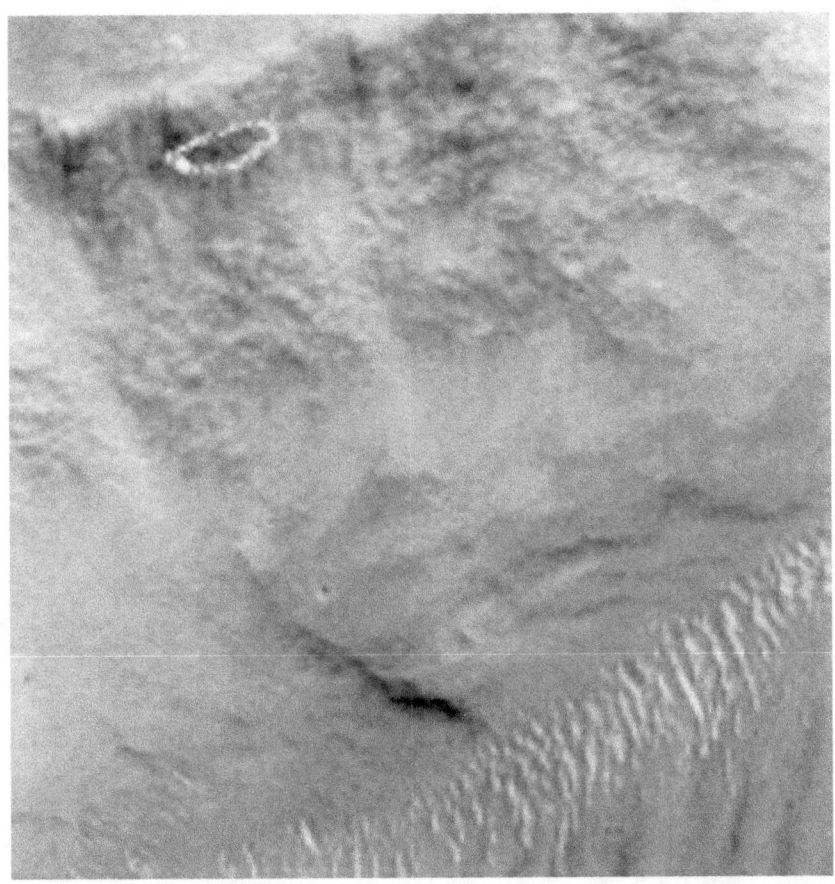

Figure 8

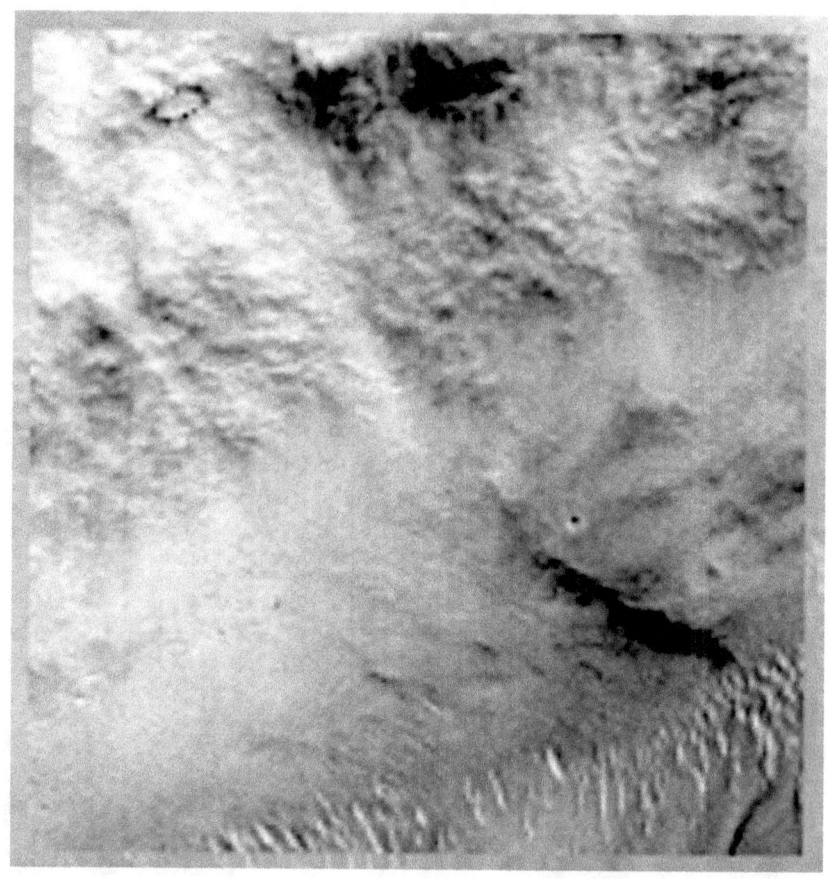

Figure 9

In Figure 8 there is a small oval shape on Face Two marked in white, in Figure 9 there is a small oval in approximately the same position on Face One. The oval on Face One is not a crater and it doesn't look to be caused by a piece of debris having fallen out of it, so it would be unlikely to have formed naturally. The odds against this similarity occurring by chance are at least 10 to 1 so with the previous similarity this gives 100 to 1 against chance so far.

The next part is where the two faces join. This line is also very close to where the right hand face joins. The line extends up from the top of the nose line in both cases.

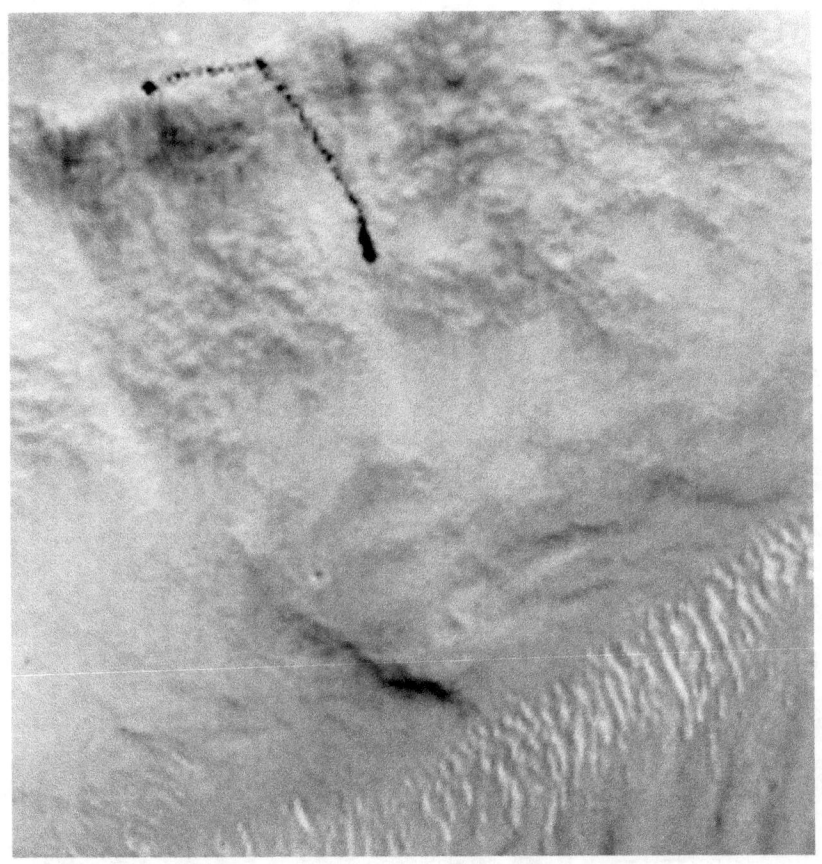

Figure 10

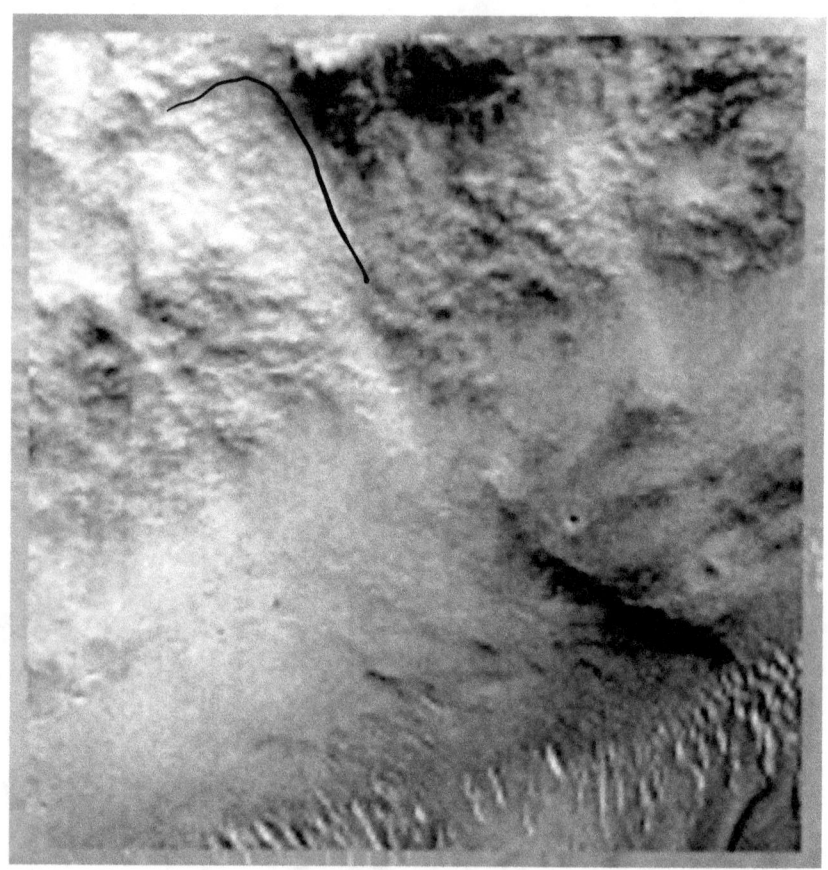

Figure 11

There is a line here where the two faces join, in Figure 10 it is marked and is the dividing line between Face Two and Face Three. In Figure 11 the marked line shows the boundary between Face One and Face Two. The shape of this line is very similar in the two images. This similarity is again 10 to 1 against chance so this makes 1,000 to 1 against these three similarities so far occurring. This is still not an impressive result because as said earlier there are at least 10,000,000 places the size of a Crowned Face in the Mars Orbital Camera photos so there are plenty of places for some similarities to occur. I will now introduce a notation to make it easier to keep track of the size of the numbers, 10^3 means 1 followed by 3 zeroes or 1,000, this is much easier to write and understand than large numbers with hundreds of zeroes in them that occur later.

Next the crown can be seen outlined in Figure 12, the border of a darker area. In Figure 13 there is another darker area at the same place. In both images there is a V from the right side, one of which helps make an eyebrow on Face Two.

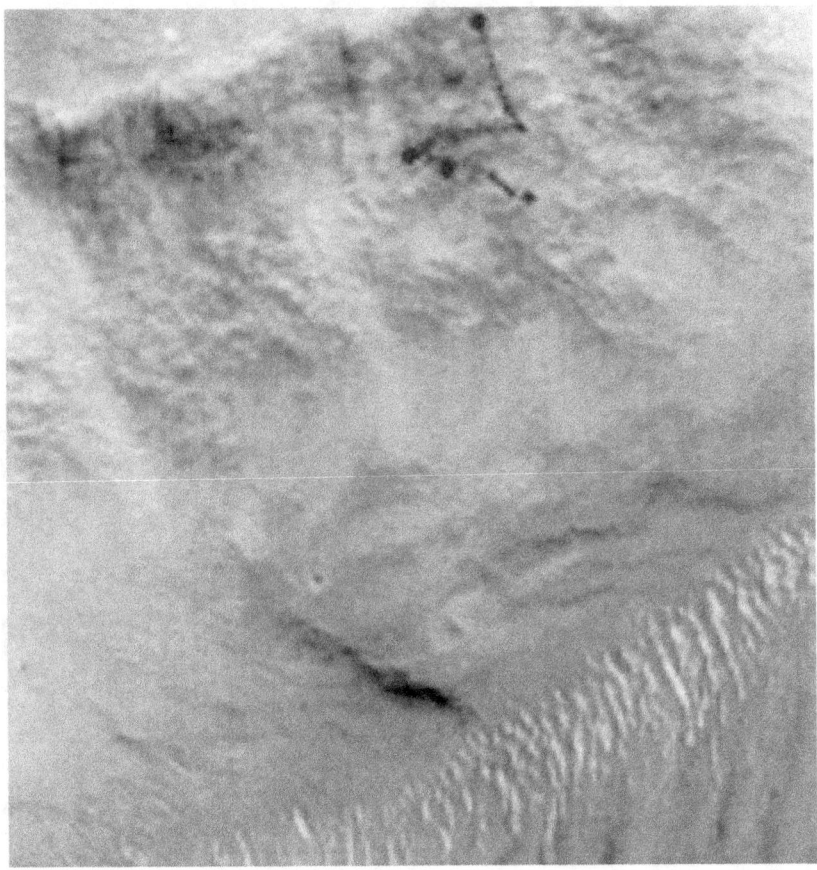

Figure 12

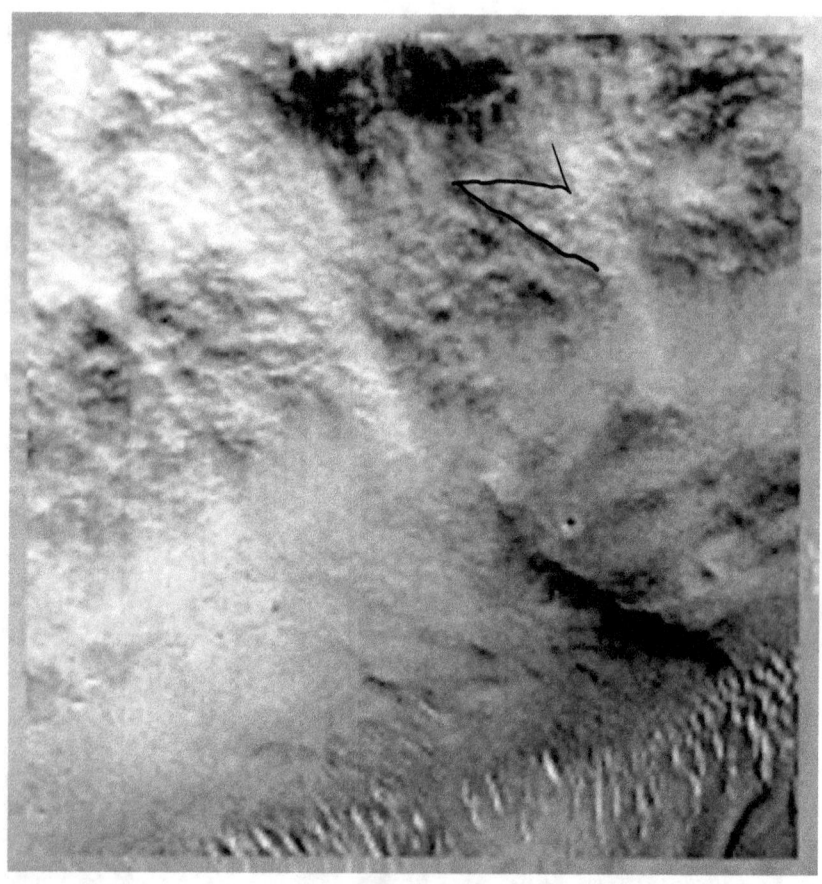

Figure 13

In Figure 12 there is a zigzag shape marked on Face Three, and another in Figure 13 on Face Two. This shape is unlikely to occur naturally because a sharp point like this surrounded by a fault should have broken off with this high level of erosion. This indicates that they are probably not faults or cracks, but cuts in the rocks made in construction. This is 10 to 1 against occurring naturally so this makes 10,000 to 1 so far or 10^4.

The next part is an area above the Face Two eye. This piece actually fits in very well with a raised area on Face One, a third photo is shown below with the two morphed together. The similarity of pieces like these implies some kind of ornamentation.

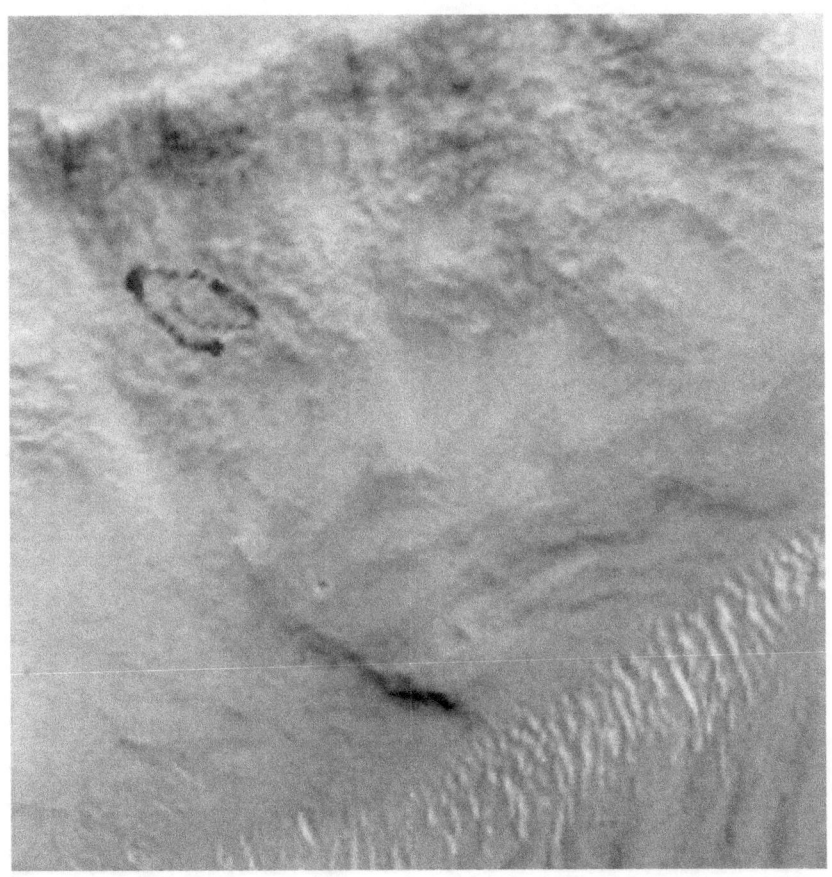

Figure 14

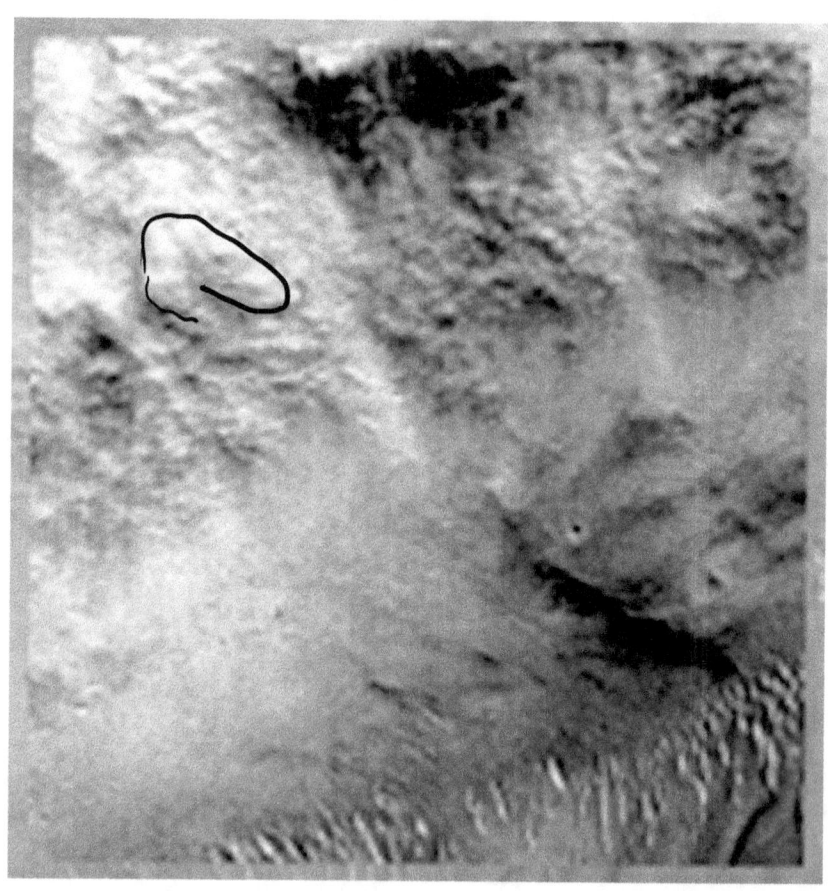

Figure 15

The marked line in Figure 14 shows a shape on Face two, the marked line in Figure 15 shows a similar shape on Face One. Erosion tends to make a surface smoother so convoluted shapes like this should have eroded away like the smooth cliffs on the other side of the valley, shown elsewhere in this book. This similarity is at least 10 to 1 against occurring naturally so this gives new 100,000 to 1 or 10^5.

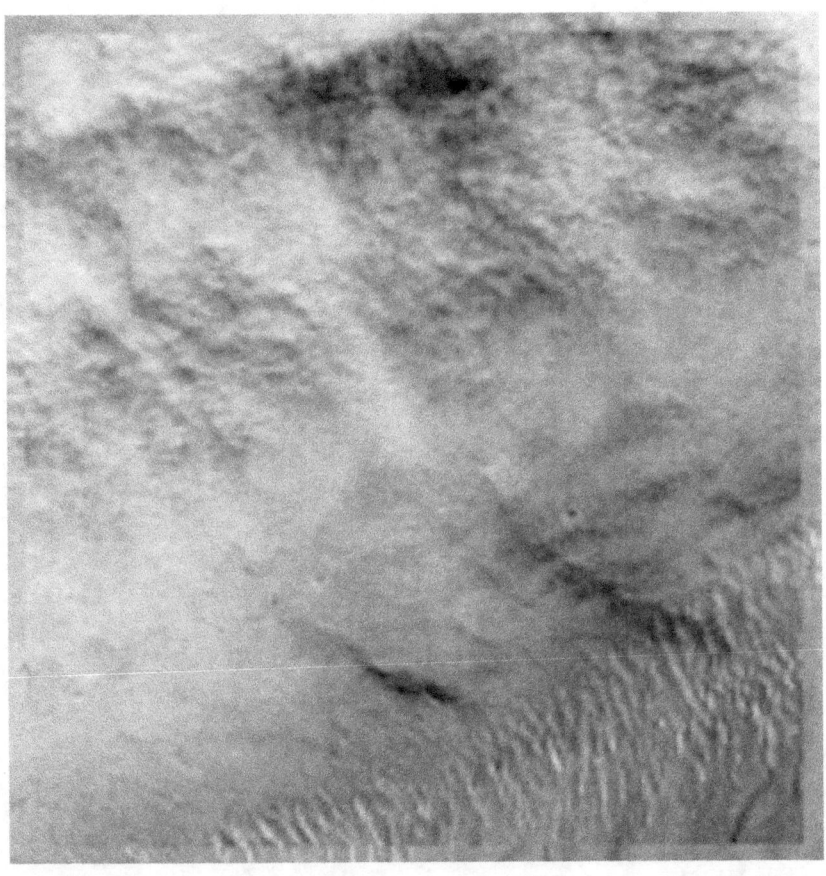

Figure 16

When Face One and Face Two are overlaid it gives another possible insight into what these aliens looked like. If these Faces are similar but of different people then like doing an overlay of two similar human faces this creates a more generic alien Face. As can be seen there are many similarities between these Faces and so the odds against chance must be very large. People sometimes see faces in clouds, but not the same face over and over. Cloud faces are very different looking from each other though we tend to recognize them as faces in a process called pareidolia. Here however the issue is not whether faces are being seen but why they are so similar to each other.

Here the left eye is outlined; note how it extends to the left in both Faces.

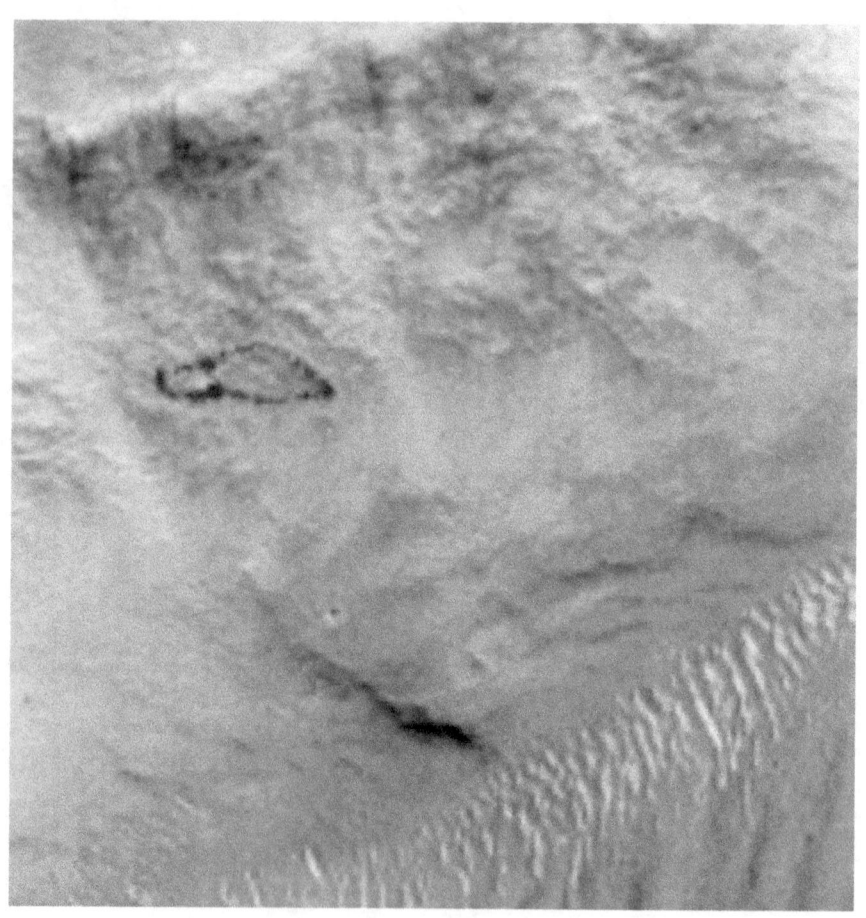

Figure 17

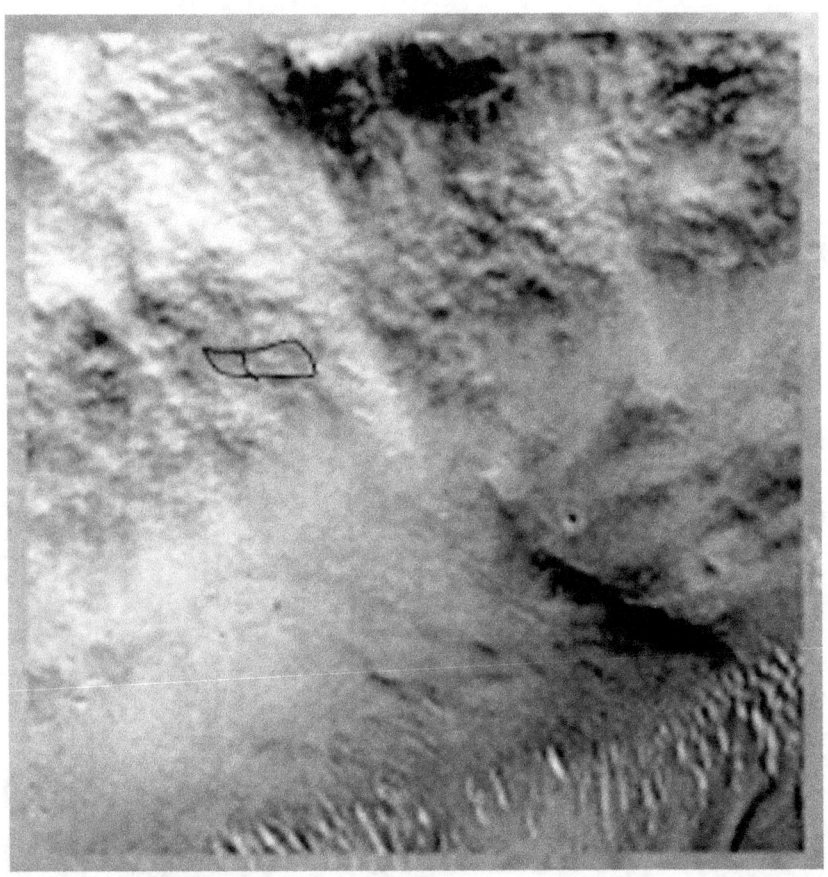

Figure 18

This repetition of the eye pattern implies a decorative purpose; it is unlikely a Face so like our own would be this asymmetrical. These eye shapes are difficult to create naturally, an impact crater tends to be circular and without an iris in the middle of the eye. The ridges around the eye are very thin so they would tend to break off with erosion, but they haven't. Figure 17 shows the eye on Face Two and Figure 18 on Face One. This is another 10 to 1 against chance so this gives a total improbability so far of 1,000,000 to 1 or 10^6.

Here is an intermediate shot where Face One and Two are overlaid again.

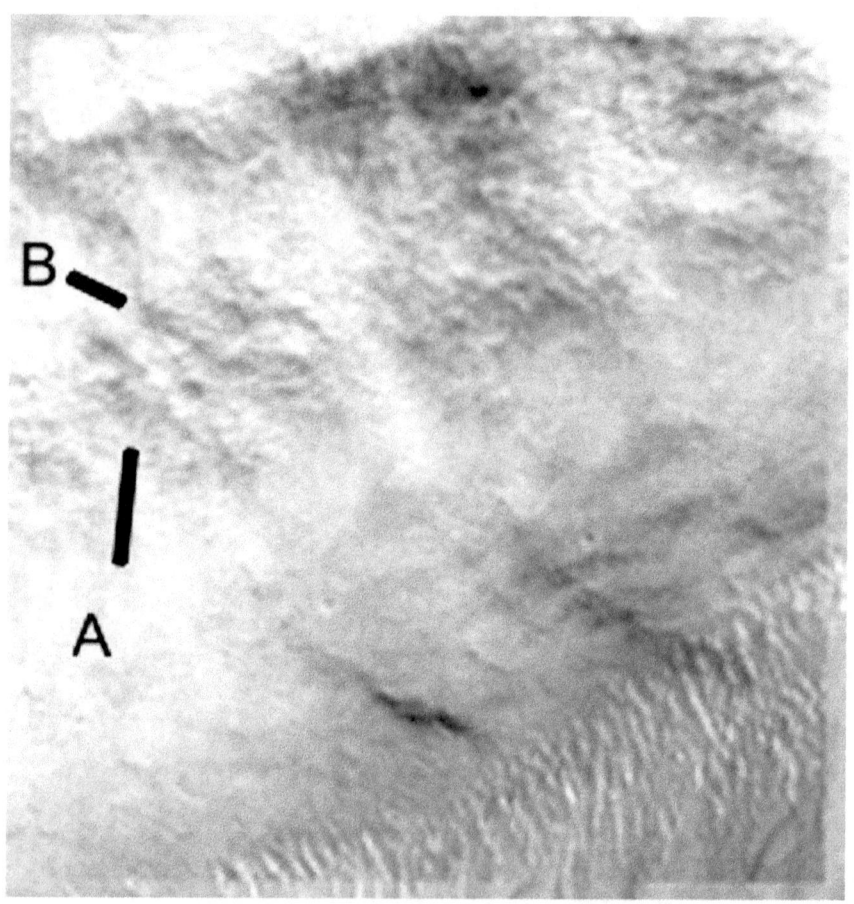

Figure 19

*Each time I overlay one photo on the other I change the relative percentages of the photos. For example one might have 30% of Face One and 70% of Face Two while the next might have 70% of Face One and 30% of Face Two, the former will look more like Face Two than the latter. This gives a slightly different appearance which gives more insight into what these aliens would have looked like. Note how the left ear at **A** is more prominent and there is a line at **B** pointing up from the left eye like a decorative feature.*

On the left of Face Two there is an ear shape and another separate ridge. This matches very well to the left of Face One. The third photo is another overlay of the two Faces.

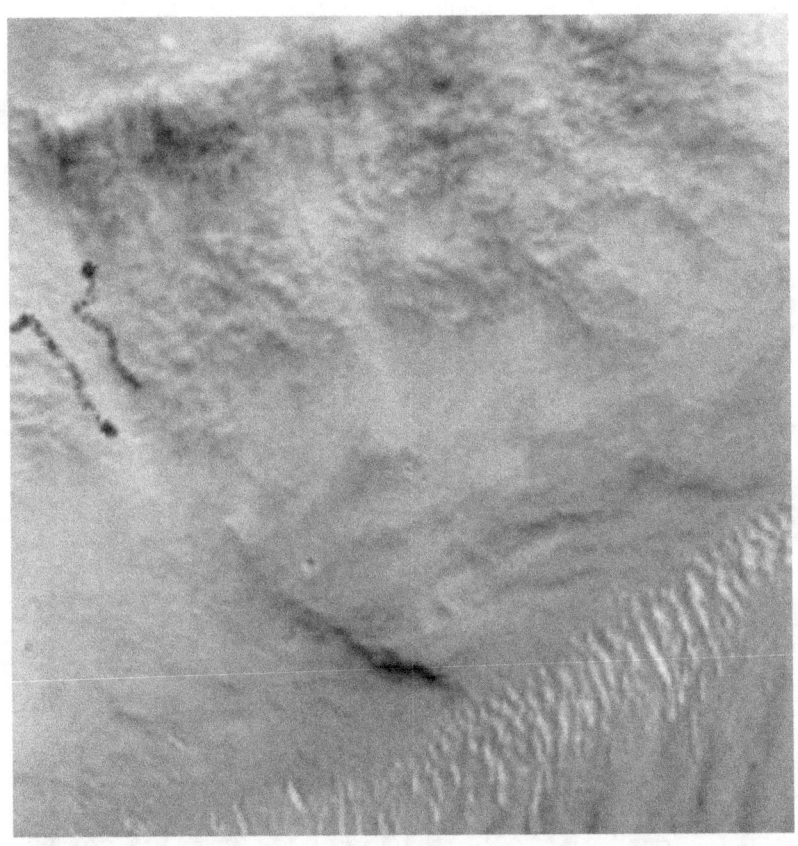

Figure 20

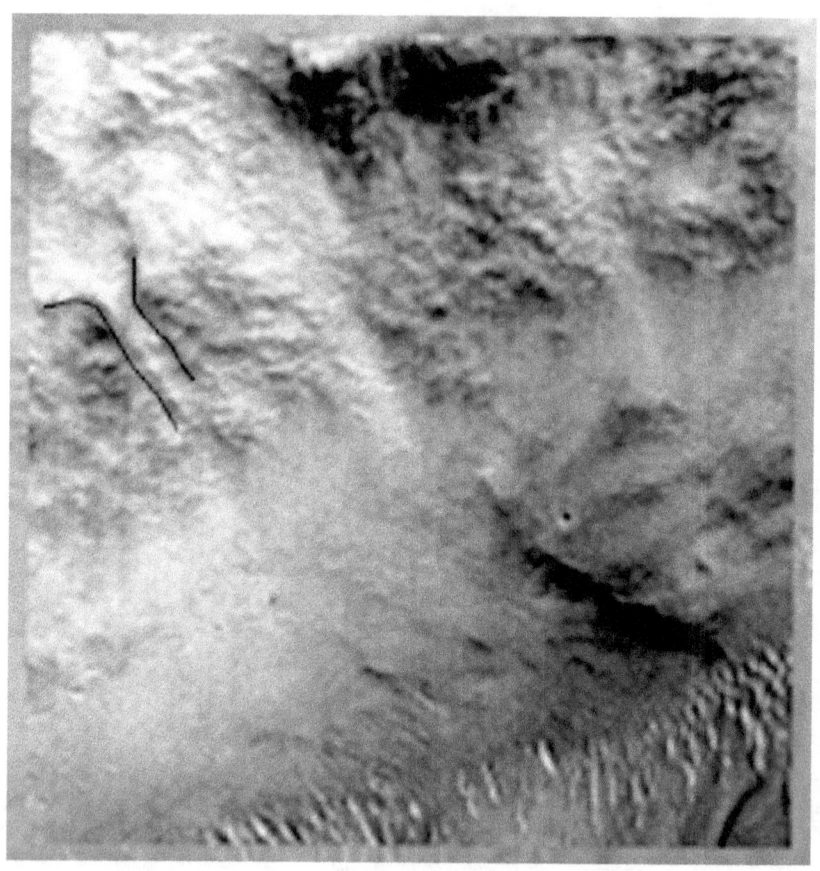

Figure 21

In Figures 20 and 21 this small depression is marked. This could have been dug to give a shadow defining the edge of the Faces but also perhaps the left edge of this depression would catch the sun and appear like part of the ear. It also doubles as the area just above the nose on Face Two for Face Three on the right, which will be shown later. This is 10 to 1 against occurring naturally this so that gives 10^7 to 1 against chance so far.

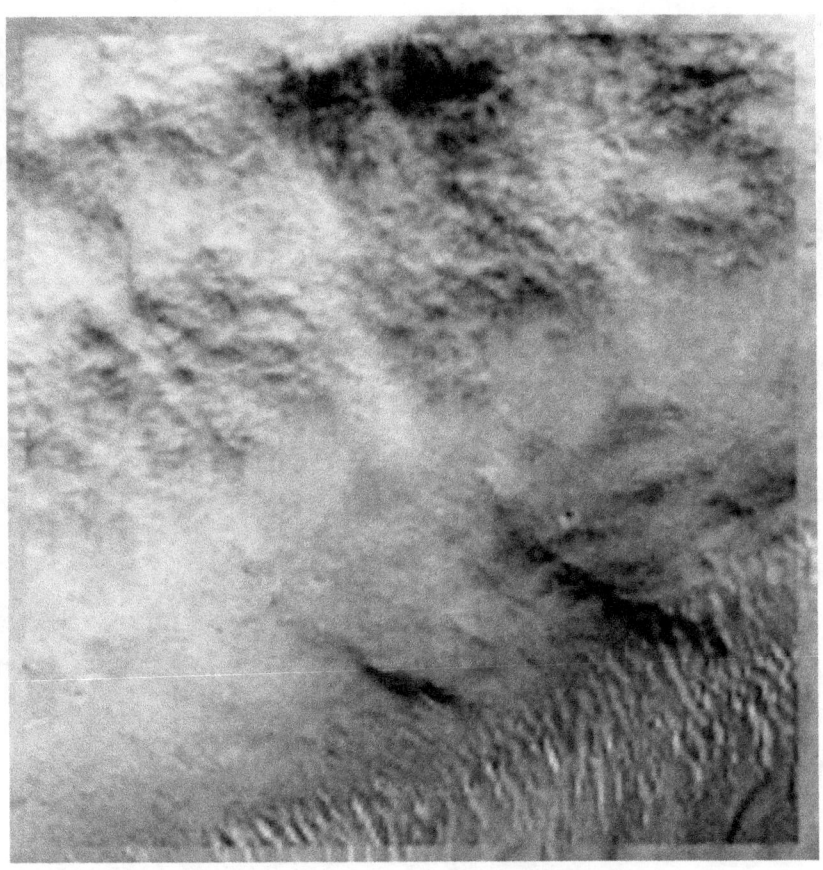

Figure 22

In this overlay the nose from Face Two fits very well onto Face One as well as the mouth looking natural. There are many different parts of the nose which could be argued to be similar, the length is about the same, they have a similar curve to the left, the width is similar, this would in itself give an additional $10^\wedge 3$ or 1,000 to 1 against chance for a total so far of $10^\wedge 10$ or 10,000,000,000 to 1 against this occurring randomly. The nose is also unlikely to occur naturally particularly on Face Two because it protrudes and should have eroded away, the nostrils in the tip of the nose should have been even more likely to have the nose break off as it probably did on Face One. The Face One nose is shorter; this may be because the end has broken off but as shown below the tip of the Face One nose appears in the HiRise image. This gives the noses the same approximate length; also the nose tip has the same approximate shape and places for two nostrils.

Figure 23

This is Face One; the area is highly eroded so this face must be extremely old.

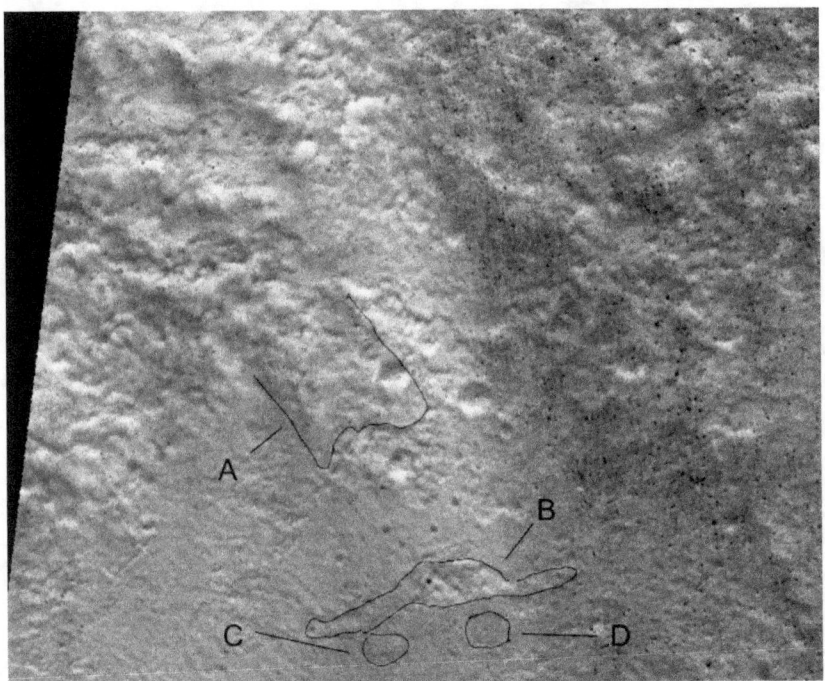

Figure 24

A shows how the nose is broken off about half way along its length. **B** shows the nose tip which is still there but much of it has broken off. **C** and **D** show holes in the right place for nostrils. Note how the nose tip is at an angle to the nose bridge giving it a similar bend to the left as in Face One.

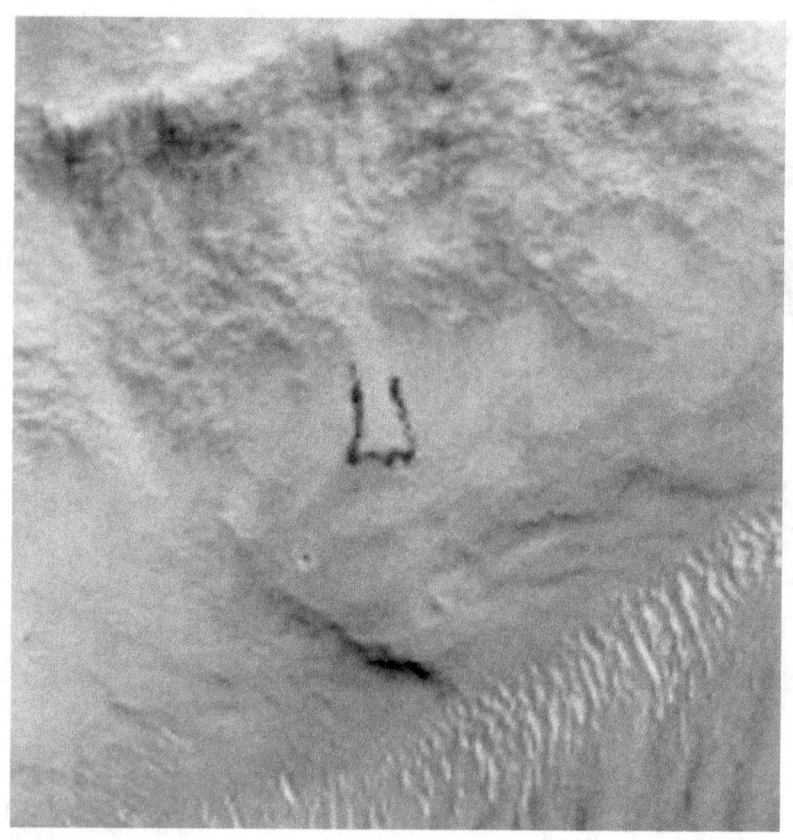

Figure 25

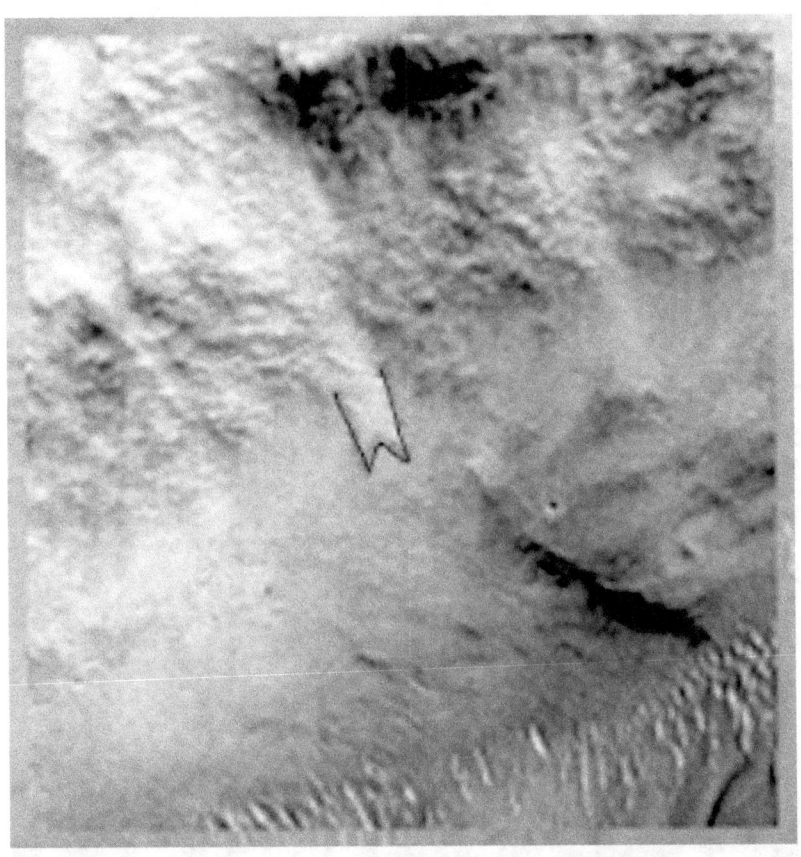

Figure 26

In the two images above the noses are marked.

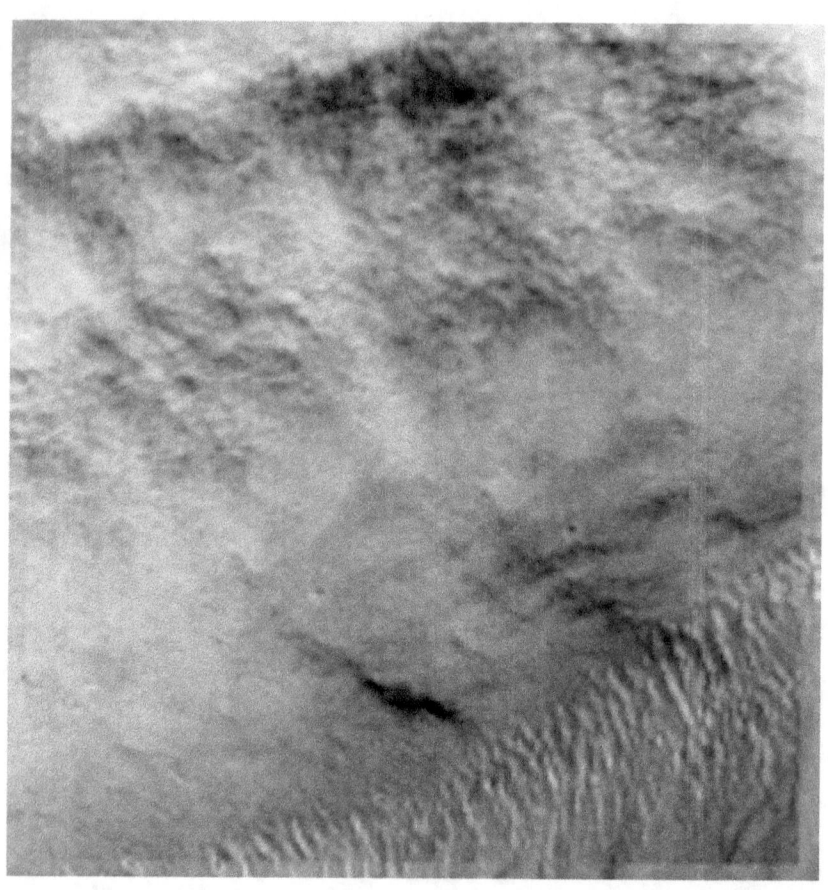

Figure 27

Note how well the crown fits in this overlay above, it has a symmetrical shape around the features of Face One.

The right eye shape is outlined on Face Two below. The next photo shows a similar shape for an eye on Face One.

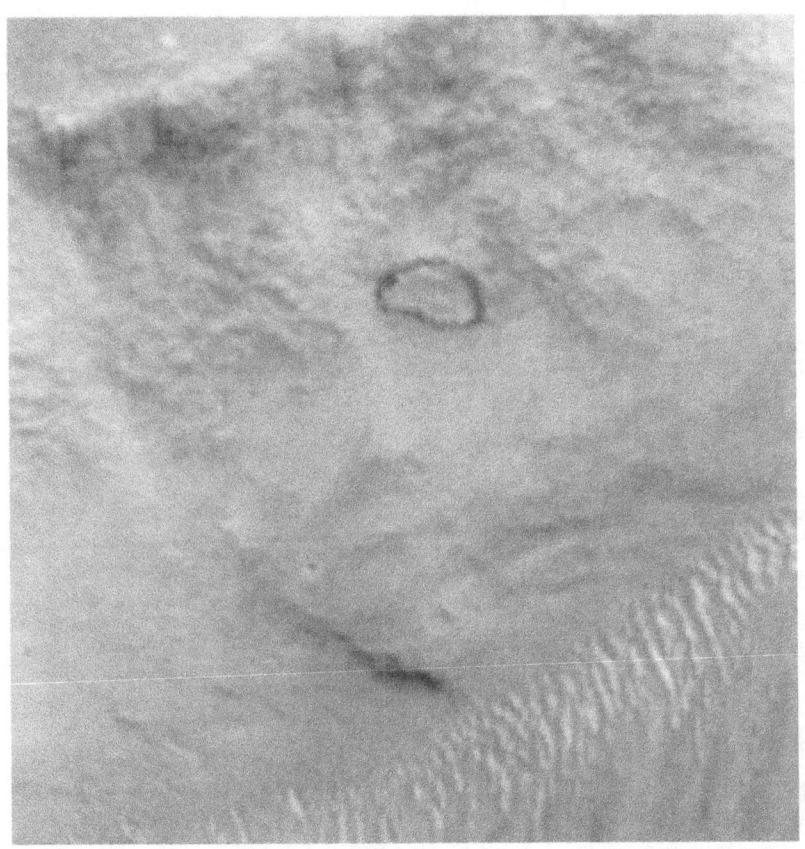

Figure 28

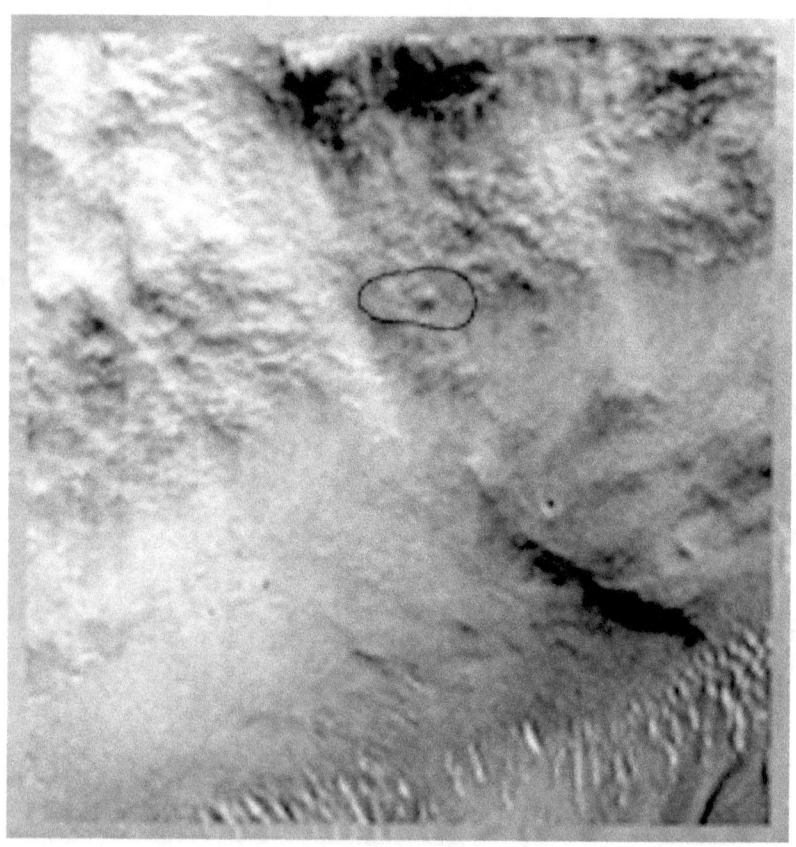

Figure 29

In Figure 29 the marked circle is the right eye of Face One, it has the pupil intact. The right eye of Face Two is seen in Figure 28 with a similar oval shape. The right eye of Face Three can also be seen in this image, it appears to be eroded or perhaps it broke off as an oval section and fell off the face. The oval shape is 10 to 1 against occurring naturally, also each has a pupil shape inside it which is an additional 10 to 1, that makes 10^{12} to 1 against chance so far.

In the overlay below the left eye looks more real as the parts of the eye of Face Two fit over it very precisely. The impression is of an eye half closed with an eyelid.

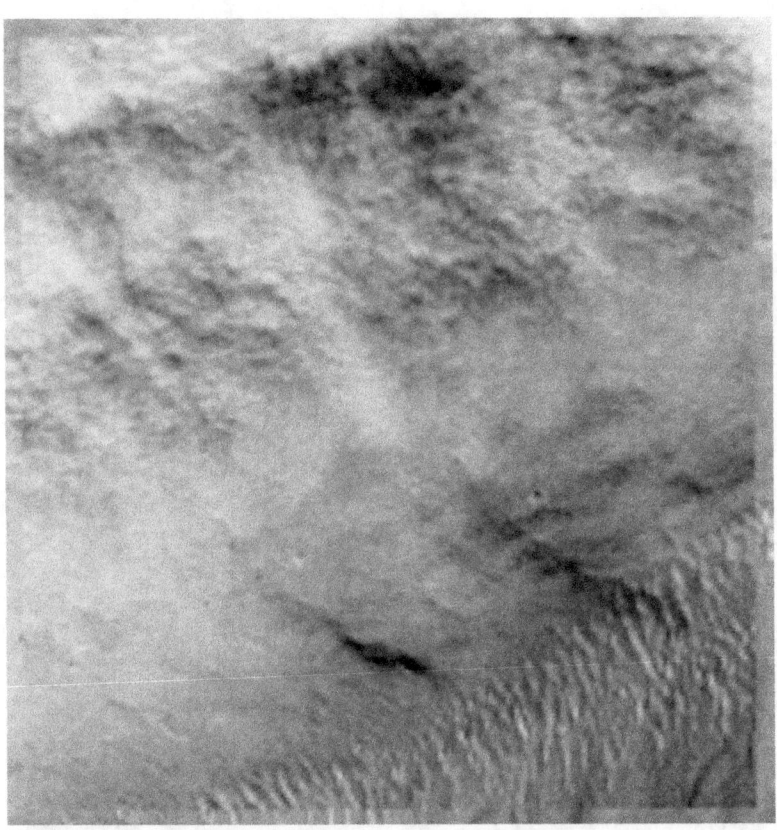

Figure 30

The Face Two chin below matches well with the edge of Face One. The separate lines marked **A** and **B** are found on both Faces.

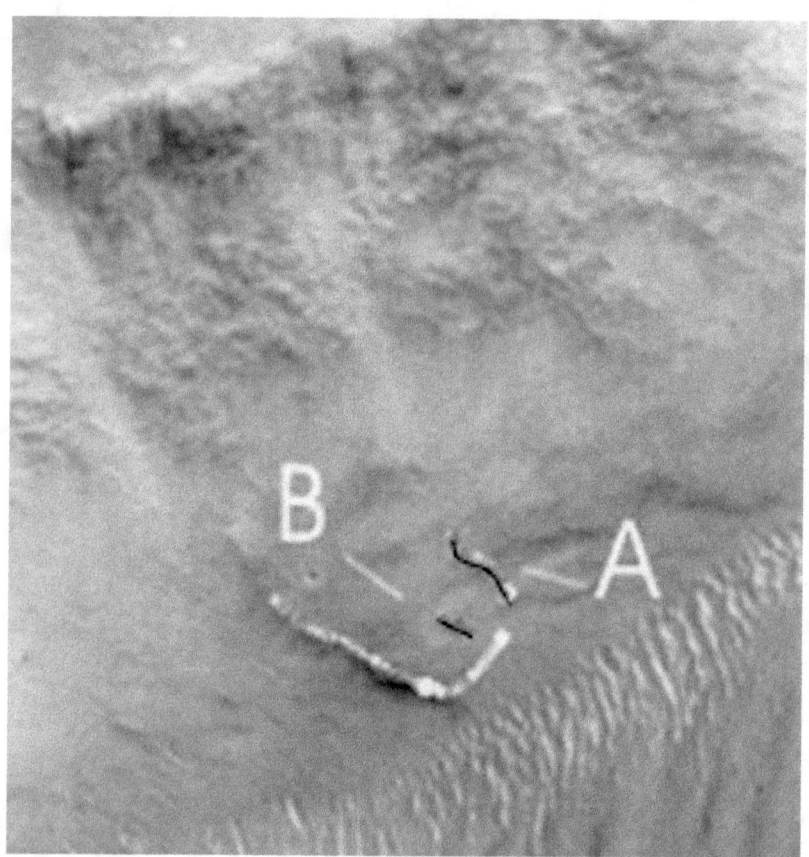

Figure 31

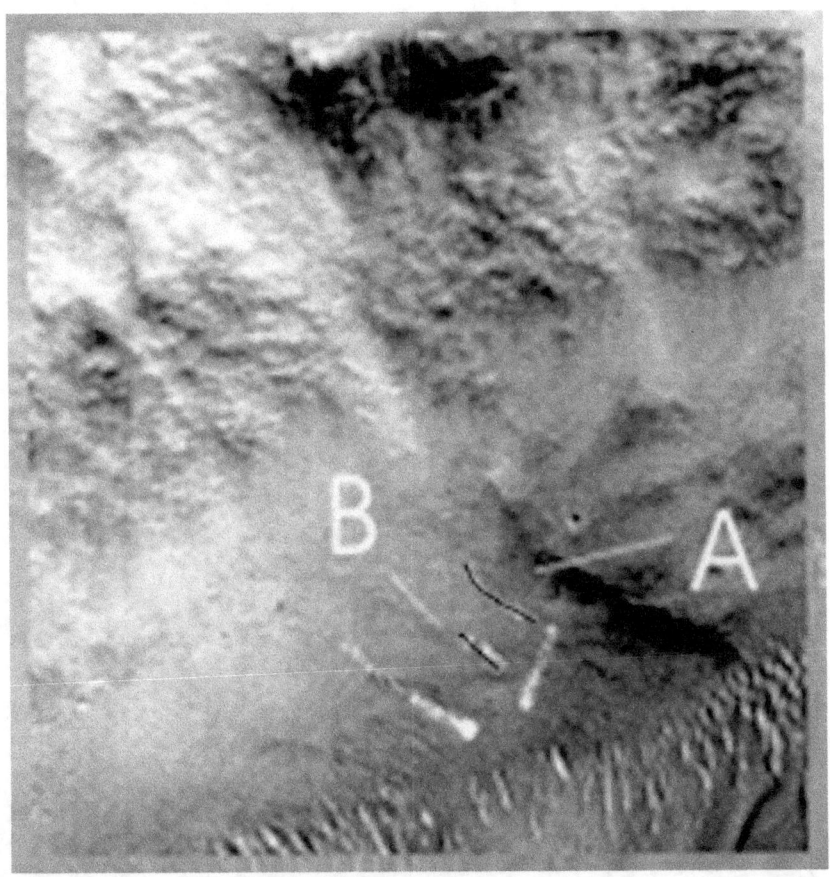

Figure 32

Note the odd shapes around the mouth of Face Two, the Crowned Face, are replicated on Face One. Elsewhere in this book the Crowned Face is overlaid on the Cydonia Face and the Meridiani Face, they also have a similar distorted chin and mouth which makes it more likely they are all of the same humanoid. Another possibility is these Faces have a more reptilian neck with folds of loose skin that gives this distorted shape. There are at least three features here which are more than 10 to 1 against occurring by chance, the general shape of the chin and the two ridges at A and B. That makes a total of 10^{15} to 1 against all these similar features on both Faces occurring by chance. The ridges A and B also seem unlikely to occur naturally because they go across the open ravine that is a mouth on both Faces, how could the ravine form with erosion while leaving these ridges intact?

There is a round shape under the Face Two left eye and a small ridge under its left eye; this is also seen in Face One below.

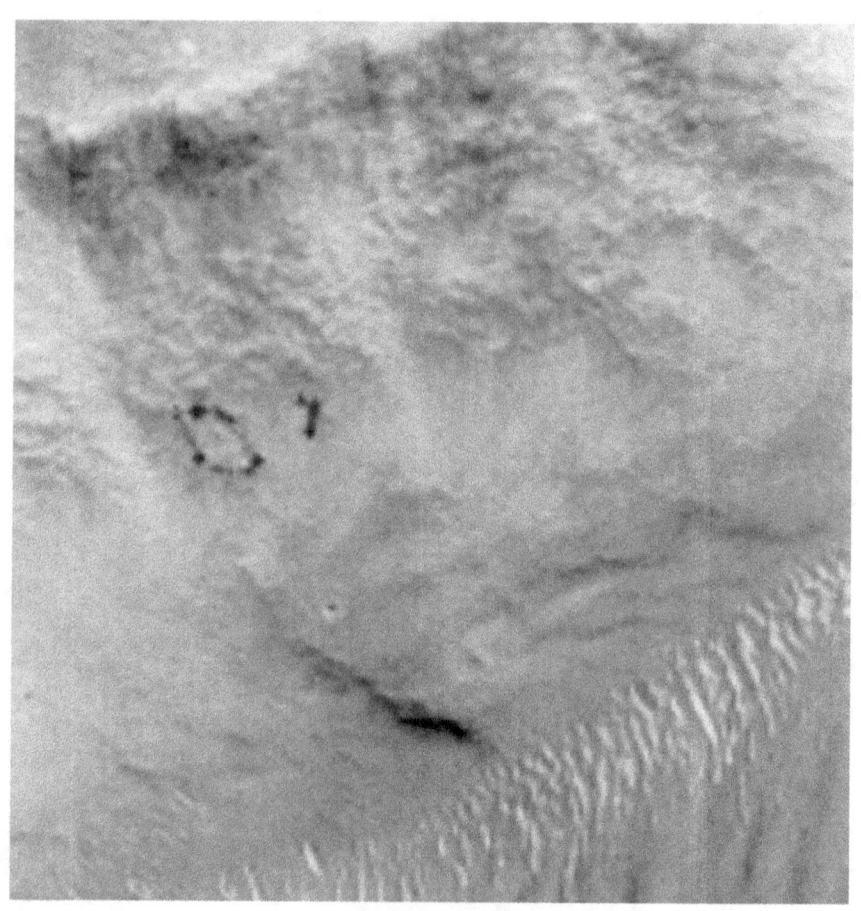

Figure 33

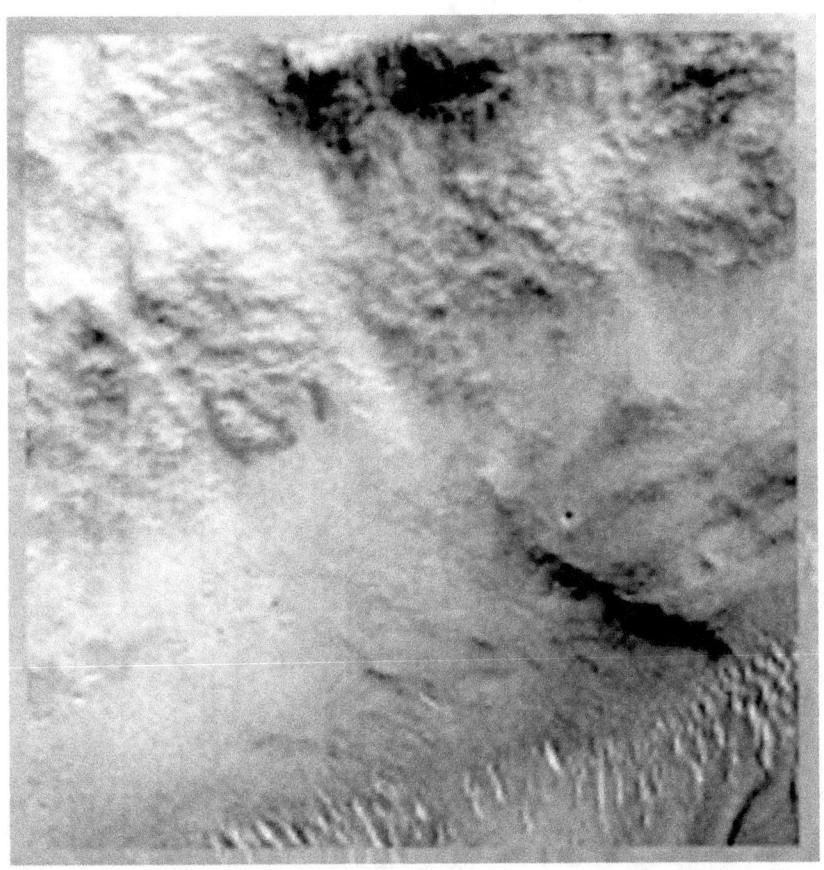

Figure 34

In Figure 33 the dark circle and line represent these features, also outlined in Figure 34. These are both 10 to 1 against chance and together are then 10^2 or 100 to 1 for a total so far of 10^{17} to 1. This oval shape on both faces also looks like an eye; it is possible that under a different lighting such as winter versus summer the different sun angle would make this eye appear to be part of another face. If so then there might be three Faces that appear sequentially through the day that also look subtly different depending on the season. This would be remarkable to see but the sun angles are different on Mars now that the poles have wandered to a new position. This theory could be tested with an accurate 3D model of the three faces however.

Chapter Fourteen

A comparison of Face Two and Face Three

Face Two is the Crowned Face which shares an eye with another face to its right called Face Three just as it does with Face One on the left. The Crowned Face or Face Two appears neutral in expression but Face Three appears angry, which further implies three seasons or aspects of the one face. It may also be because of the shadows when this photo was taken, a particular sun angle may accentuate some features which simulates an emotional aspect to the faces. On the left crown edge of Face Two there is an oval shape in Figure 35 below, there is a similar shape where this would be on Face Three in Figure 36. The Cydonia Face also has a similar notch in the crown in this position.

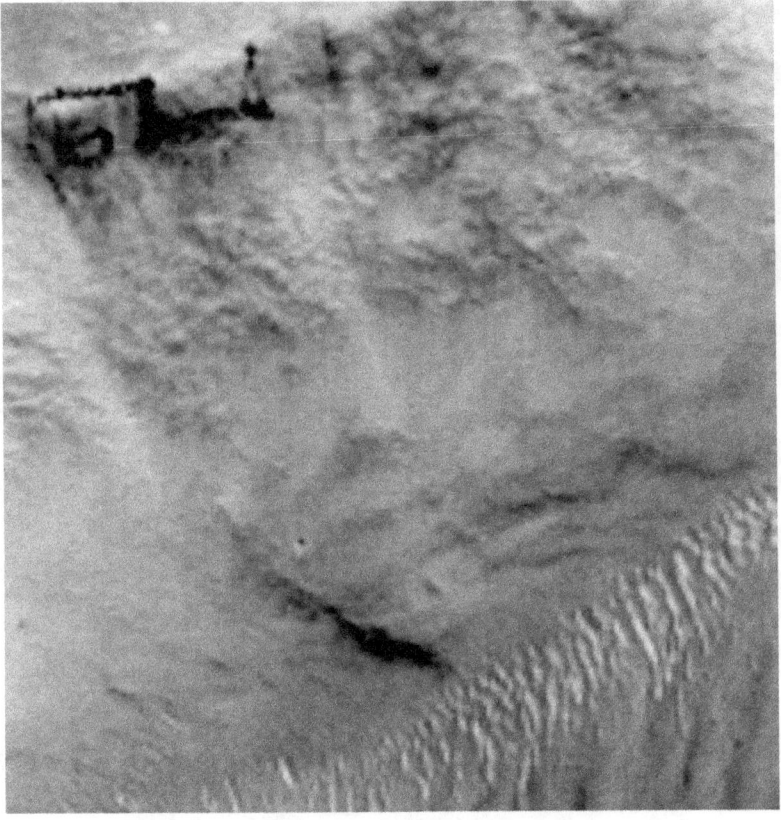

Figure 35

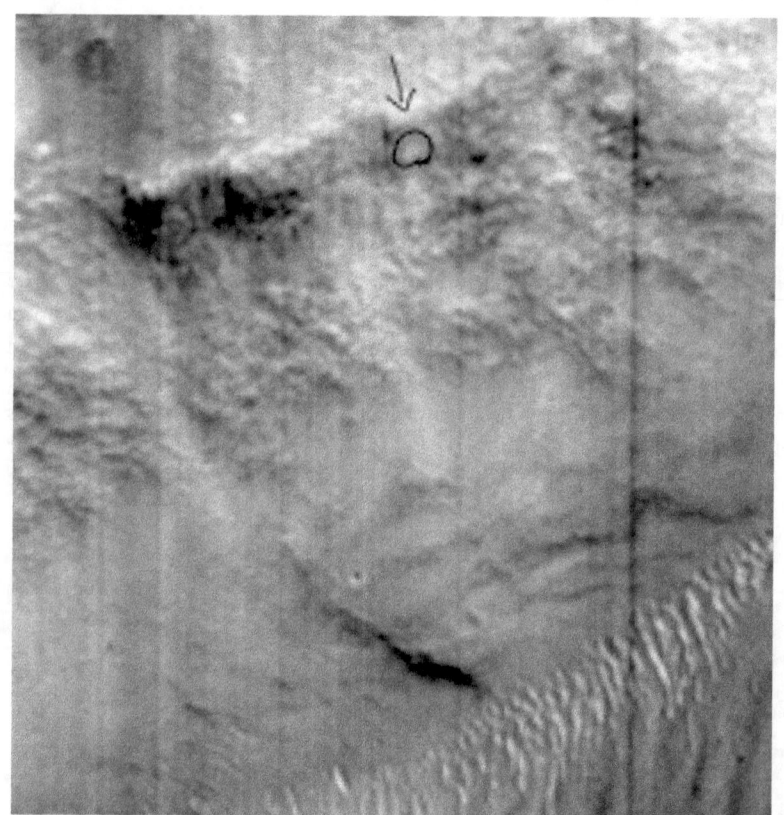

Figure 36

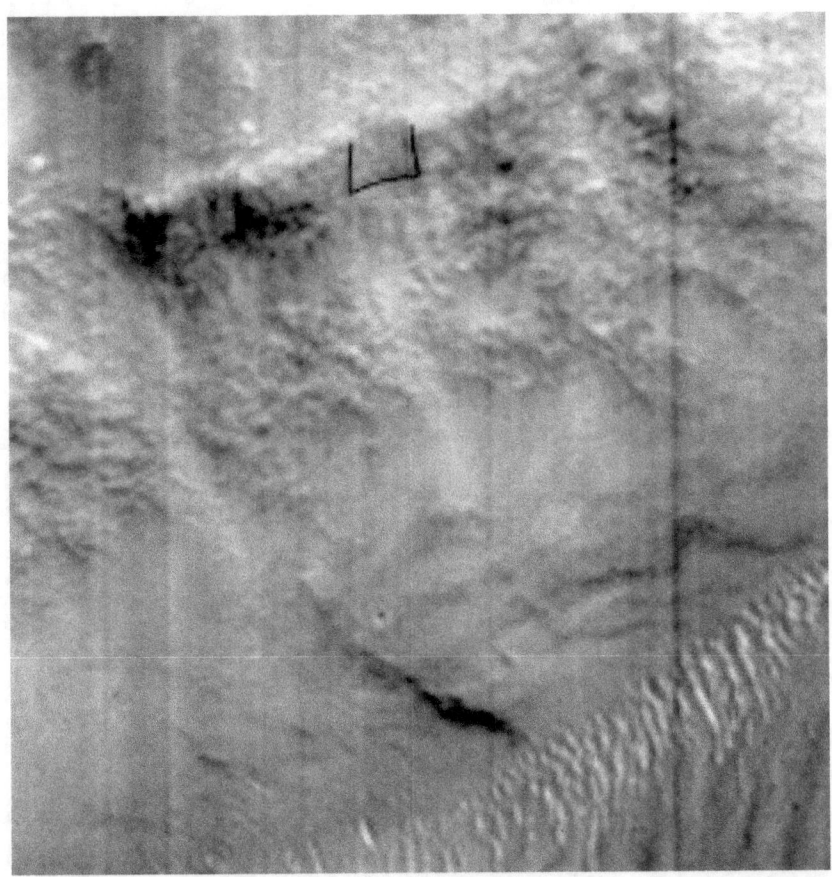

Figure 37

In Figure 35 the oval shape and notch are shown outlined in black on Face Two and in Figure 36 there is a similar oval shape on Face Three not exactly in the same position. This cliff face seems to be curved here and it may be there is a different perspective with this face which seems to be directed more to the right. In Figure 37 there is a notch exactly in the right position on Face Three but no oval, it may have eroded away. Figures 36 and 37 are more enhanced because they come from the raw Mars orbital Camera image, they also have less shadows in this area to see more detail. The streaks are from the raw image, when the photos were taken they usually had these streaks from cosmic radiation which is then processed out later. To see some finer details it was sometimes useful to look at the raw .IMG image like this. Note that while exact matches in patterns between faces are extremely unlikely to occur by chance small differences may simply mean the faces are of different humanoids or in this case are wearing a slightly different crown. The notch in the same place is 10 to 1 against chance against it occurring randomly, this takes the total to 10^18 to 1 against these being random coincidences.

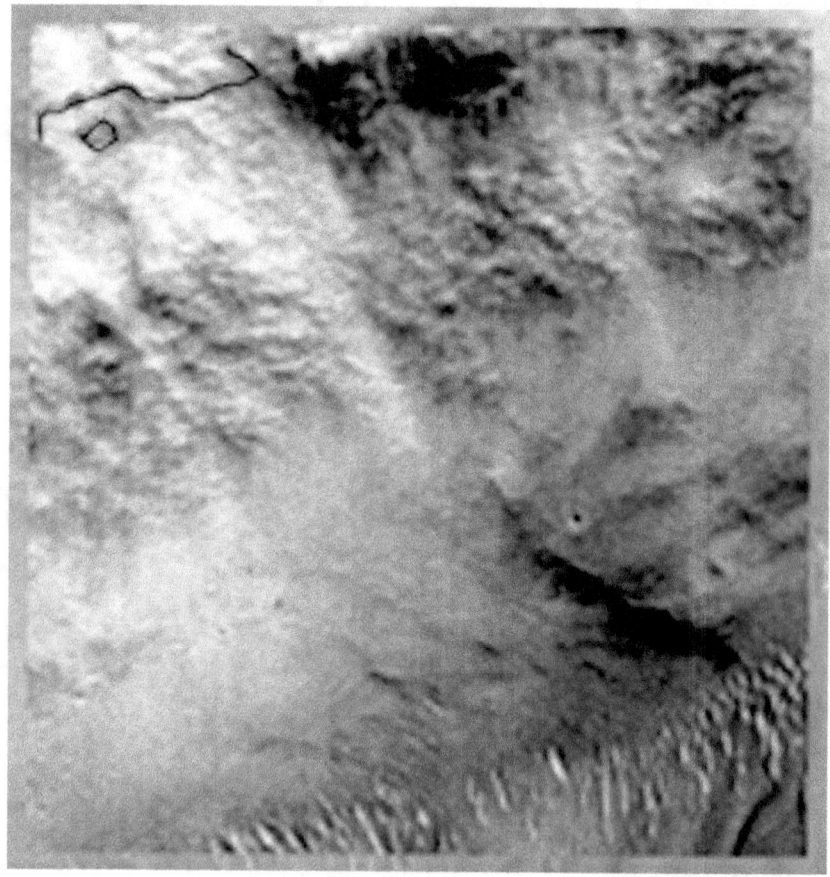

Figure 38

In the above image a similar shape appears on top of Face One, this may be where the crown would have extended to before this area was eroded. This area appears to be where the sand came from that ran down onto Face One's nose and covered its mouth. It may have been caused by water and damaged the Face if it was made of a material like mud that eroded more easily when wet. This similarity between Face Two and Face Three is 10 to 1 against chance and brings the total to 10^19 to 1.

The left hand side of Face Two is better defined than the left side of Face One. The left side of the right face, Face Three, however is partially defined by the Face Two nose. The left jawline of Face Three is very similar to that of Face Two when the Face Two nose is taken as the jawline, along with the vertical ridge seen as **A** in Figure 31.

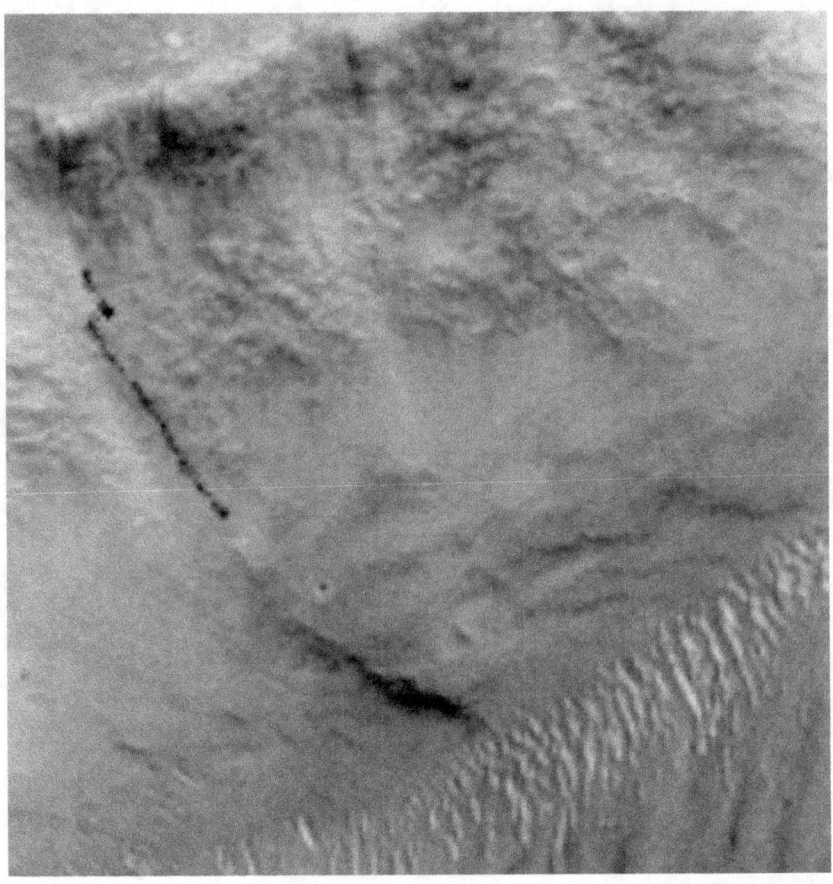

Figure 39

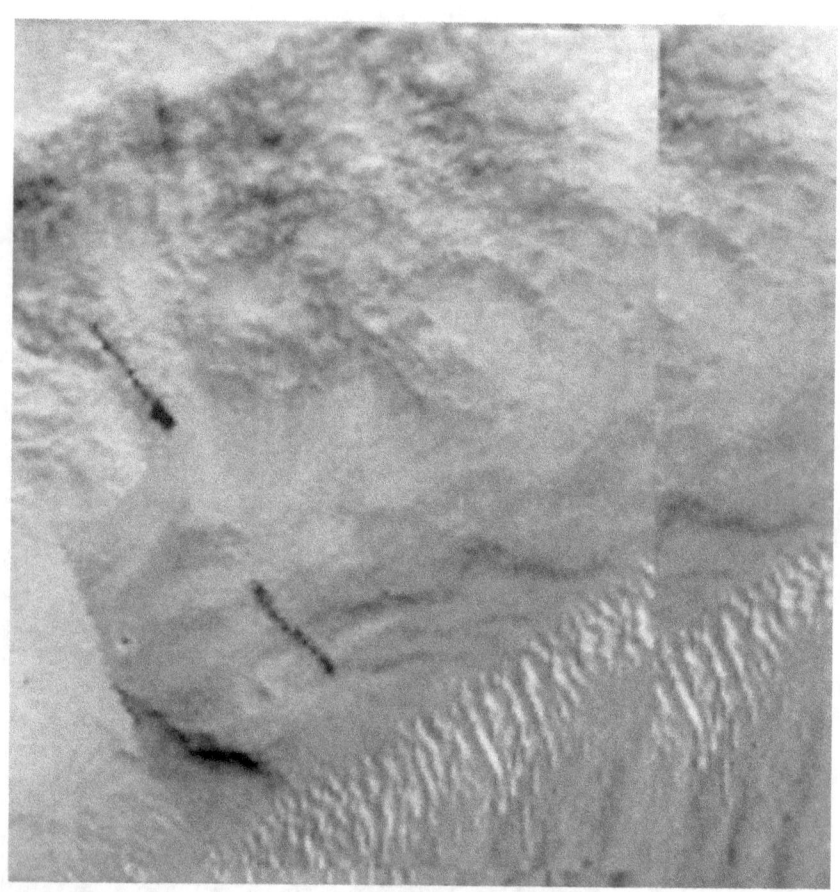

Figure 40

In Figure 39 the left side of Face Two is defined by a black outline and a shadow by the small ditch there, on the left of Face Three in Figure 40 there is a faint shadow in the same position marked in two black lines. It uses one of the vertical ridges under the nose of Face Two to define this face edge; the other vertical ridge may mean that the face edge moves with the sun. In Figure 40 I have cut out the left eye of the Crowned Face i.e. Face Two, this is to avoid distracting the eye which seems to be drawn to the Face Two features instead of Face Three because it is so life like. By removing the left eye of this Face it is easier to concentrate on Face Three. Each of these lines is 10 to 1 against happening by random chance, this comes to a total of 10^{21} to 1.

A similar shape appears in the image below like an upside down U shape at **A** which would give a similar shadow on the edge of Face Two. It is just where the black line marking the side of the face ends in Figure 39. This U shaped feature would also be difficult to explain with natural

318

geological processes because of its symmetrical shape, it appears to be mimicking the nose tip of Face Two at **B** so perhaps another Face appeared between Face One and Two with certain sun angles. The piece on the left side of Face Two is a similar same shape as the tip of the Crowned Face nose. **A** is however not the tip of the nose of Face One because that was shown to be in a different position in the HiRise image earlier.

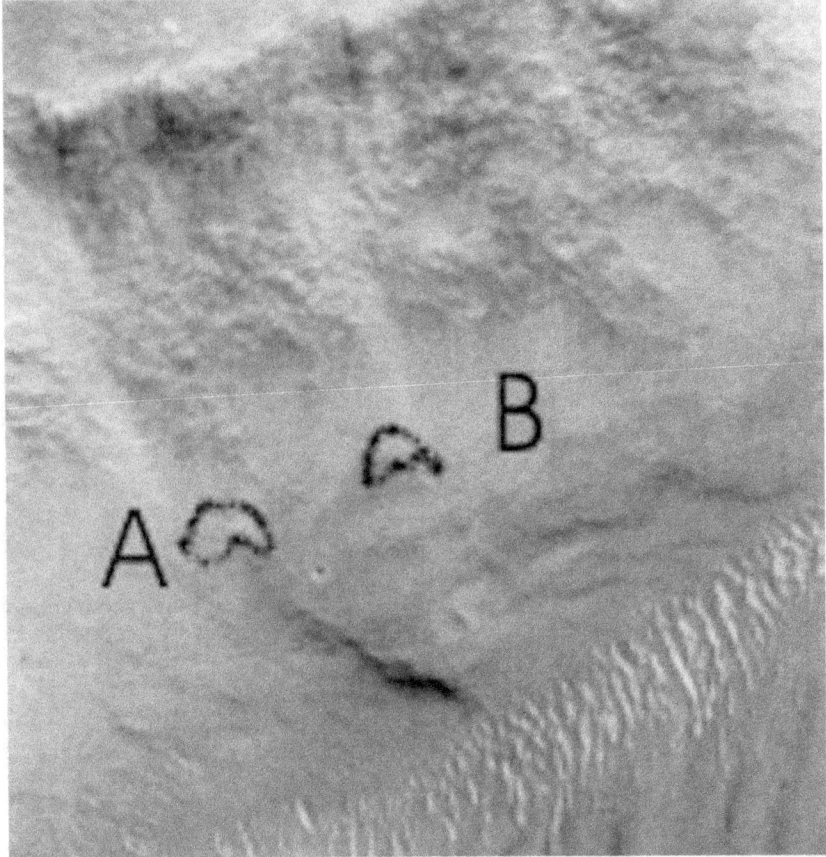

Figure 41

*This feature is 10 to 1 against occurring naturally bringing the total to 10^22 to 1. The formation at **A** might be impossible to occur naturally because the rocks around it are all eroded into a smooth surface.*

Another possibility is the marked feature shown below in Figure 42 is a kind of decorative adornment like jewelry, this may have connected to the

319

U shape but this ornamental shape might also have appeared as a nose using the left eye of Face One and the left eye of Face Two. So then the sun may have caught this adornment on Face Two at some sun angles and at other time it highlighted the upside down U shape at **A** in Figure 41. This may have used the left eye of Face One and the left eye of Face Two, with so much erosion it may be difficult to tell without a manned expedition.

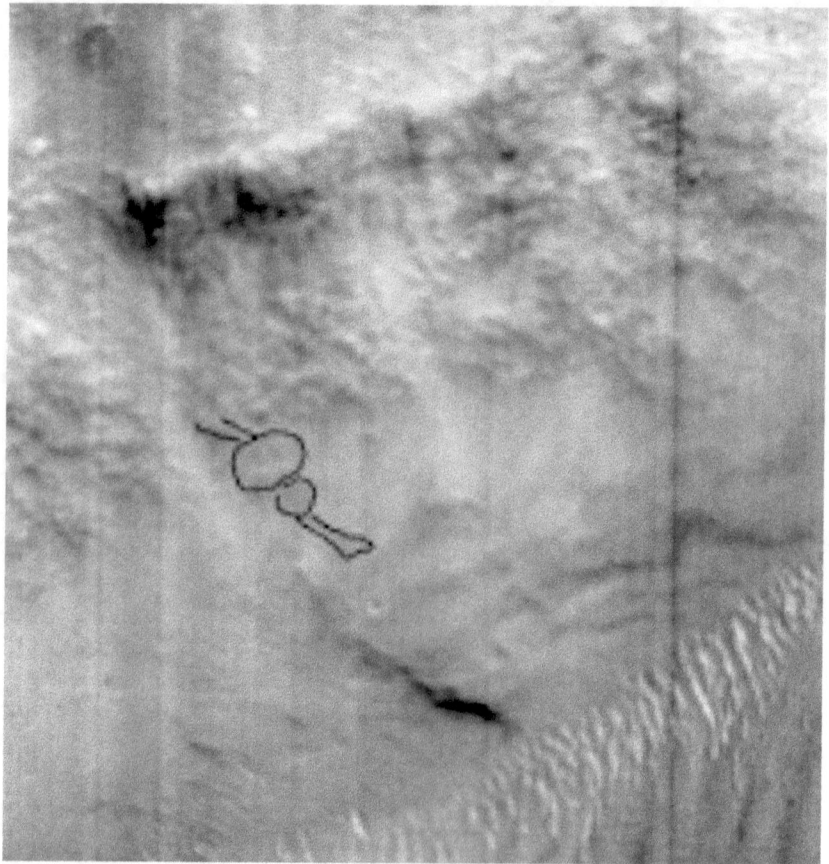

Figure 42

This jewelry shape is also unlikely to occur geologically because the cheek and other areas are highly eroded, this should have erased this feature yet it remains with many details.

There is a shape on Face Two marked in white below and the same shape is found on Face Three below also marked.

320

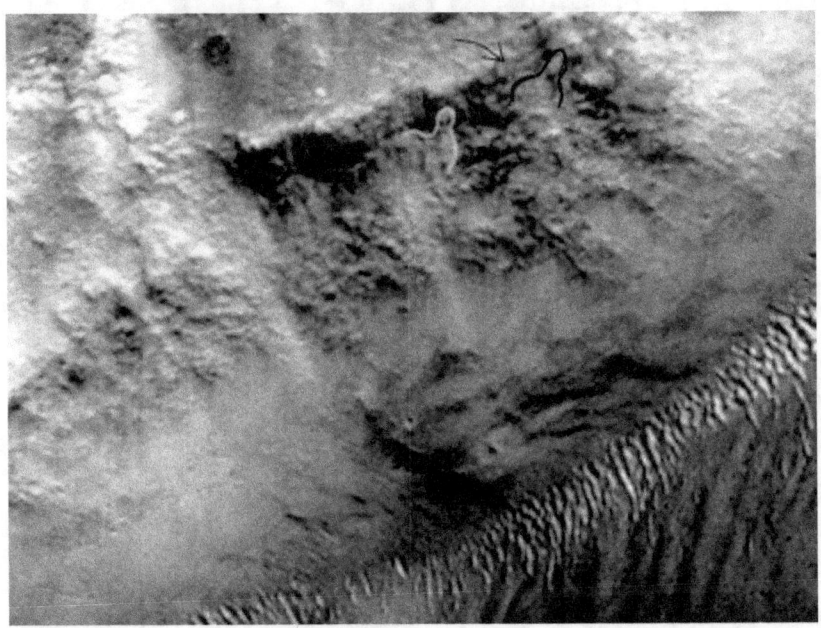

Figure 43

The shape marked in black may be partially buried by debris from the notch in the crown above it. Note that the crown of Face Two is more prominent and this probably prevented debris from rolling down the Face, damaging and burying it. Face One doesn't have this protection and has a lot of sand on it which would have come from above, perhaps sand dunes which were blown into the valley from this cliff edge. These two shapes are at least 10 to 1 against being so similar bringing the total to 10^23 to 1.

This photo below shows Face Three superimposed onto Face Two. The Face Two nose fits well onto Face Three as does the jawline.

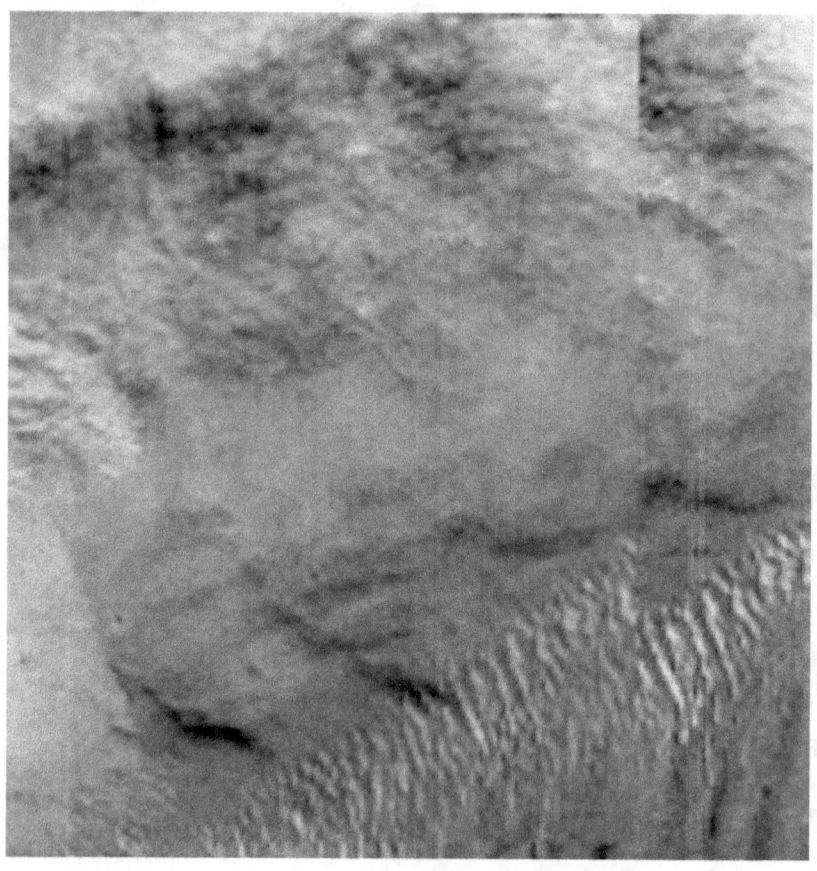

Figure 44

Compare above to the photo below where Face One and Face Two are superimposed; note how they share the same left jaw shape. While humanoid this doesn't appear to be a human face. It is also not like any other supposed alien face such as the Greys, there is enough information here to do a 3D image of these aliens in an upcoming book.

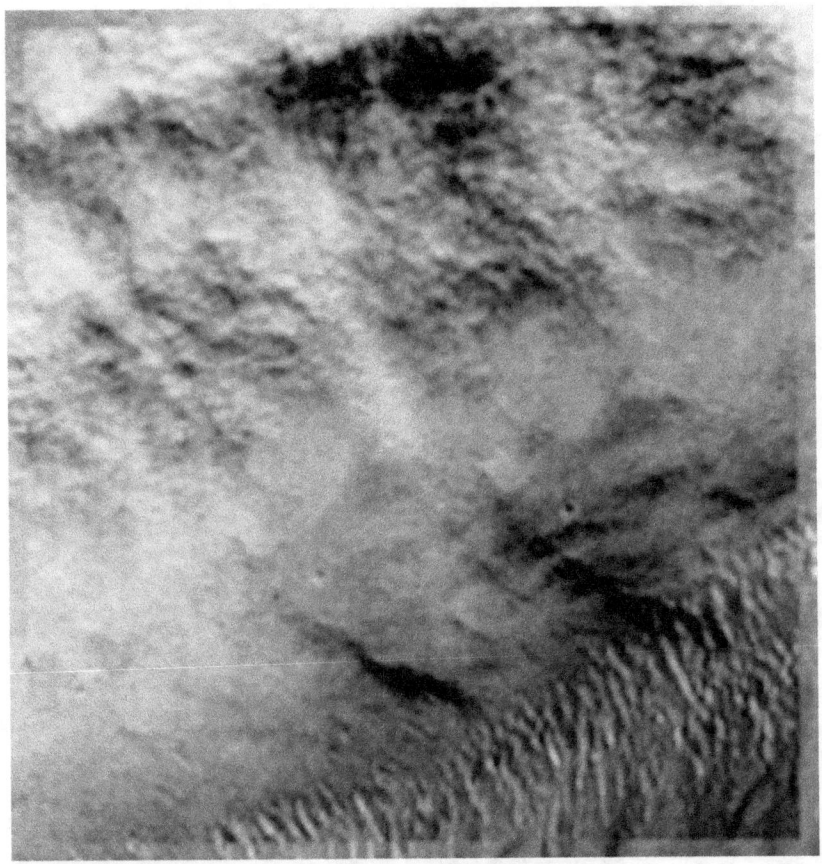

Figure 45

Examining both of these overlays gives a clearer picture of what these aliens or indigenous Martians looked like; the similarities imply that they are different facial expressions of the same humanoid; the odds against them occurring naturally are astronomical. To give an idea of how high these odds are a lottery calculator is used, playing 6 numbers out of 39 gives odds against winning of 1 in 13,983,816. That means the odds against the similarities so far occurring by chance are higher than winning this lottery three times *in a row*. If someone did this then most people would suspect the lottery was not working randomly, in the same way these similarities should not be happening with random geological processes.

There is a shape above the Face Two left eye. There is a similar blank shaped area above the Face Three left eye.

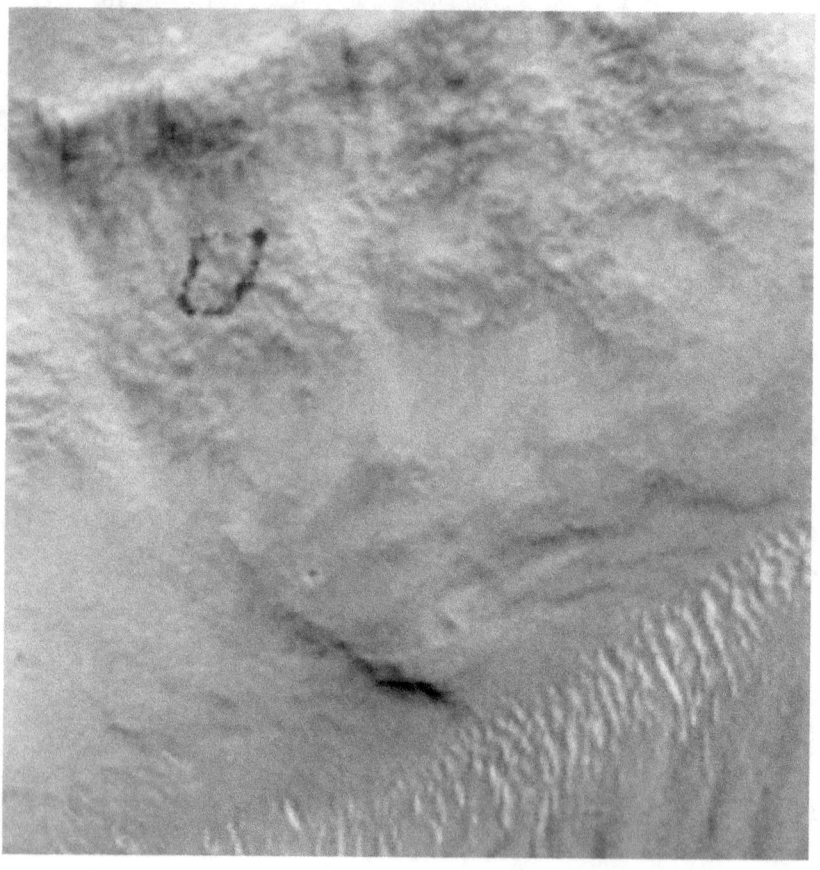

Figure 46

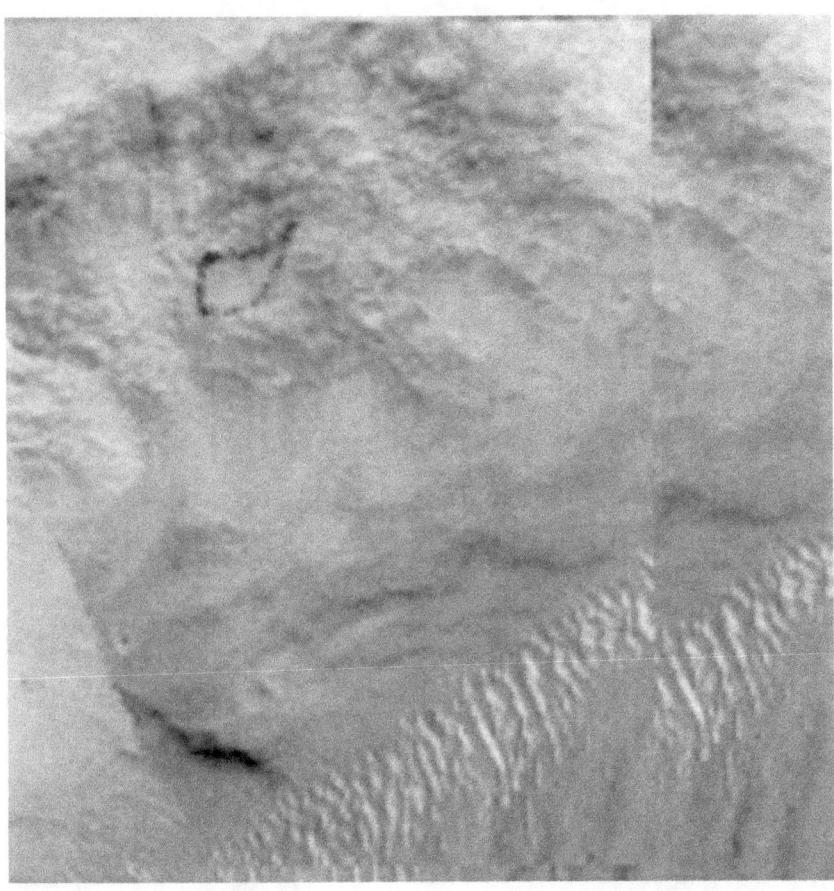

Figure 47

This is at least 10 to 1 against random chance causing this, making a total so far of 10^24 to 1. It's unlikely for this on Face Three to be natural because the cliff face around it is so rough and it is not the shape of an impact crater.

Here is the similar shape above the Face One left eye.

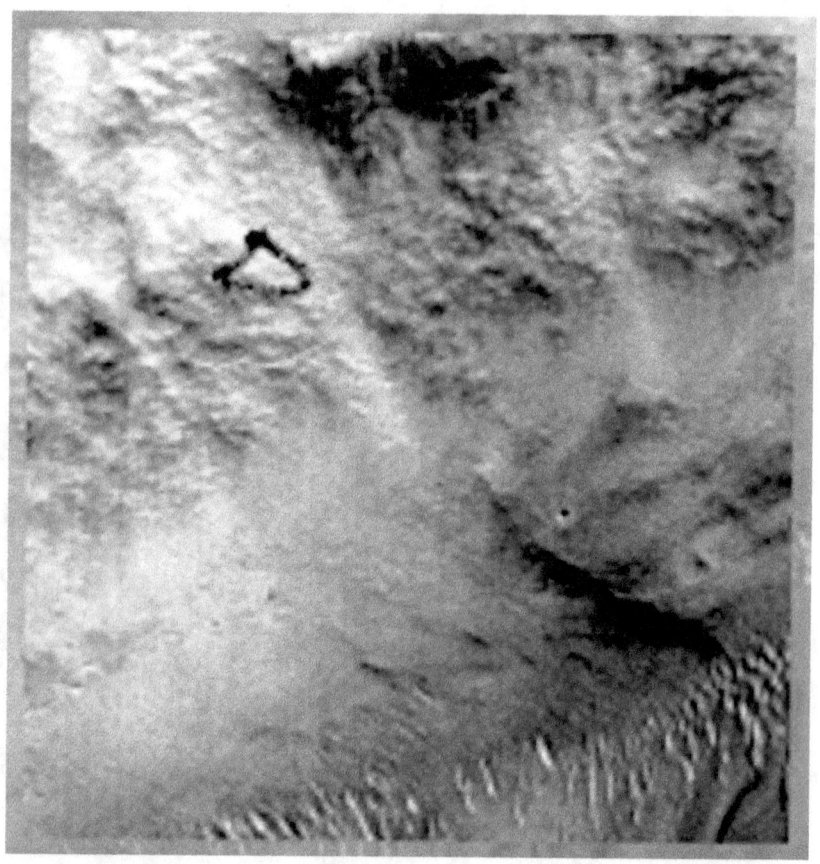

Figure 48

The shapes are not identical but there is a depression in all three positions, this might move slightly in relation to the eyes forming a different expression like a mark above a human eye would if a person frowned, laughed, etc. This is an additional 10 to 1 against chance for a total of 10^{25} to 1.

Below is the Face Two left eye and the Face Three left eye, and then there is an intermediate shot of them together, matching very closely. Previously the Face One left eye and the Face Two left eye were also very similar to each other, much of this is necessary to have the faces overlap like this. The small difference in shape could be from the different facial expression. The eyes seem larger perhaps because of ornamentation or cosmetic markings on the outer left edge of the eye.

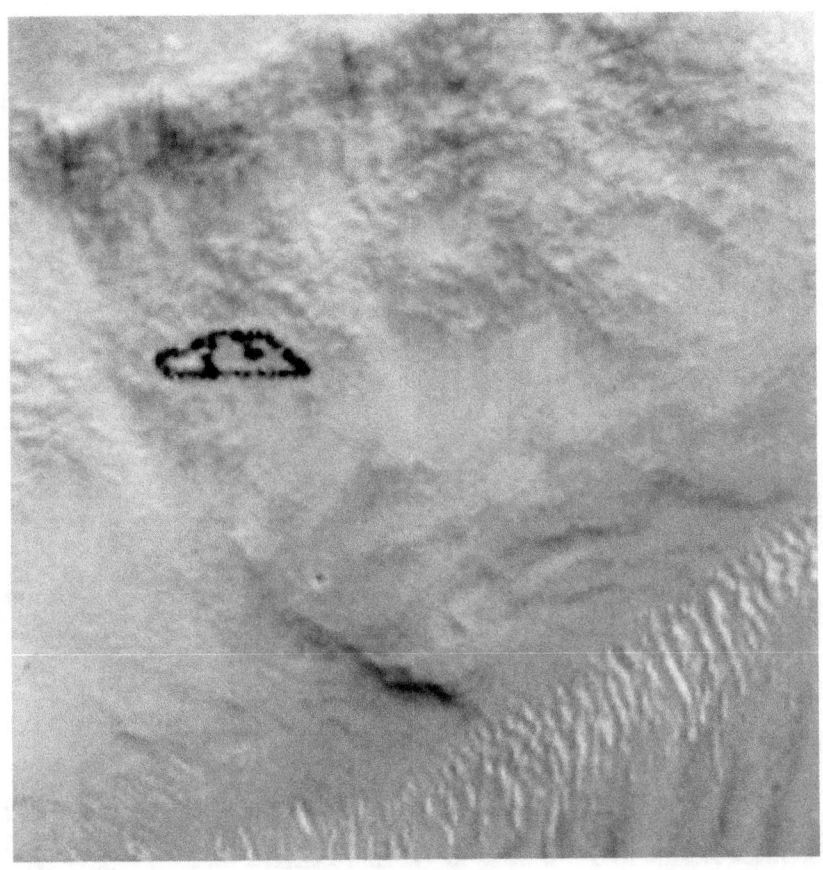

Figure 49

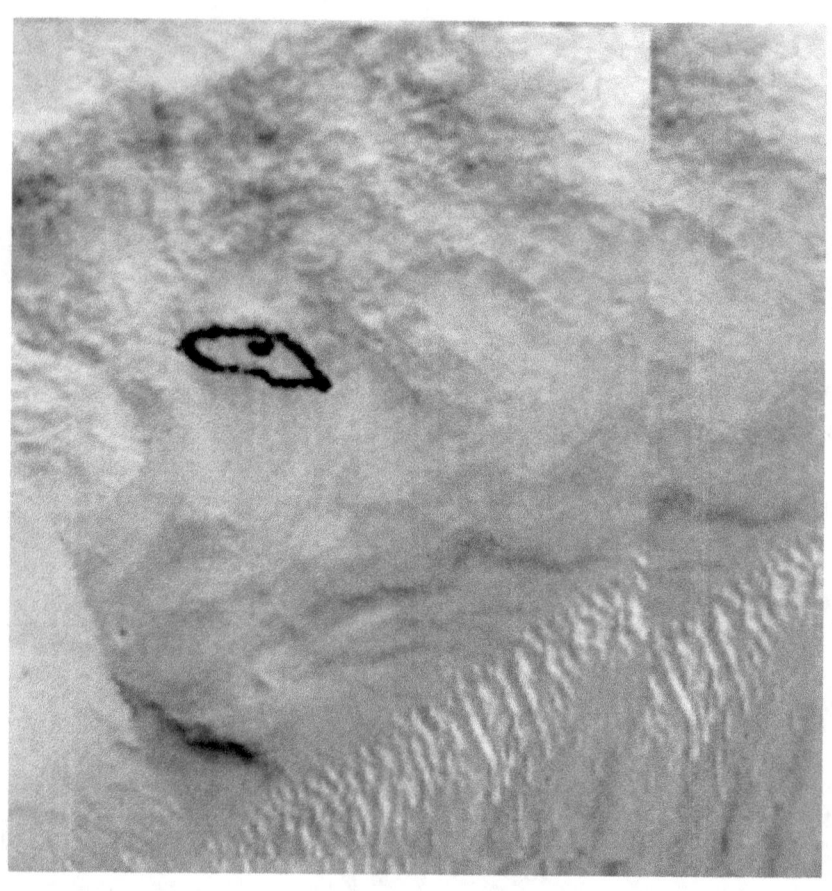

Figure 50

This has at least two aspects which are 10 to 1 against random chance, the length of the eye, the slant of the eye, and the pupil in it. This is 100 to 1 or 10^2 to 1 against chance giving a total of 10^27 to 1.

Below is an overlay of Face Two and Face Three.

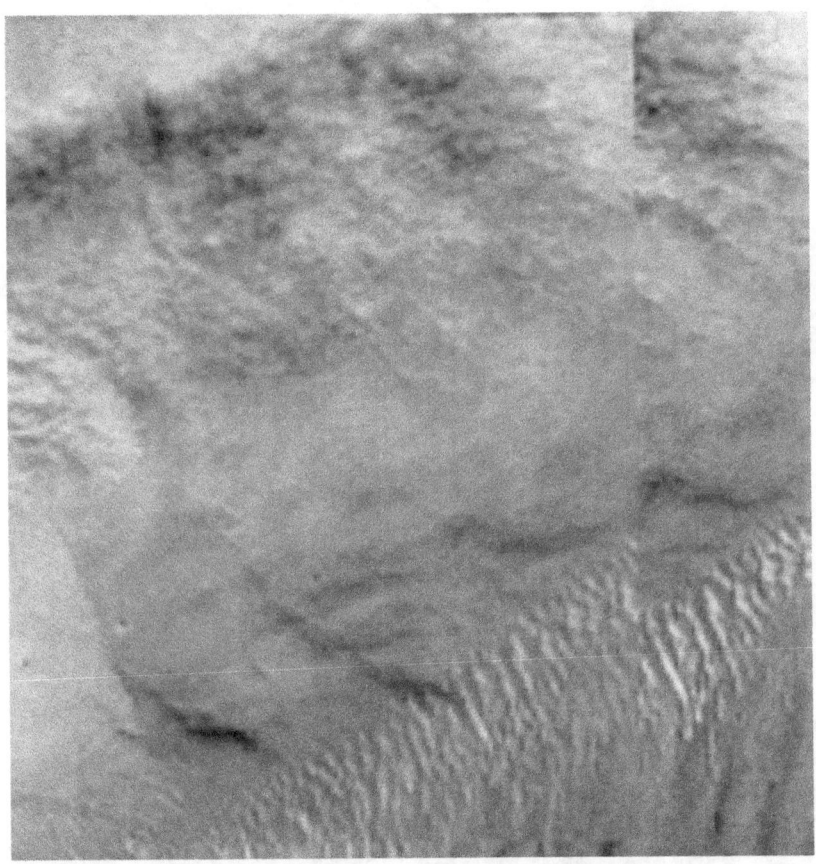

Figure 51

Superimposing the crown onto Face Three fits well, it probably also looks like this in certain sun angles by itself because of the shadow from the Profile Face's crown to its right. With the right sun angle the Face Three crown would appear and perhaps the shadows on Face Three would be accentuated. Emotionally it looks much more stern and angry than Face Two; this may have stood out with shadows more in the Northern winter. This assumes of course that the emotional expressions of these faces are like our own, on Earth there can be variations in how facial expression portray emotions even in animals but anger seems to be recognizable in nearly all of them. It may be that anger would give a similar expression in a humanoid race but there is no real evidence for this.

The Face Two nose is shown below with the Face Three nose, and then the Face One nose. All are very similar. This is unlikely to occur by chance because shapes like this nose are rare geologically; this can be seen by looking at other photos in the King's Valley as well as the areas in the Controls chapter of known natural formations. Usually a nose like this would be worn flat with erosion like the cheeks to their side.

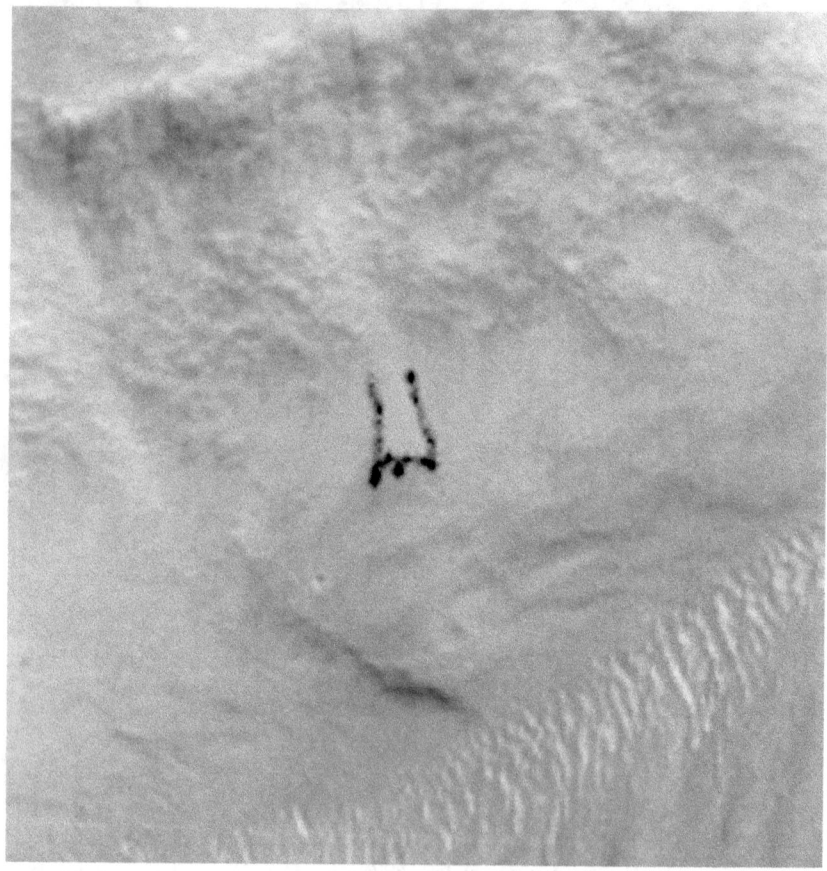

Figure 52

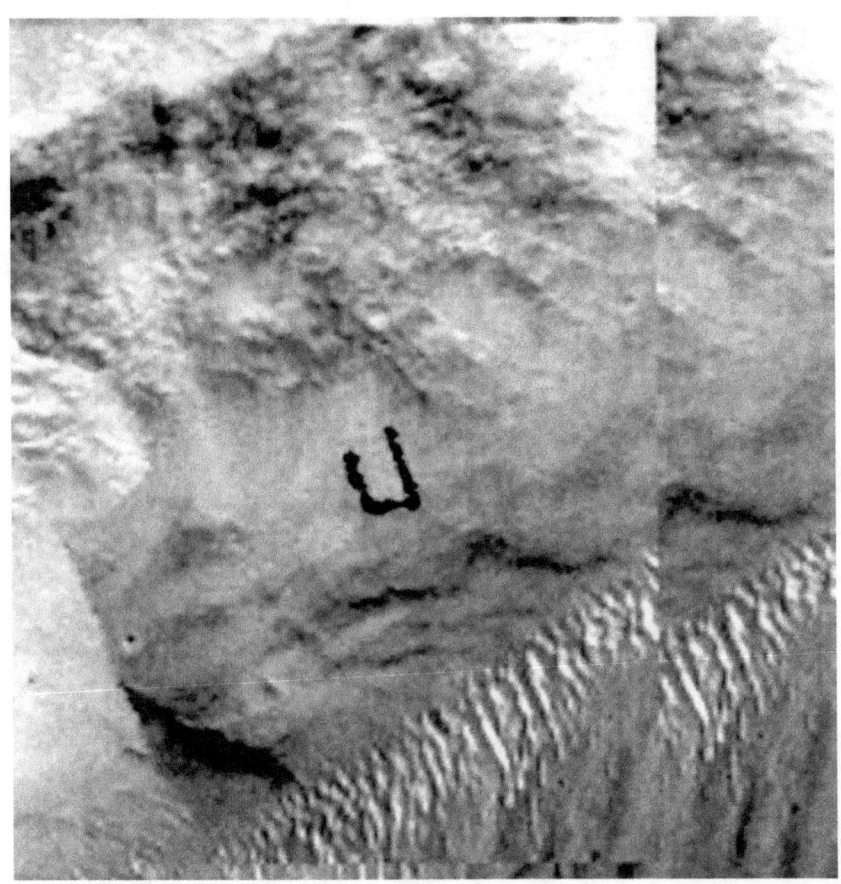

Figure 53

As shown below the nose tip of Face Three has two indentations where the nostrils would be, very similar to the Face Two nose tip. The shape of the nose, the curve to the left, and the two nostrils in the tip make 4 similarities for 10,000 to 1 against chance for a total of 10^{31} to 1.

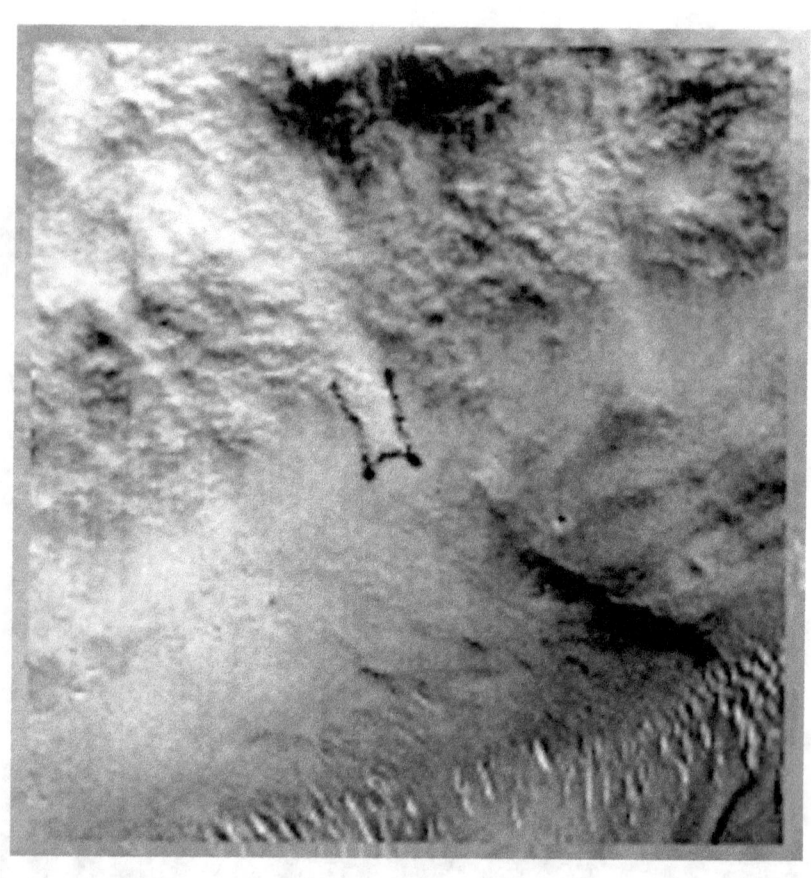

Figure 54

This shows the similarity of the Face One nose; compare it to the noses in Figure 53.

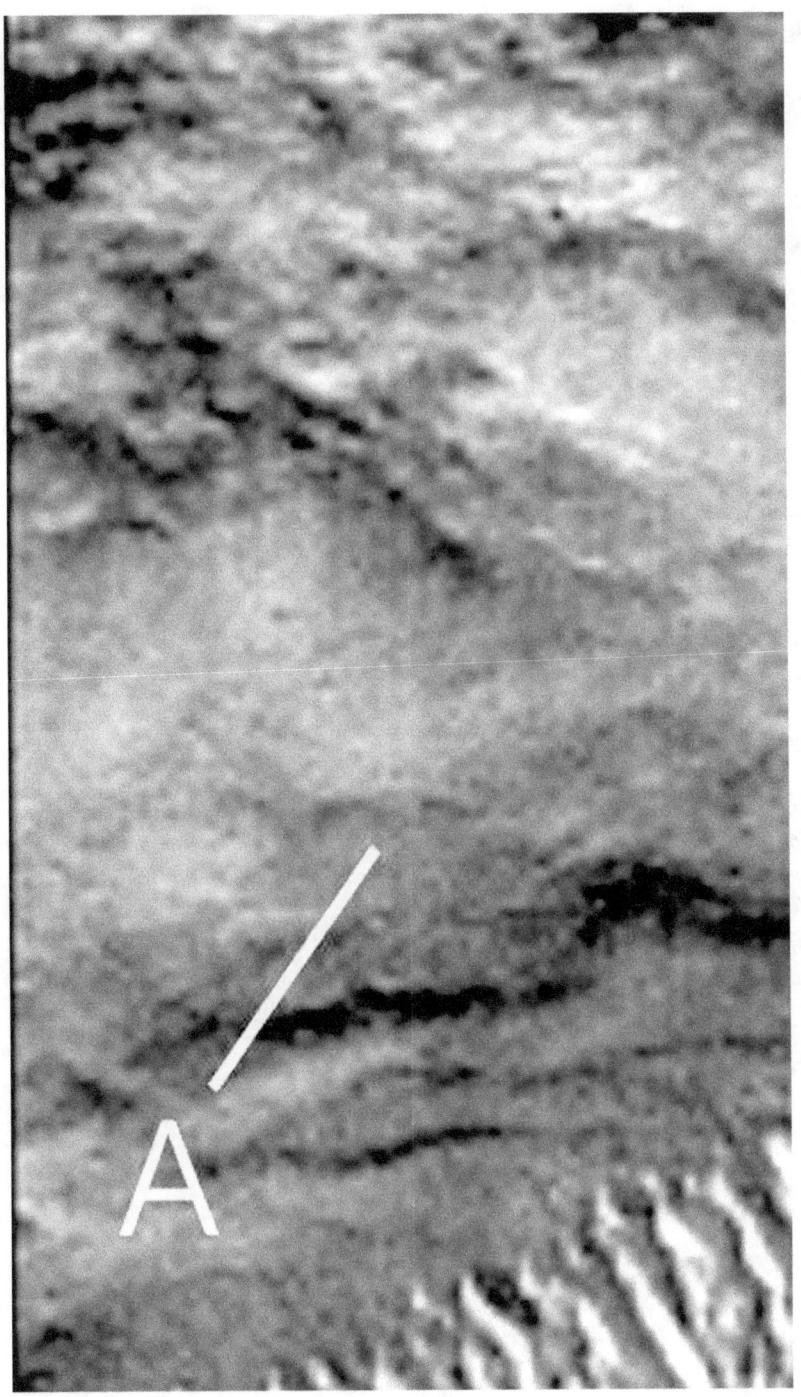

Figure 55

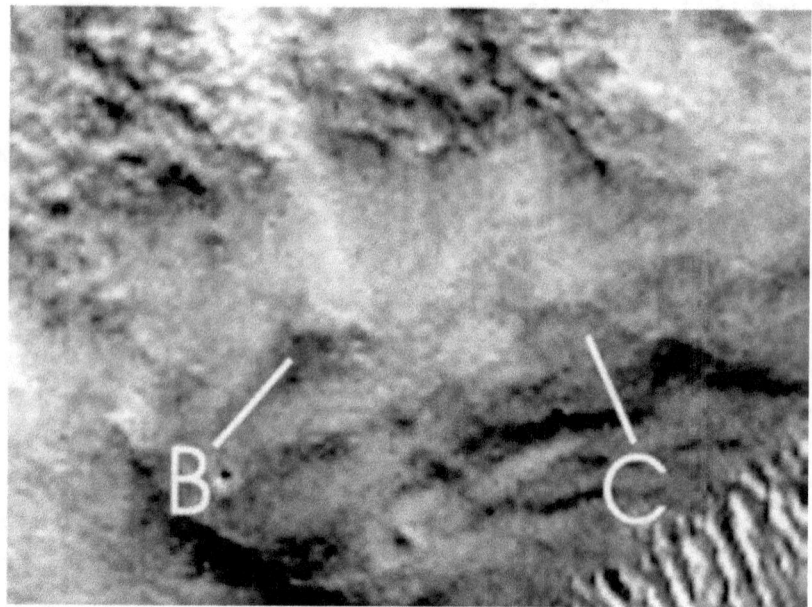

Figure 56

*A in Figure 55 shows the Face Three nose in photo M0303483, this is much clearer than it was in the photo M0203051 with **B** and **C**. Note that the left nostril is bigger than the right in both noses so there is an asymmetry in these noses as they curve to the side.*

Next is another overlay of Face Two and Face Three, note how well this fits and how it gives a subtle angry expression to the Crowned Face.

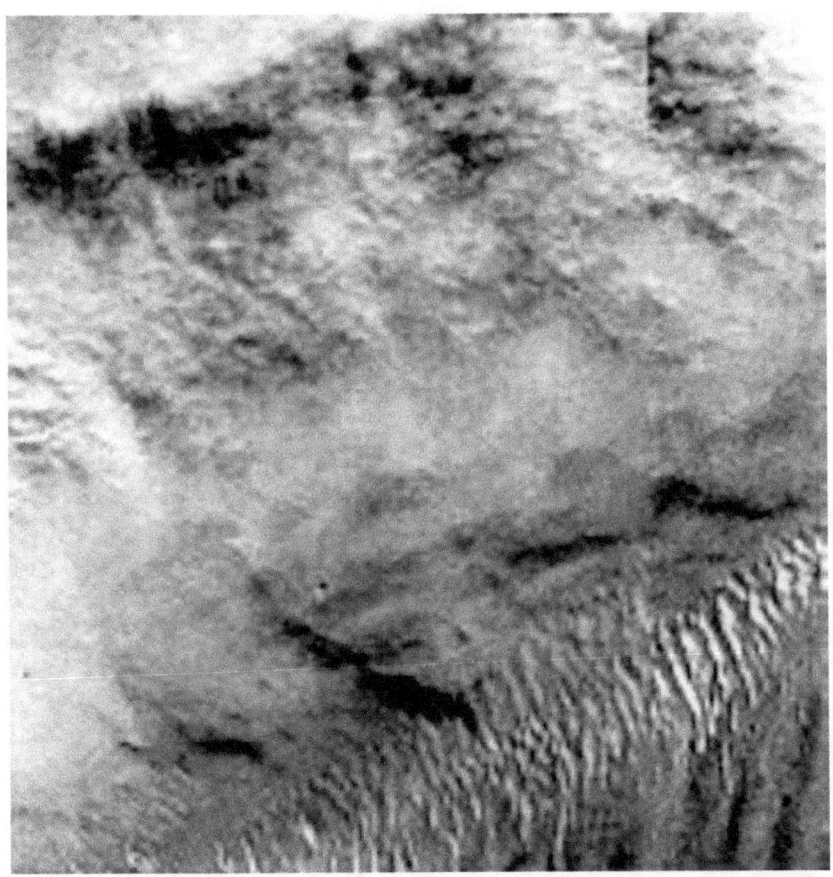

Figure 57

By varying the relative amounts of each photo in these overlays all kinds of subtle emotional expressions are created, this is highly unlikely to occur by chance with random rock formations. For example rough faces might be sometimes found in clouds, mountains, even pieces of toast but they don't change with different sun angles and they would rarely be able to be superimposed on each other to give an extra face like formation.

The right eye of Face Three below appears to be missing in a section that may have had a landslide, the eyeball of Face Three still seems to be there when reimaged by HiRise. This and other parts of Face Two and Three are analyzed in the new HiRise images elsewhere in this book.

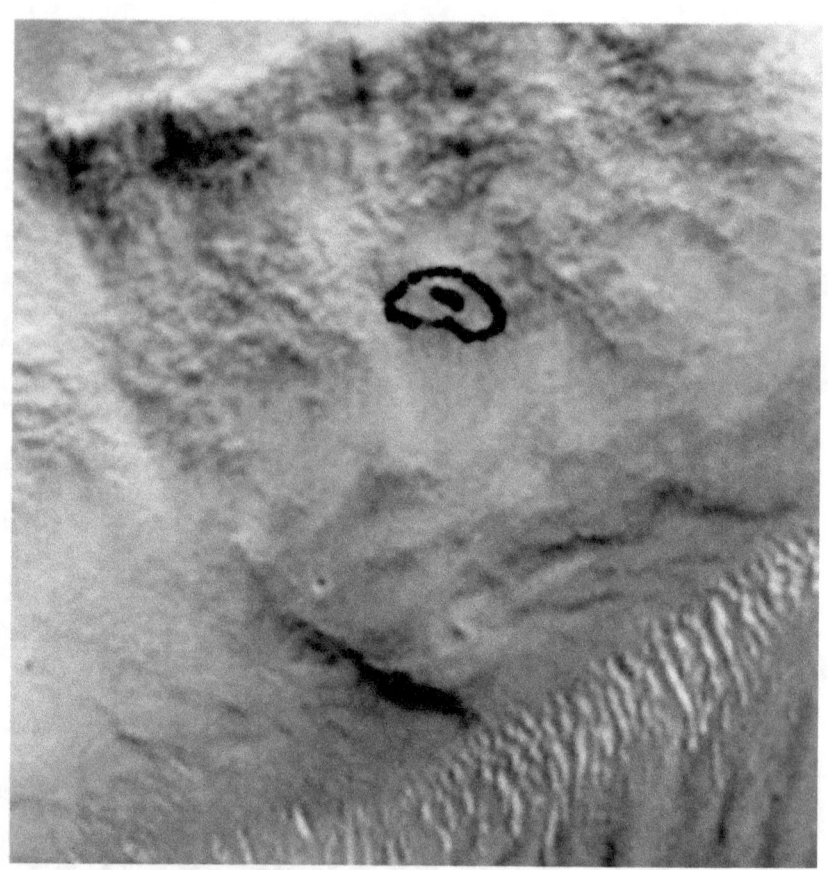

Figure 58

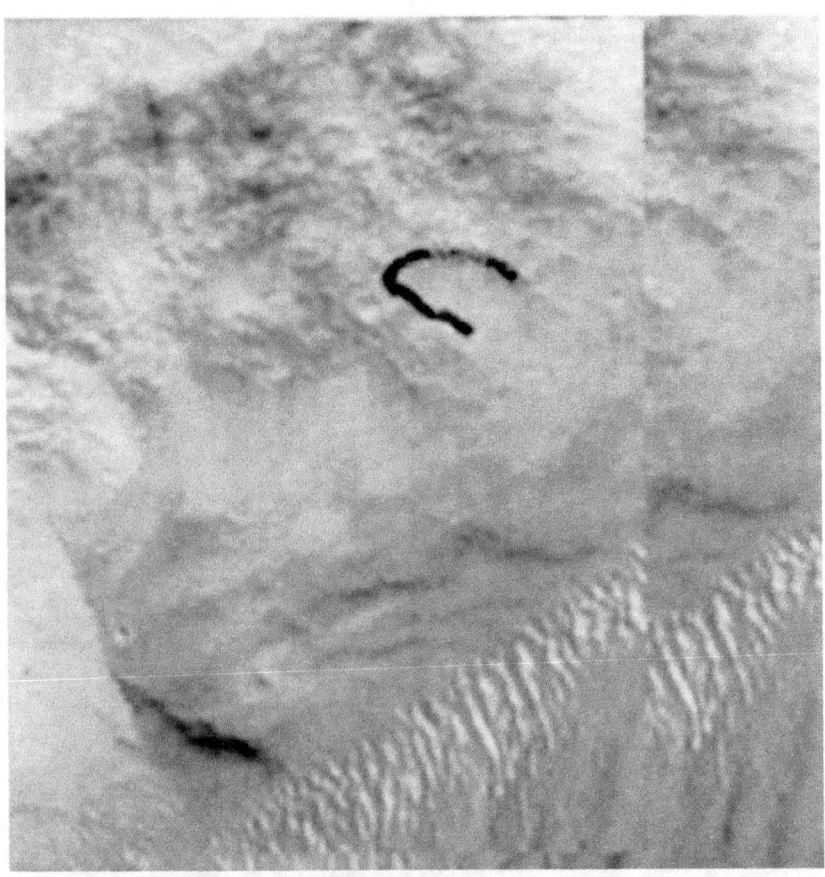

Figure 59

Figures 58 and 59 show how the two right eyes are very similar. Below there is another overlay of Faces Two and Three, note how perfectly the Face Two right eye fits into the hole making it more likely the whole eye section broke off. The channel of debris from above may have damaged it though there is no debris from the eye visible on the valley floor below, there is so much sand there it may be buried. This sand should be investigated in an expedition because many artifact pieces may be buried down there as they fell off the faces. This is 10 to 1 against chance making a total of 10^32 to 1.

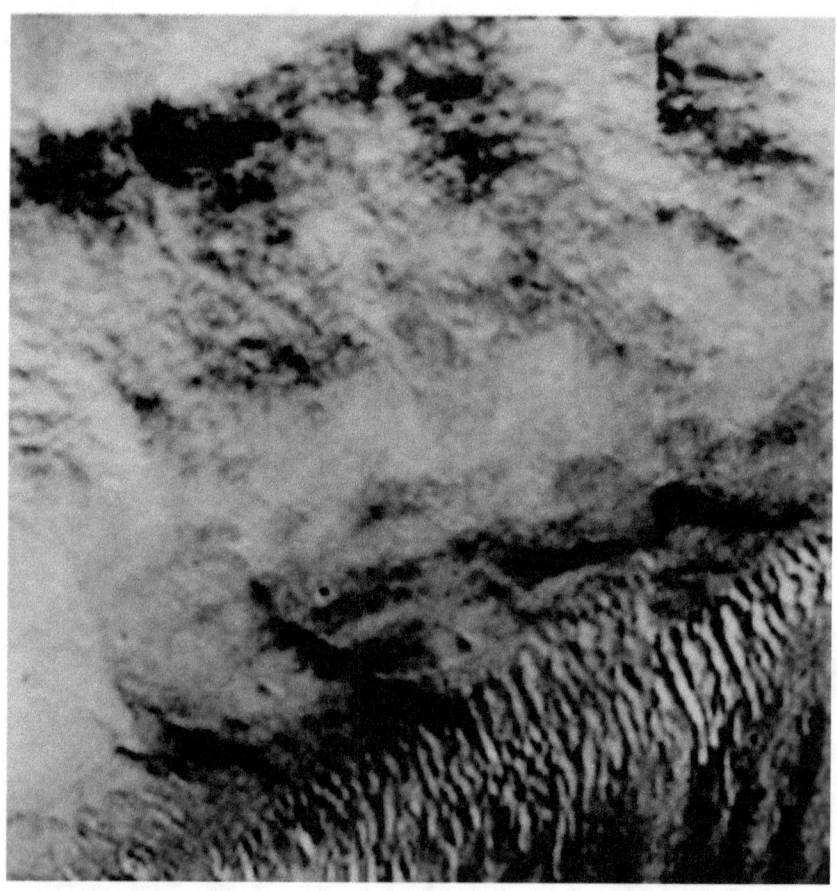

Figure 60

Note how the angled eyebrow of Face Two fits well onto Face Three making it appear more angry or indignant, another different emotional expression.

There may also be another right eye that services all three Faces. Face Two below has the normal left eye but also a hollow under the usual right eye. This fits in more with the face shape looking rightwards and may indicate one right eye merging into another as the sun moves. There is a matching hollow on Face Three and Face One consistent with an eye. This is also like the hollows above the Profile Crowned Face nose mentioned elsewhere in this book.

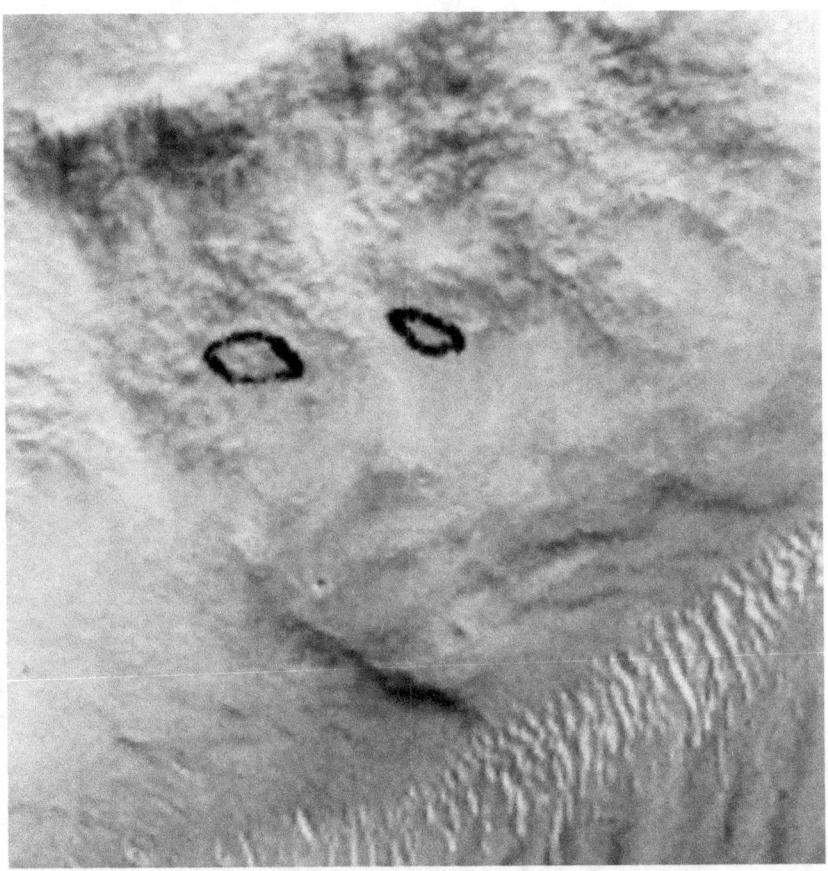

Figure 61

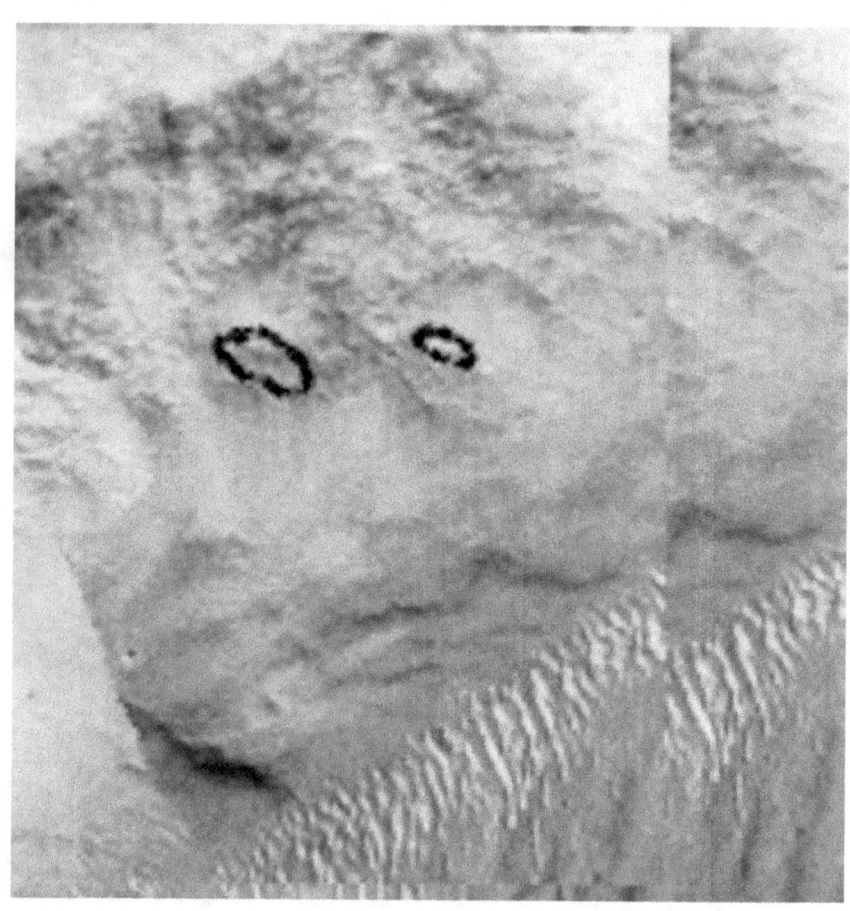

Figure 62

These eye positions give the impression of the eyes looking to the left, as if the faces might have their eye move as the sun angle changes. This is 100 to 1 against chance giving a total of 10^{34} to 1.

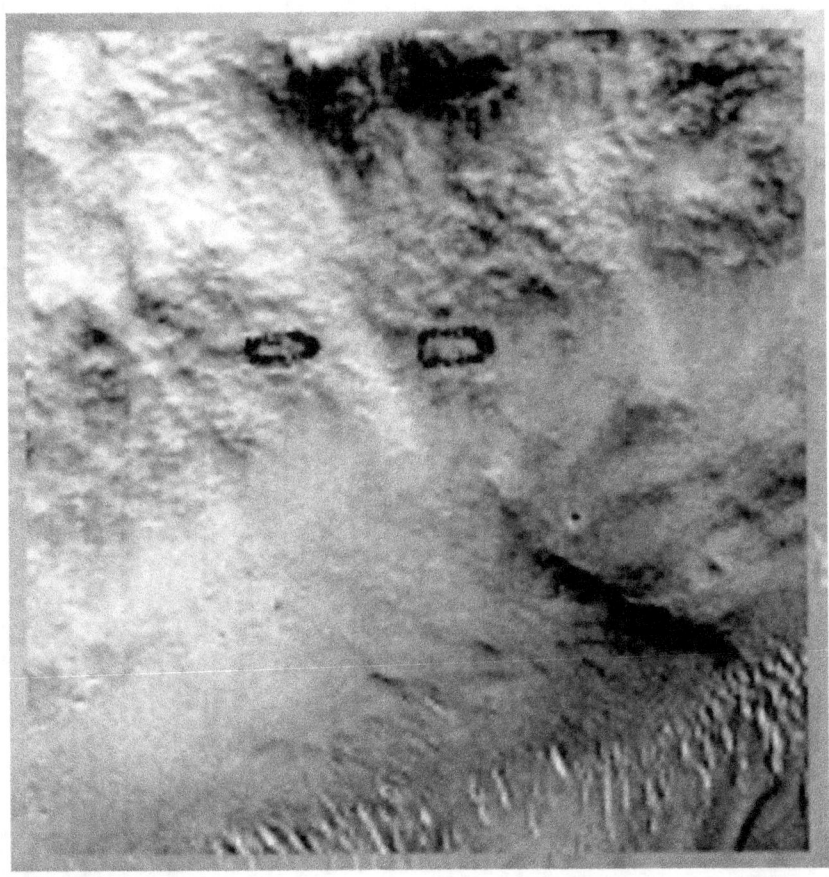

Figure 63

There is another possible pair of eyes for all 3 faces. While it is hard to evaluate how this would look for a different angle or shadows this is present in all 3 faces. It may be ornamental or cosmetic like a face tattoo.

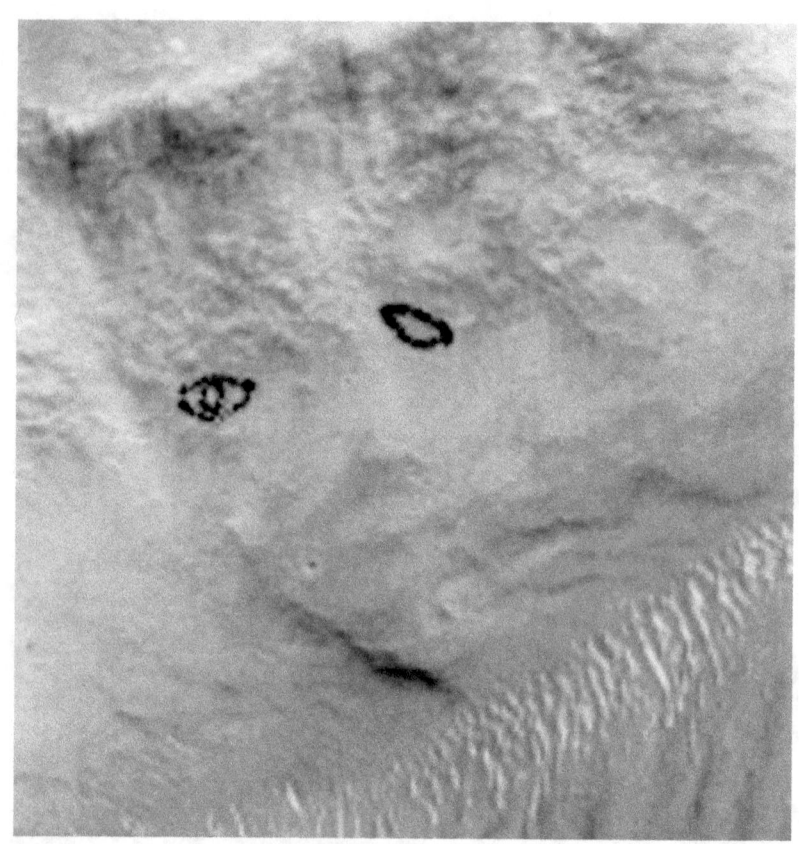

Figure 64

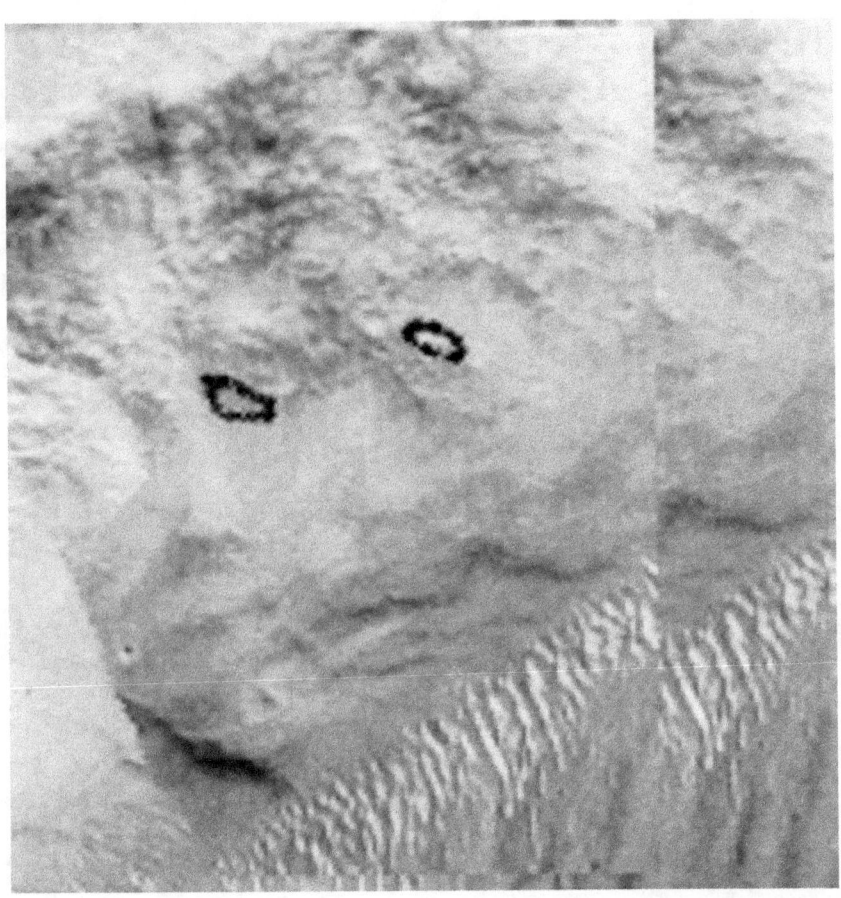

Figure 65

As shown above there might be yet another set of eyes moving to look at different things as the sun moves, there may be something on the other cliff face it would be looking at or perhaps it is intended to be looking at a star or place.

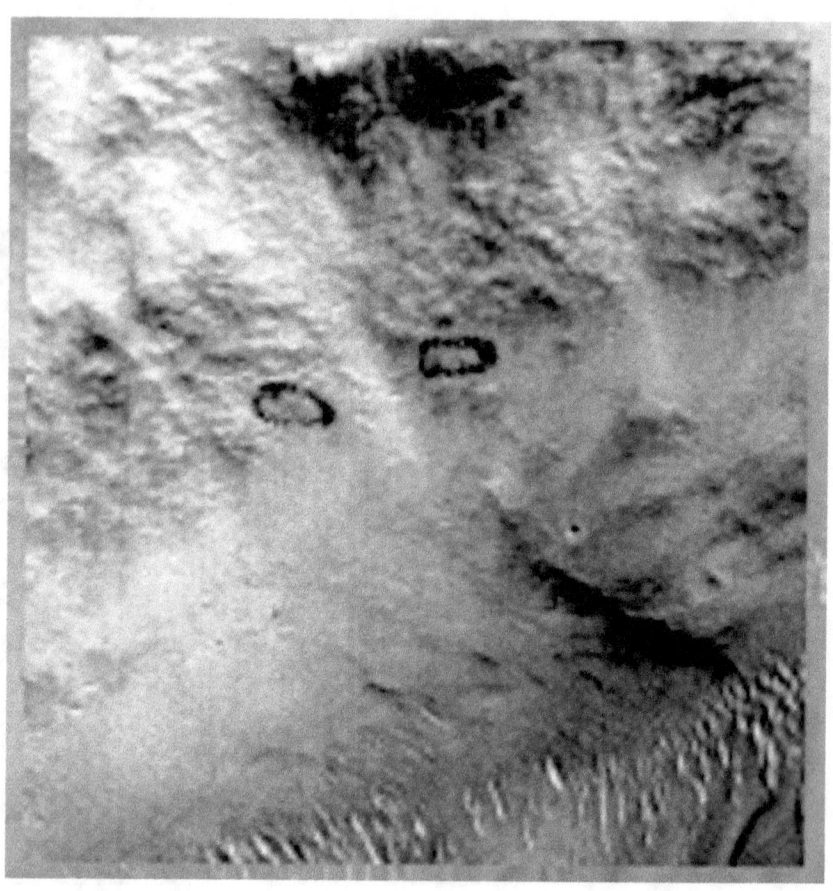

Figure 66

The eyes may also move to the side with different sun angles on Face One.

There is a difference in the mouth positions which could be accounted for by different facial expressions, or the expression may change as the sun angle changes. For example the sun may outline the upper part of the mouth onto the lower part of the mouth ravines. Because these are curved this may give a different expression as the sun moves.

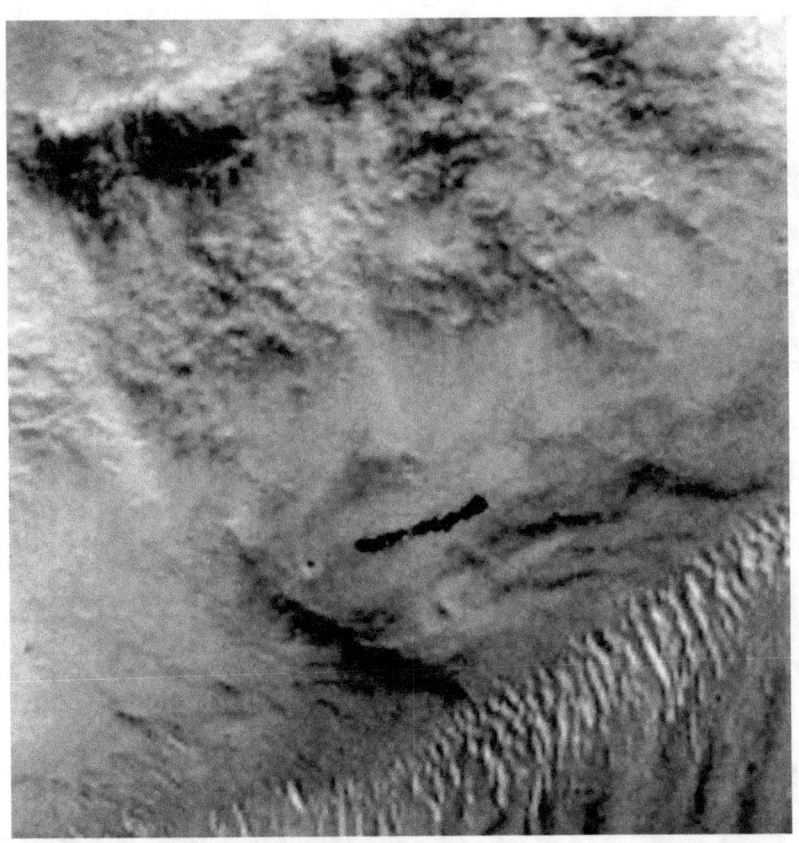

Figure 67

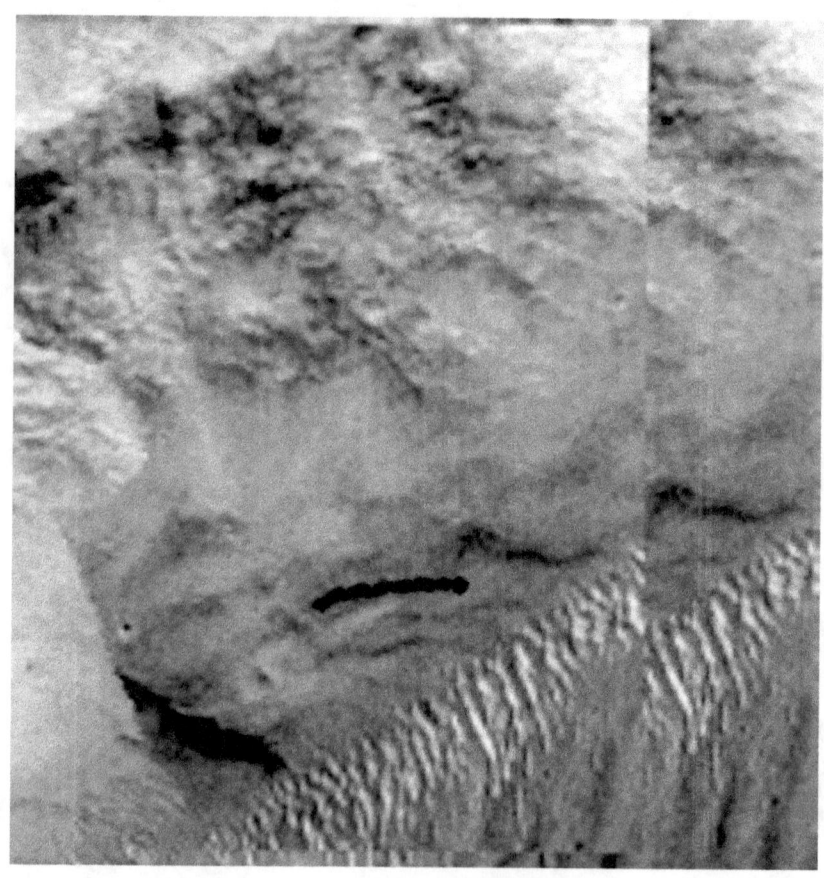

Figure 68

Note above how a slight upturning of the mouth gives a different facial expression to Face Two. The similarity in the moths is another 10 to 1 against chance making a total of 10^{35} to 1.

In the image below superimposing Face Three to its right also fits in with a rounded area on the cliff, there may have been another face here as well.

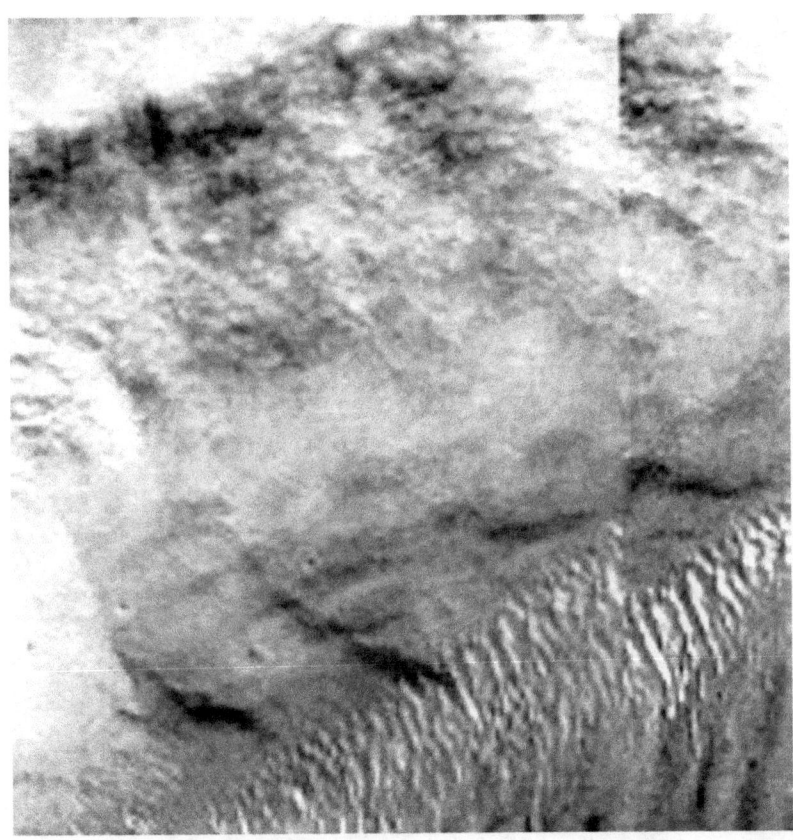

Figure 69

There is an eyebrow like line on Face Two and on Face Three as well as on Face One below.

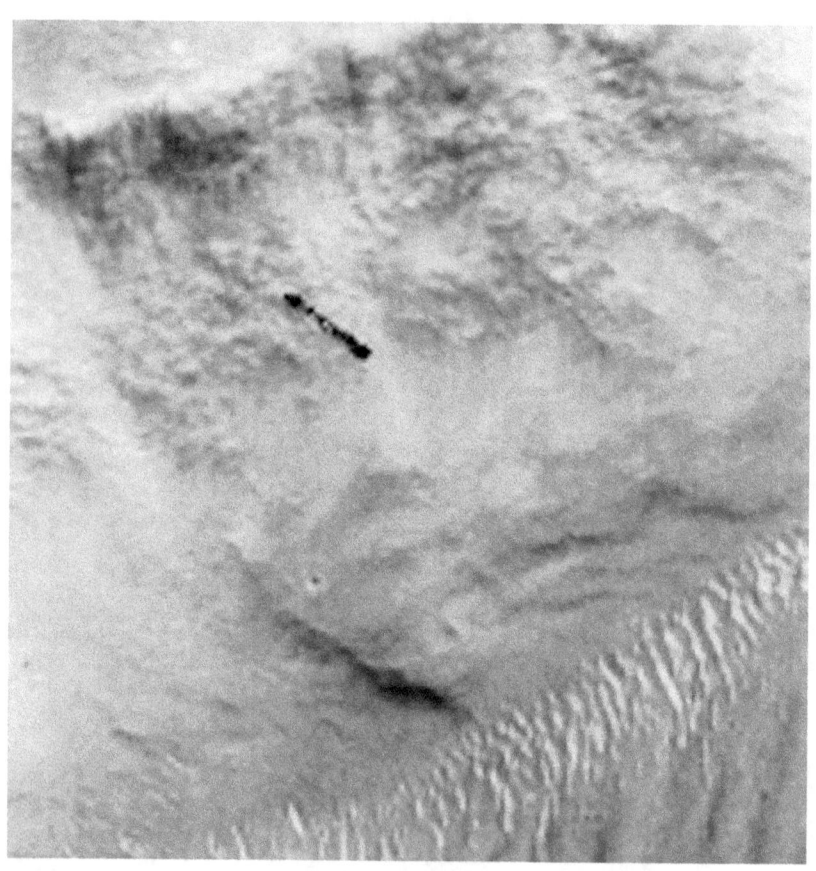

Figure 69

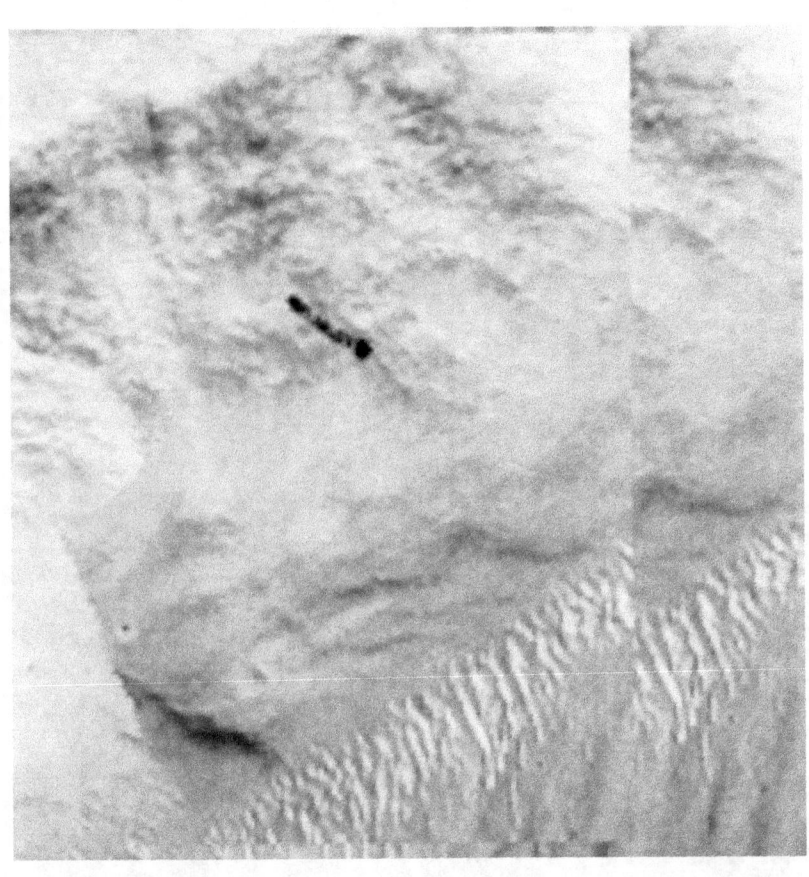

Figure 70

This is another 10 to 1 against chance making a total of 10^36 to 1.

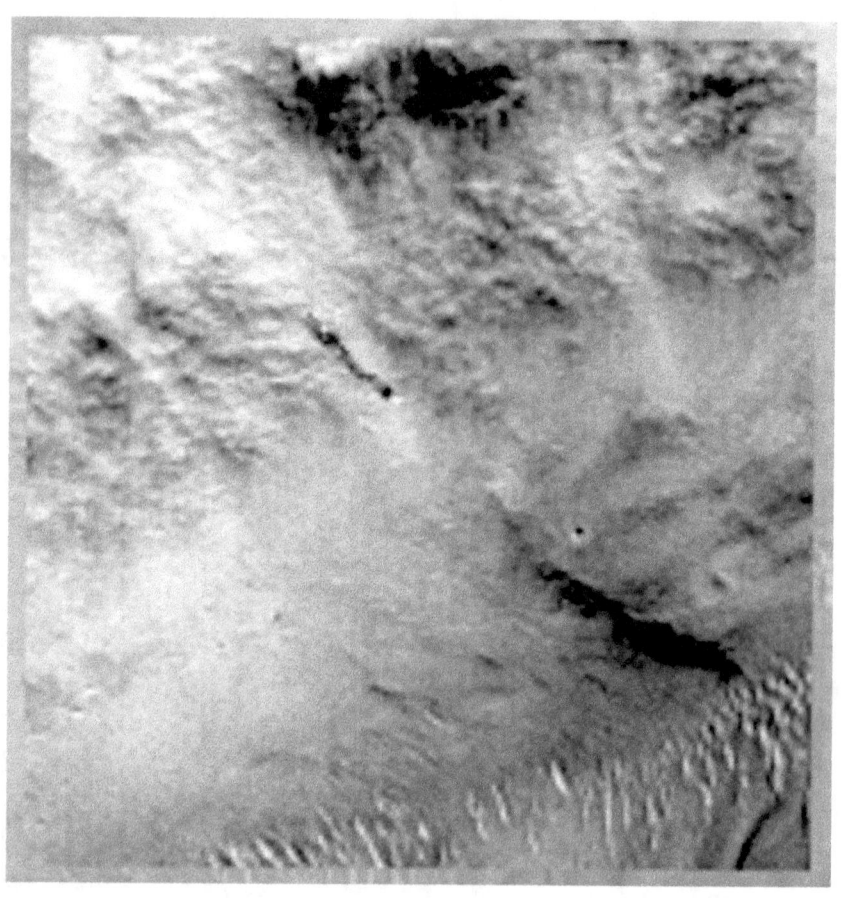

Figure 71

There is a similar line on Face 1 which is 10 to 1 against chance for a total of 10^37 to 1.

This is getting to be a large number, written out in full it is 10,000,000,000,000,000,000,000,000,000,000,000,000 to 1. To put this number in perspective there are only thought to be 3 to 7 times 10^{22} stars in the observable universe[cxii] so the odds against these similarities in the three faces being a coincidence is indeed astronomical. As mentioned earlier the number 10 to 1 is almost certainly too low, for this to be the correct odds 10% of all the area of the photos taken so far would have to have almost the same faces with a total number of faces on Mars of around 10 million. However the number 10,000,000 could arguable be substituted for 10 which would give 10,000,000^37 or 10^222 to 1 against these faces being so similar by random chance. By comparison there are

350

about 3 times 10^79 atoms in the observable universe. Pick any atom in the observable universe and someone else picks any other atom. The chance you both picked the same 1 is still far less than the chance these Faces are coincidentally so alike. Somewhere in between these two numbers, 10^37 and 10^222 is the most likely odds against chance these three faces formed randomly. In any scientific experiment such odds would be regarded as proof. There is still more evidence however, not only are these three faces so similar to each other they are also very similar to the Cydonia Face and Meridiani Face which are analyzed below. Then those odds are multiplied by the other faces in the King's Valley also having crowns and the result is surely far beyond what could happen by random chance.

As mentioned earlier one problem with this analysis is that a certain number of these similarities would need to occur before an area would look enough like a face to be evaluated. For example an eye shape sitting by itself would not be noticed and so the odds cannot take its existence into account. Therefore taking 10 of these similarities looking face like as a baseline to an area means that the odds would be reduced by between 10^10 and 10,000,000^10 or 10^60 in the higher estimate. This would give adjusted odds against chance of between 10^27 to 1 (instead of 10^37 to 1) and 10^162 to 1 (instead of 10^222 to 1). These odds are still larger than the number of stars in the visible universe, and as said earlier much higher than winning lotto 3 times in a row.

Chapter Fifteen

Comparing the Crowned Face, the Cydonia Face and the Meridiani Face

Like with the three Crowned Faces a statistical argument is used here, that 3 pieces of ground should not be so similar to each other. That they also look like Faces is useful in some ways to the argument but not really necessary. In fact it makes it more difficult because of the unsupported assertion that aliens or indigenous Martian would not build Faces on Mars. In fact we have no idea what these hypothetical builders would choose to create, we don't even understand the motivations of the different races on Earth. It is much easier to find evidence that most faces on Mars are of the same humanoid rather than being all different faces. If all the faces are different to each other then one can argue that these variations could be random, there is no real way to prove they are related to each other unless they have some similarities such as being of the same species or ideally of the same humanoid.

Geology should form faces randomly, making them look different from each other but also randomized in all different ways. For example some might have one eye, others three, the eyes might be on the chin, on the forehead, etc. A nose might be upside down or sideways, a mouth vertical, on the forehead and so on. On Mars there are indeed some highly distorted faces that have been found, this are likely to be pareidolia because there is no reason why random formations on Mars should not produce some faces just as geology does on Earth. One problem with looking for faces is that our imagination has become accustomed to cartoon faces in the media so we tend to also see faces with impossibly large noses, eyes out of proportion, and faces that look nothing like each other. However there is a difference with some more plausible faces in that they have so many similarities to each other, also there are no faces like them with distorted variations such as extra eyes, etc. That is, the more plausible faces seem to be a in class by themselves, there are no variations in between them and the more obviously random faces. This makes them less likely to be formed by random chance. For example if we found many variations of the Crowned Face on Mars but with extra eyes, a smooth area where the mouth should be not explainable by erosion, a Face much longer and thinner, etc. then this would imply these are random variations and the Crowned Face just happened to be one where these variations produced a less distorted Face. However this doesn't happen, any face like the Crowned Face doesn't have these defects.

If some of these faces are artificial then it is most likely they are related to each other in some way, either a single or limited number of alien races visited Mars or a limited number of indigenous races evolved on Mars. The main faces seem to be all of one species and arguably all of the same humanoid, the only exception to this are the two Spaceman images which will be analyzed in a later book.

The Meridiani Face compared to the Crowned Face

The Meridiani Face was found by Terry James.

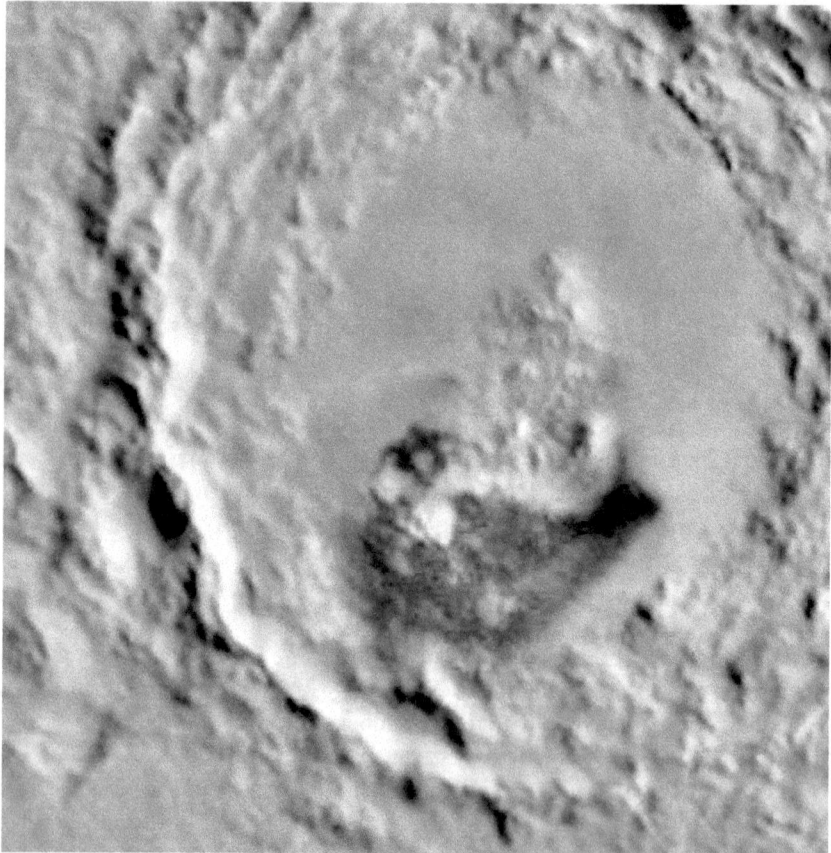

Figure 72

This is the Meridiani Face, so large it is seen in the original Viking images of Mars. It is too large to be seen in one piece in the Mars Orbiter Camera images.

In the image below the Crowned Face, namely Face Two is overlaid on the Meridiani Face. It appears the Meridiani Face may also be a dual face, sharing one eye as with the Crowned Faces.

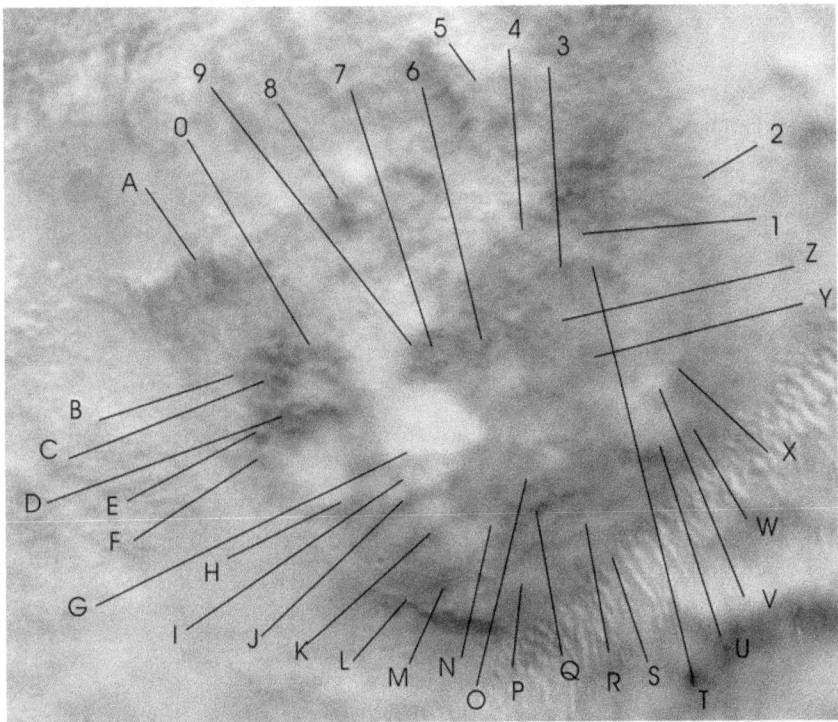

Figure 73

Here I have overlaid the two faces and compare the similarities. Assuming each of these is 10 to 1 against occurring by chance then a similar argument can be made as with the overlays of Faces One, Two, and Three. Since there are 36 similarities this gives an odds against chance of 10^{36} to 1. If we again assume that an area would have to have 10 points of similarity to be considered a candidate for this comparison then that leaves 10^{26} to 1. That is, this comparison is made because the Crowned Face, the Cydonia Face and the Meridiani Face look somewhat like each other. When we see a face on Mars it has to have some similarities to even try this overlay so assuming 10 points of similarity as a candidate face means the first 10 similarities are excluded. That then leaves 10^{26} to 1. Assuming again that 10 to 1 is far too low for the initial odds, otherwise 10% of Mars would be covered in faces, with 10,000,000 to 1 as the base odds that gives $10,000,000^{26}$ to 1 which equals 10^{156} to 1. The comparisons of Faces One, Two, and Three gave odds against chance of between 10^{27} and 10^{162} making a total so far of between 10^{53} and 10^{318} to 1.

A is a cleft on the Crowned Face Crown and there is a dark patch just above it. Another dark patch is over a similar cleft at **8**.

B shows how the dark area curves around an eye like feature on the Crowned Face.

C is another dark area that curves around the feature to its right, which also matches the mesa above the left eye of the Cydonia Face.

D shows the iris of the Crowned Face eye and how it falls exactly on the rounded area in the Meridiani Face eye. This also overlays exactly on the Cydonia Face eye.

E shows a small raised area on the Crowned Face which is also dark on the Meridiani Face.

F shows a dark line on the Meridiani Face which moves down through the jewelry like shape on the Crowned Face cheek.

G shows a lighter area on the Meridiani Face which is in the same position as the tip of the nose on the Crowned Face.

H shows the end of the jewelry like shape on the Crowned Face, also a dark area on the Meridiani Face and a raised area on the Cydonia Face.

I shows a lighter area on the Meridiani Face which is a similar shape to the bottom of the nose on the Crowned Face.

J shows a dark shadow like area (from the end of the arrow line sloping to the left and downwards) from the Crowned Face nose, and a similar dark area on the Meridiani Face which may also be a shadow.

K is the Crowned Face mouth and is a dark line in the same position on the Meridiani Face.

L shows how the edge of the dark area lines up well with the jaw line of the Crowned Face.

M shows a dark area on the Meridiani Face corresponding to a ravine on the Crowned Face.

N shows the right hand edge of the Crowned Face mouth and a small ridge there, corresponding to a dark area on the Meridiani Face.

O shows a small raised ridge on the Crowned Face corresponding to a lighter area on the Meridiani Face.

P shows a widening in the dark area on the Meridiani Face where the ravine on the Crowned Face ends.

Q shows a darker area on the Meridiani Face corresponding to the mouth on the right hand face on the Crowned Face.

R shows a ridge on the Crowned Face corresponding to a lighter area on the Meridiani Face.

S shows how the edge of the Crowned Face along here matches the edge of the dark area on the Meridiani Face.

T shows a ridge on the Crowned Face which may be seen on the Meridiani Face.

U shows a ridge on the Crowned Face which also appears as a lighter area on the Meridiani Face.

V shows a ledge on the Crowned Face which also appears lighter on the Meridiani Face.

W shows a dark ravine like area on both faces.

X shows another darker area common to both faces. W and X may be shadows.

Y shows an eye like shape common to both faces.

Z shows an eye like shape, like the right eye on the right face of the Meridiani and Crowned Faces.

1 shows the upper side of the ridge pointed out at **T**.

2 shows a similar shape on both faces.

3 shows a dark area on both faces under the ridge between **1** and **T**.

4 shows where the ridge on the Crowned Face seems to curve around, matching the edge of the dark area on the Meridiani Face.

5 shows similar shapes on both faces.

6 shows a similar shape just to the right of the eye on both faces.

7 shows how the iris on each face matches almost exactly.

8 shows the dark area which matches up with the cleft on the Crowned Face like with **A**.

9 shows the left edge of this eye is very similar on both faces.

0 shows the eyebrow on the Crowned Face is thicker on its right hand end, like on the Meridiani Face.

The Cydonia Face compared to the Crowned Face

Here I overlaid the Cydonia Face with the Crowned Face. I did not distort the Crowned Face in any way; I shrunk the size of the Crowned Face to fit and then reduced its height by about 3%. This shrinking is not distortion because images are usually from slightly different perspectives which can make them foreshortened in different ways. Altering the height of the image is equivalent to a slightly different viewing angle.

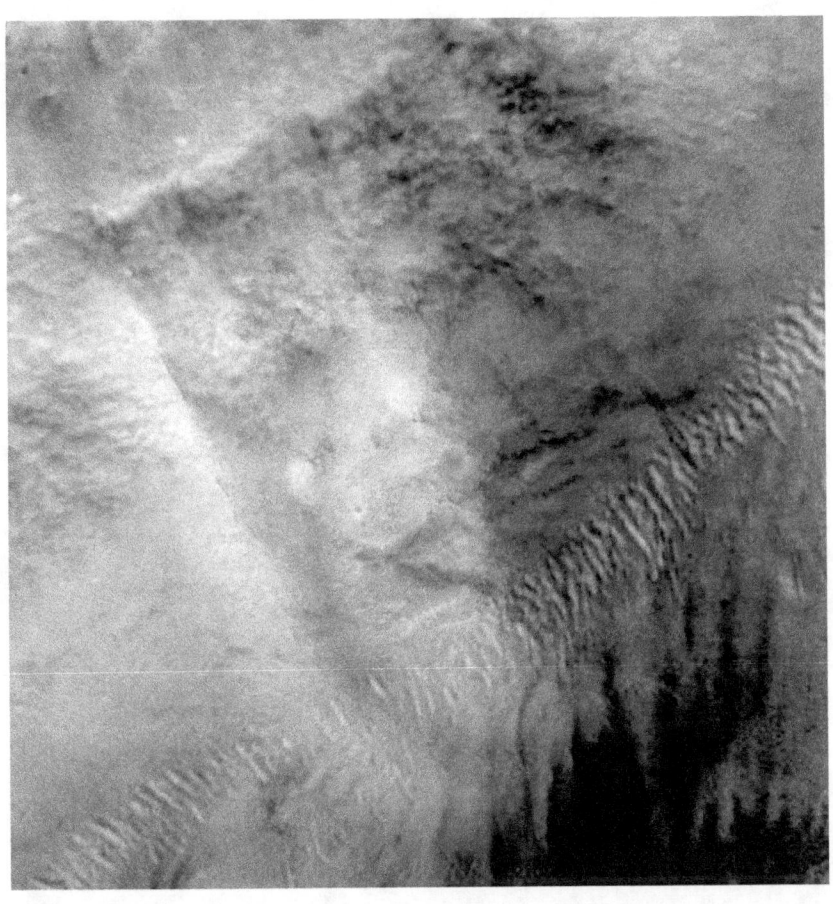

Figure 74

This is an overlay of the Cydonia and Crowned Faces; you can see the larger rounded part of the Cydonia Face mesa protruding under the Crowned Face.

It may be that the Cydonia Face is a heavily eroded version of the Crowned Face. These results are significant not only because they are two Faces with some plausible claim to artificiality, that they are so similar to each other is highly improbable as well because they were formed in a completely different way geologically. The Crowned Face is an eroded slope in a valley and the Cydonia Face is an eroded mesa. This study shows that two of the major faces on Mars (even if only faces in our imagination) are highly similar formations though any similarities should be random only. It is also important to realize that these overlays do not depend on these areas looking like faces, they simply show that there are five very similar looking areas and that geology cannot explain all these similarities. If they looked like an animal, an alphabet, geometric shapes, etc. the result would still be equally improbable to occur naturally. That they are all faces only makes it more overtly artificial looking, but it also casts some doubt on the evidence because it is seen as unlikely someone would have built faces instead of something conventional wisdom would have preferred to find as artifacts.

Figure 75

This is an overlay of the Cydonia and Crowned Faces. The letters point to areas similar to each other in the overlay.

"**a**" shows a rounded curve that is similar on both faces. I assume the faces are wearing slightly different hats, one is crown shaped and the other is like a head dress or helmet. Both have close to the same line where the hat starts.

"**b**" shows the vertical line on the edge of the Cydonia face lines up well with the edge of the Crowned Face.

"**c**" shows how the two left eyes are almost identical here.

"d" shows the rounded mesa like shape above the left Cydonia eye is almost identical to a dark oval shape above the left Crowned Face eye. It may be the Crown Face area is more eroded in this section than on the Cydonia Face.

"e" shows a shape on the Cydonia Face, but a nearly identical shape is seen on the Crowned Face and is like a long earring.

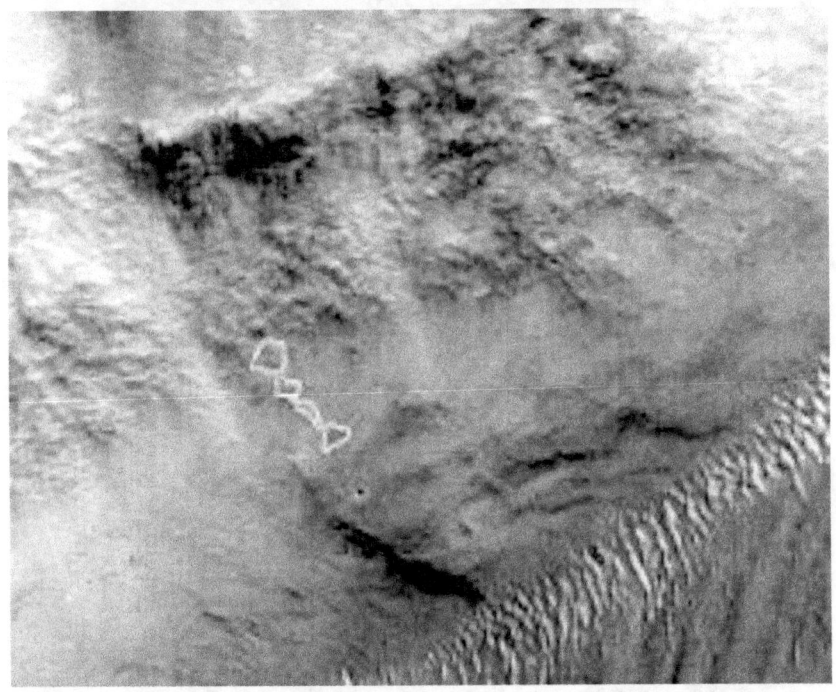

Figure 76

A similar shape on the Crowned Face above also occurs on the Cydonia Face.

The bottom of the earring is at the same position as the shape on the Cydonia Face at **"e"**. Also further up the earring shape there is a rounded shape, and there is a similar dark shape on the Cydonia Face shown below.

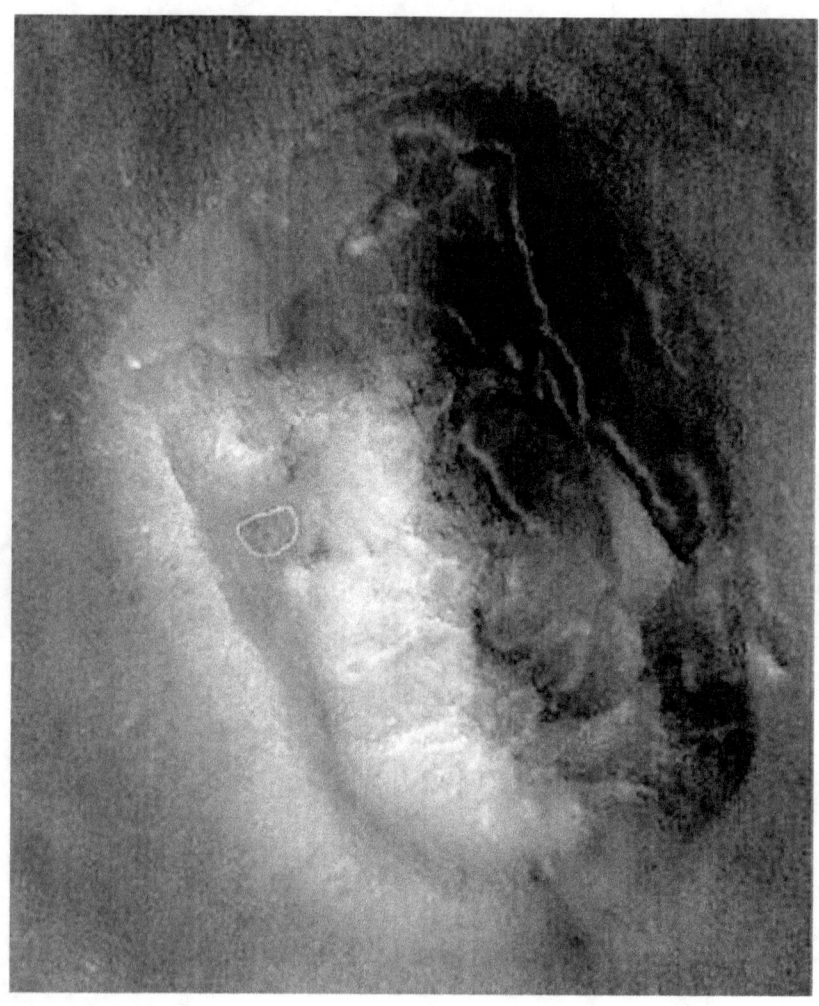

Figure 77

The dark area shown above is similar to a feature on the Crowned Face.

"**f**" shows the end of a nose shape which lines up well with formations on the Cydonia Face.

"**g**" shows this cliff face lines up perfectly with a ravine on the Crowned Face. The lip like shape above this on the Cydonia Face is nearly exactly like the mouth on the Crowned Face with two depressions or dark patches on it in the same positions.

"h" shows the lower edge of the large mouth like ravine son the Cydonia Face, which also lines up well with ravines on the Crowned Face.

"i" shows how the pointy chin and lower left jaw of the Crowned Face fits in well with the shape of the Cydonia Face here.

"j" shows an eye shape where the right eye of the Crowned Face would be.

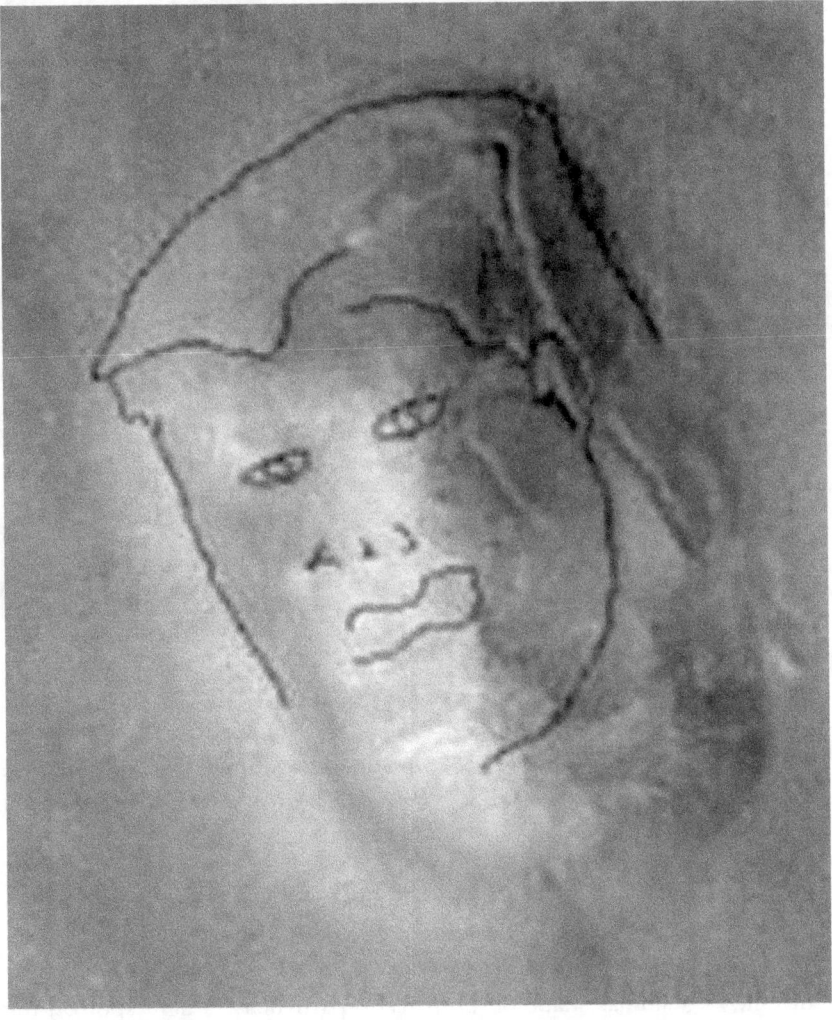

Figure 78

There is a right eye shape on the Cydonia Face implying it is also a Dual Face.

"**k**" shows an eye like shape on the Cydonia Face that is similar to a dark shape on the Crowned Face. This lines up very closely with the right eye of the "evening face" on the Crowned Face. The next image shows this formation on the Cydonia Face:

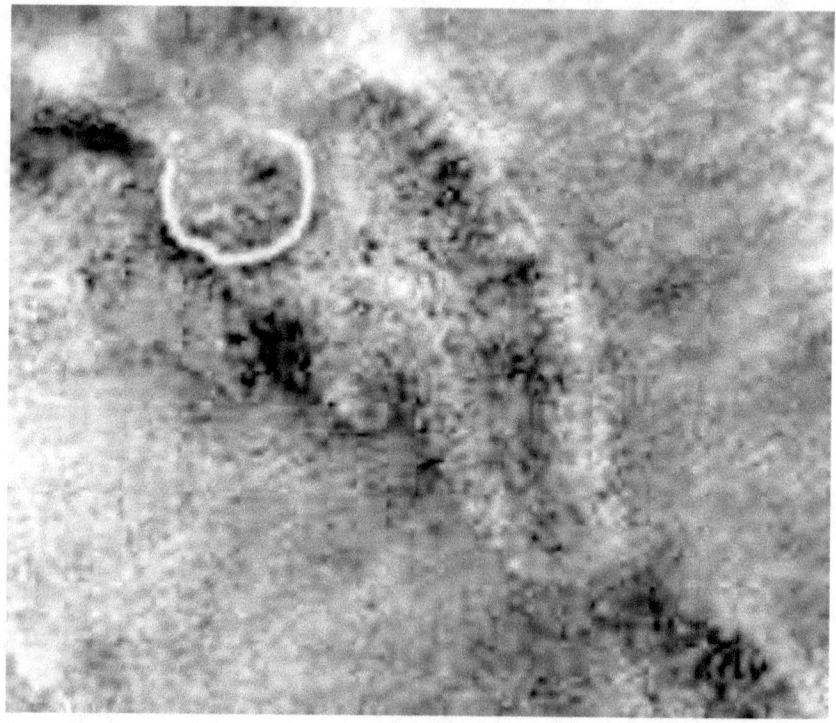

Figure 79

A round shape on the Cydonia Face like an eye matches up with the right eye of Face Three in the King's Valley. There is a lot of software compression distortion in this photo but the eye shape stands out from this.

"**l**" shows a dark spot on the Crowned Face which is at the same position as an arc like dip in the headdress on the Cydonia Face.

"**m**" shows a similar shape on both faces, particularly a light spot on the Crowned Face marked in the image below. You can see this same light spot in the Cydonia image above where the hairline I drew ends above the right eye. The two are close to each other but higher on the Crowned Face, which may be because of the dissimilar head gear.

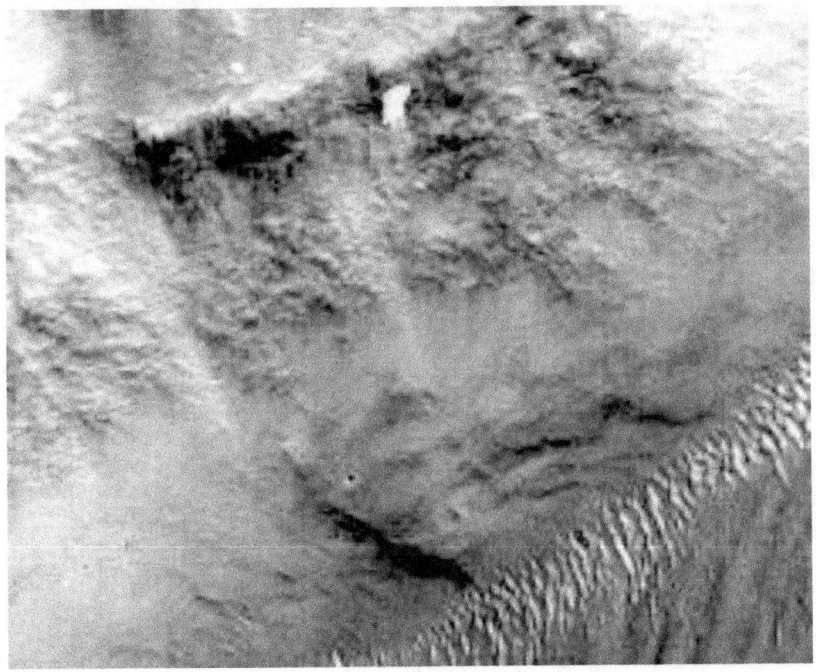

Figure 80

There is a similar shape on the Crowned Face marked above to that on the Cydonia Face.

"**n**" shows another similar area. This is shown on the Cydonia Face in Figure 9.

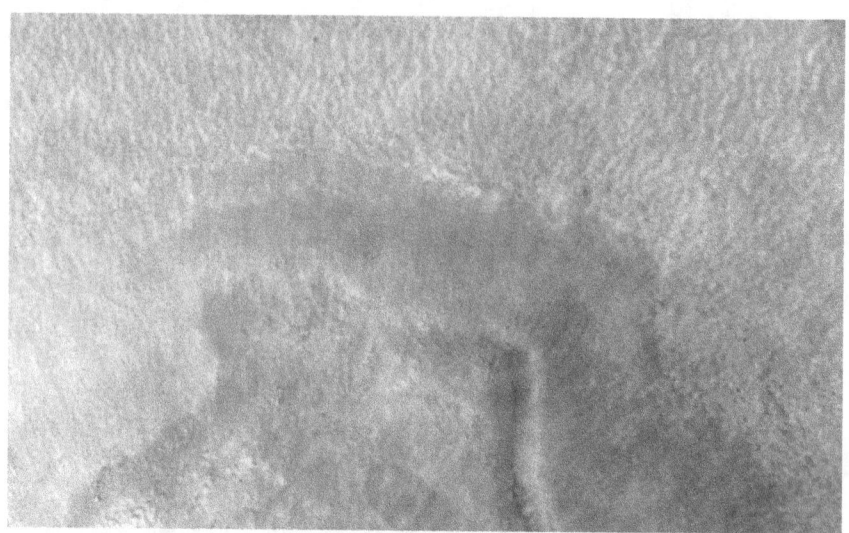

Figure 81

The edge is similar to part of the Crowned Face. It may be that the Cydonia Face was meant to show a dual face like the Crowned Face, since the edge of the whole headdress seems to have a partial correspondence in the Crowned Face area. Below the two are superimposed.

Figure 82

This shows the similarity of the right edges of the Cydonia and Crowned Face when overlaid on each other.

And here the Crowned Face area is shown:

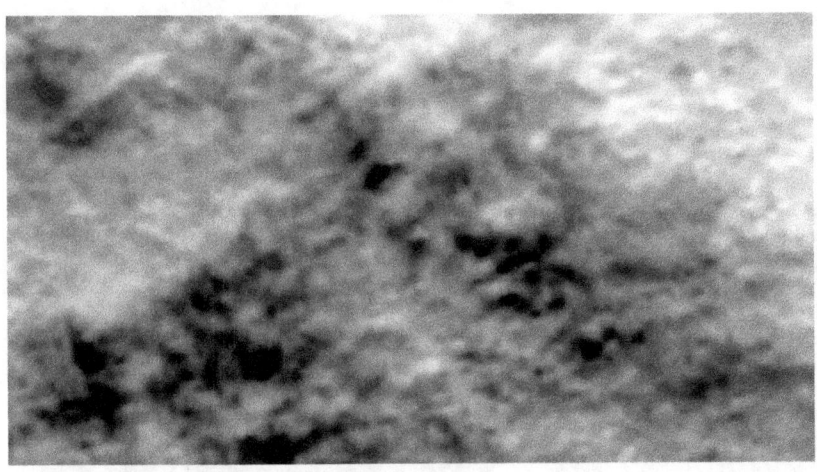

Figure 83

This shows the right edge of the Crowned Face's crown.

Another similarity between the two faces is shown below. The line marked can be seen as two sections on the Crowned Face image, I've drawn a line connecting them.

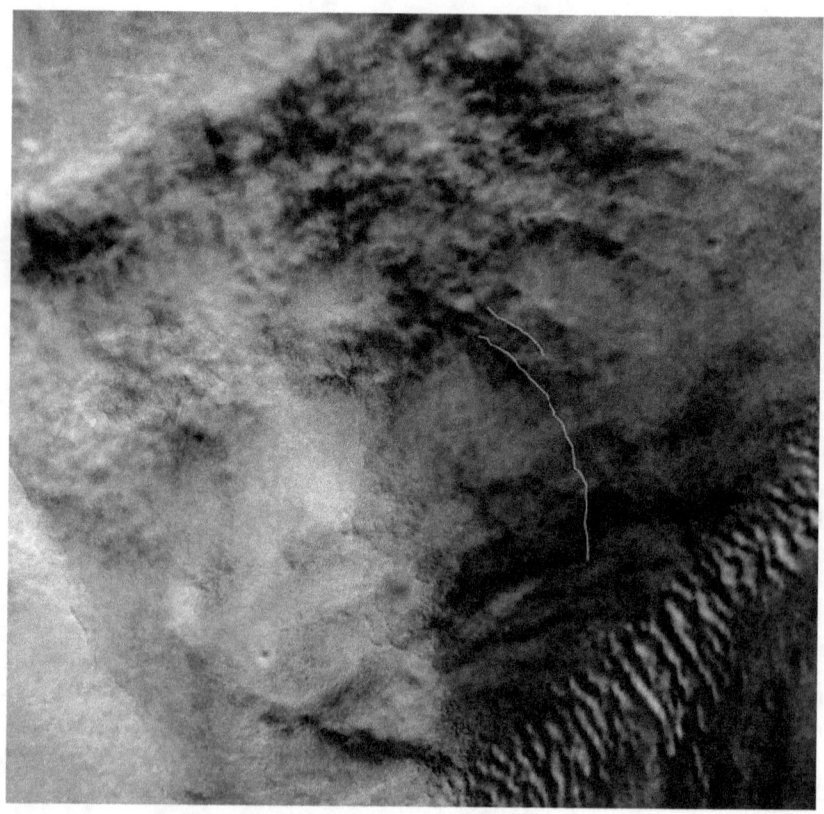

Figure 84

This is nearly the same as this line on the Cydonia Face seen below. That is, in the overlay they are in the same position.

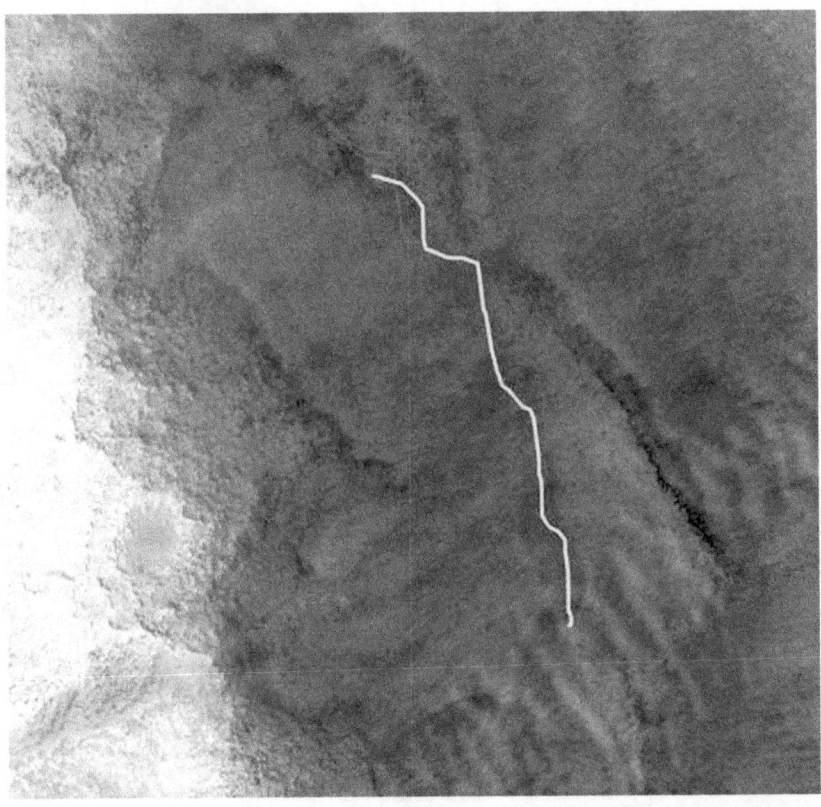

Figure 85

A section of the Cydonia Face.

This is surprising because these shapes seem to serve no purpose in the Cydonia Face, though they are part of a second face on the Crowned Face.

The part shown below is a cliff face on the Cydonia Face and represents the bottom of the Crowned Face as it is buried under sand dunes. You can see the outline of these dunes in the overlay below.

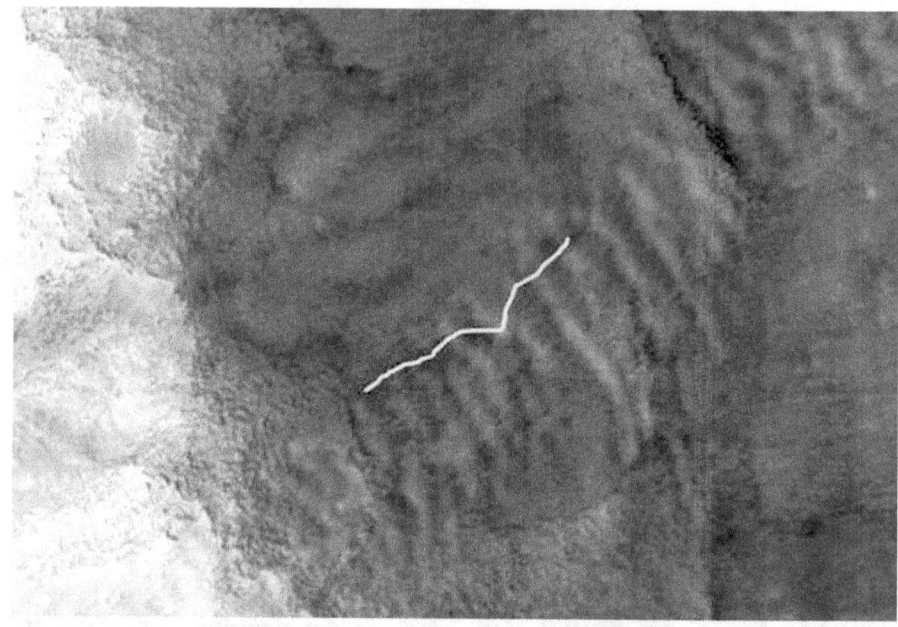

Figure 86

Both Faces have a similar chin.

Another part of the crown is very similar on both faces. The Crowned Face below has these lines on the crown.

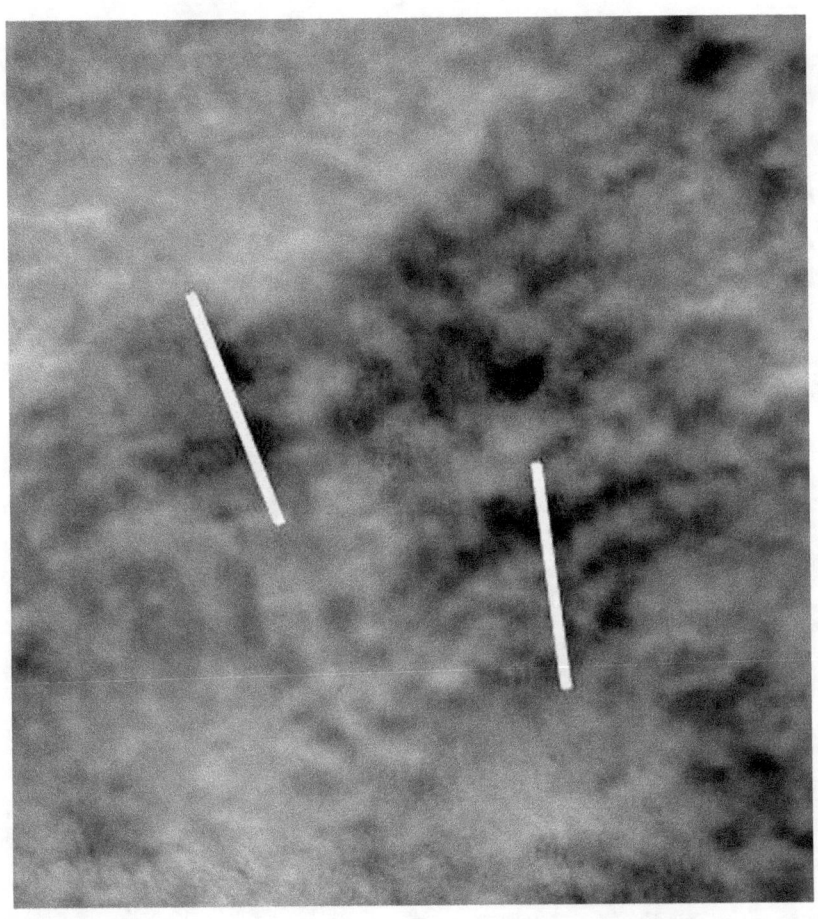

Figure 87

This is on the Crowned Face.

Compare these to the lines below on the Cydonia Face, they line up nearly exactly.

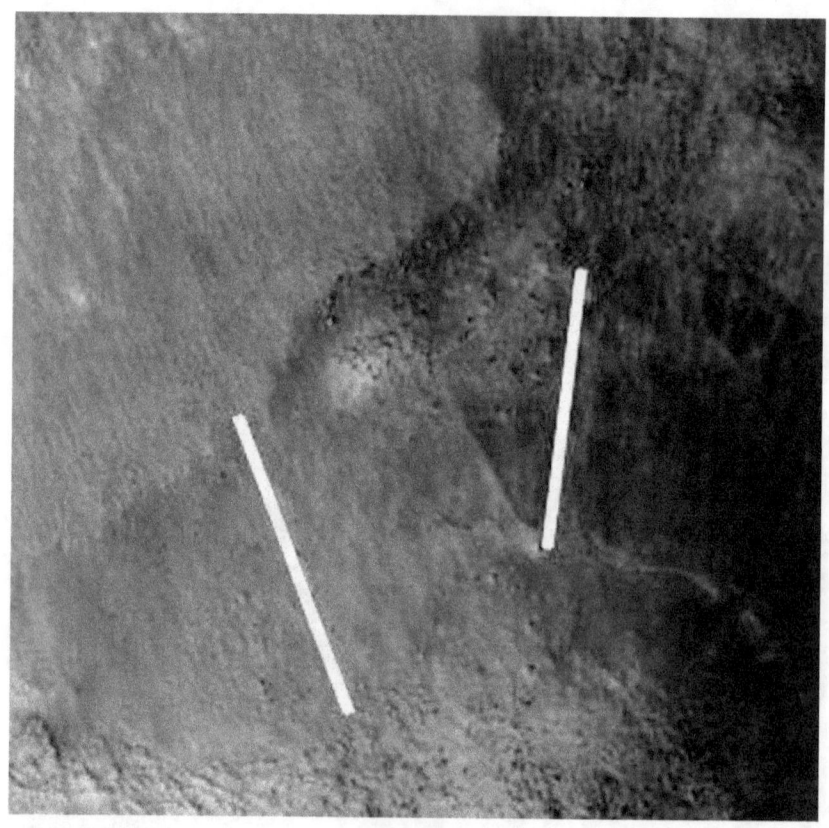

Figure 88

This is on the Cydonia Face.

The part marked on the Cydonia Face is almost exactly the same shape and position as the mouth on Face three, to the right of the Crowned Face.

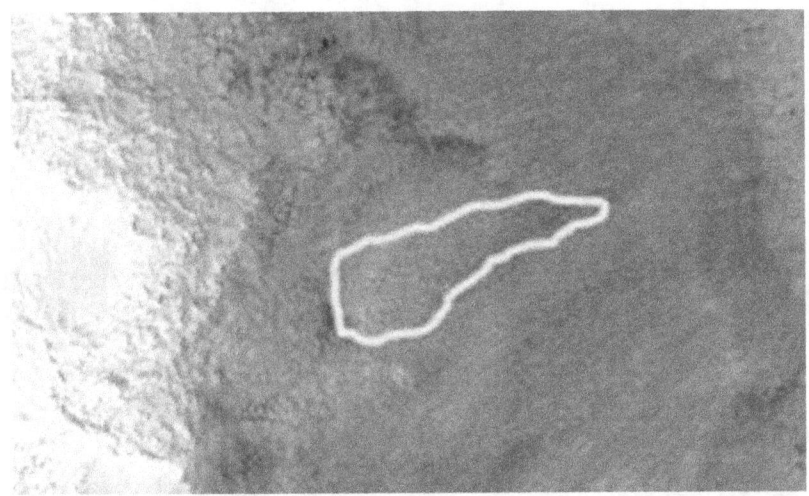

Figure 89

This is a shape on the Cydonia Face.

Compare this to the mouth on Face Three.

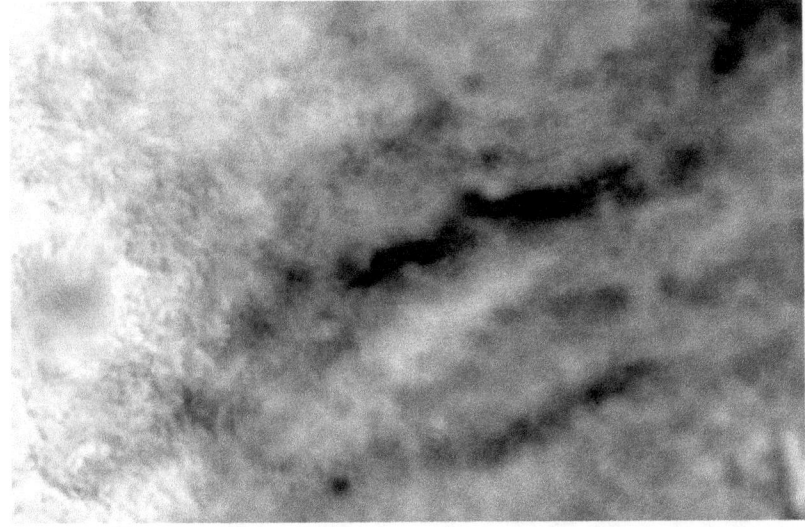

Figure 90

This is on the face to the right of the Crowned Face, Face Three.

The line of shadow across the Cydonia Face nose is shown below:

Figure 91

This is on the Cydonia Face.

This lines up virtually perfectly with a line on the Crowned Face from the nostrils in the upper left corner:

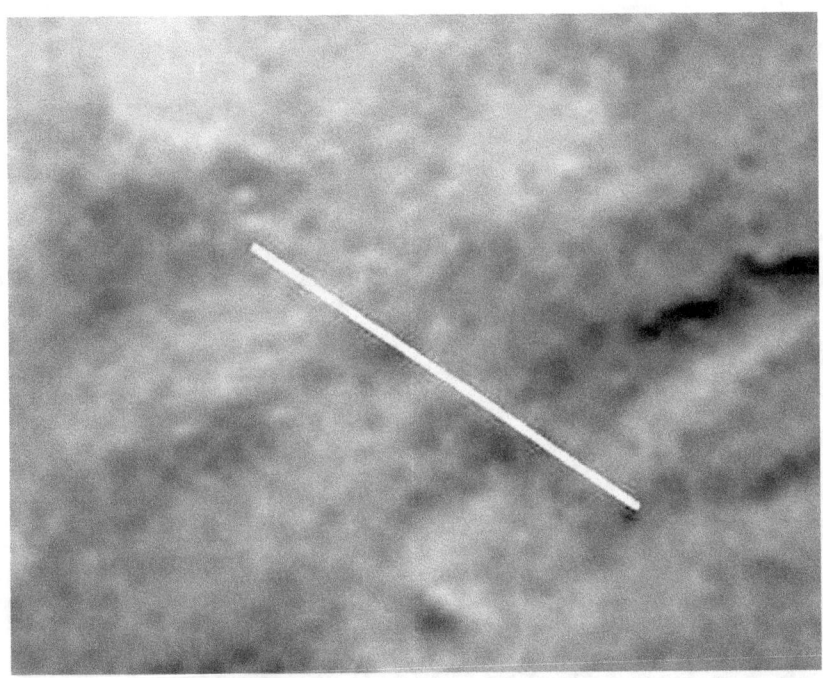

Figure 92

This is on the Crowned Face.

A fault or crevice appears to come from the Face Three mouth mentioned earlier.

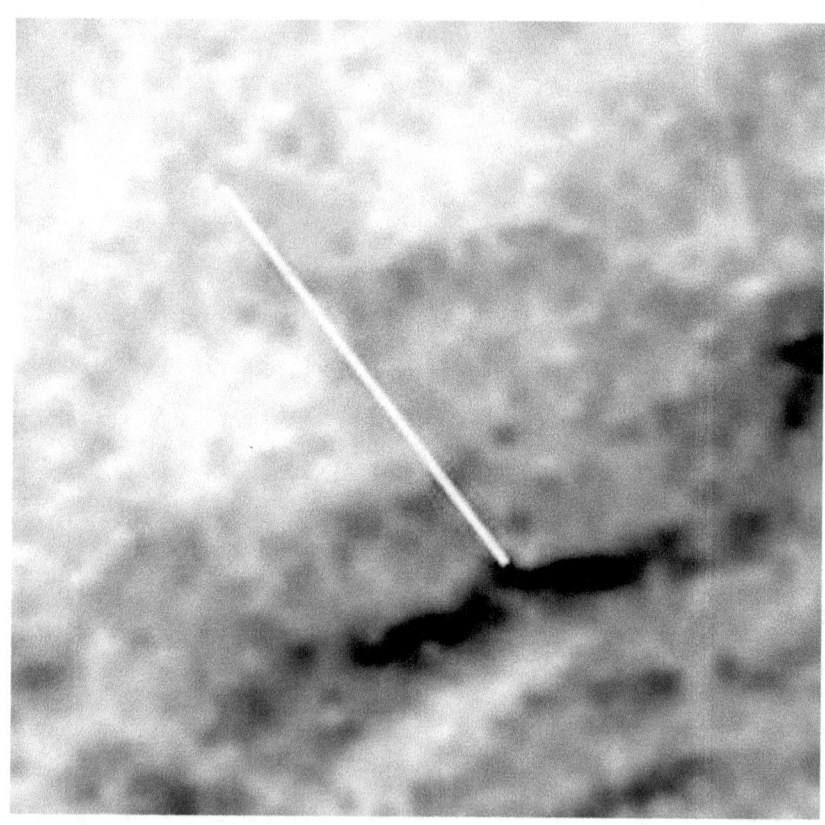

Figure 93

This is on Face Three.

This is exactly aligned with the cliff face on the Cydonia Face in the overlay:

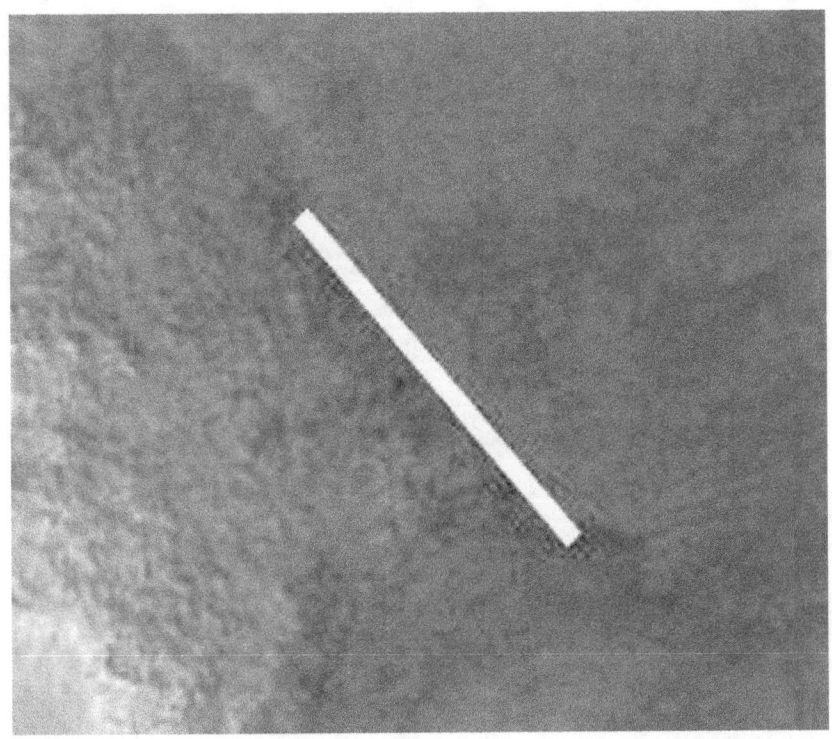

Figure 94

This is on the Cydonia Face.

There is an eye like shape on the nose on the Cydonia Face:

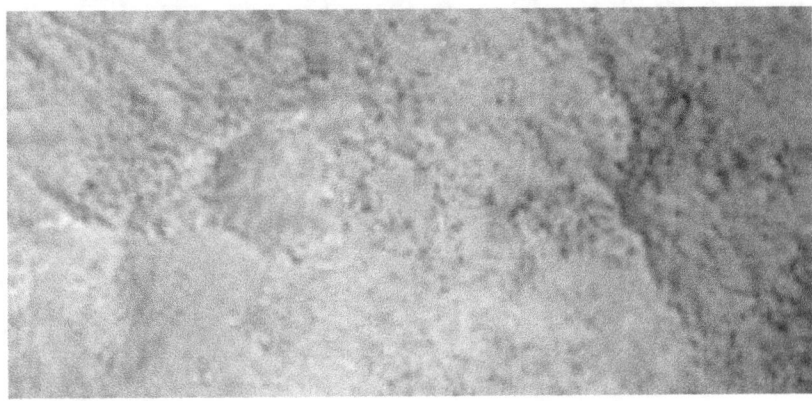

Figure 95

This is outlined below.

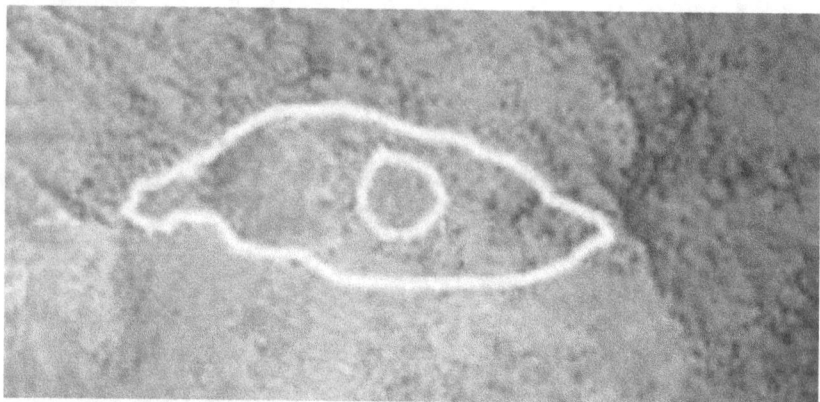

Figure 96

This is nearly exactly in the same shape and position as a formation on the Crowned Face:

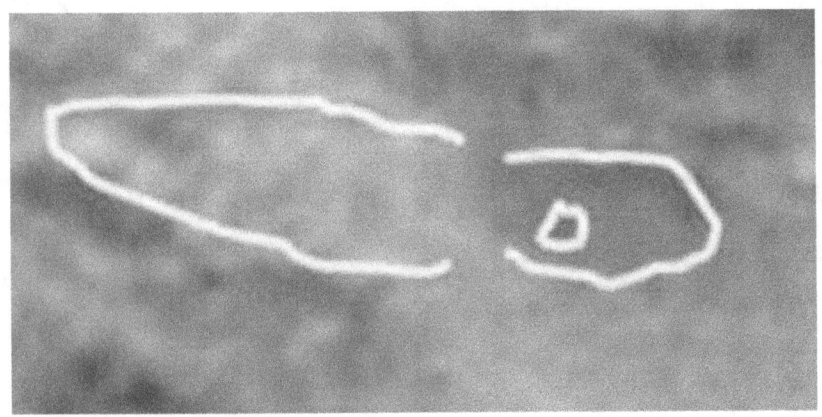

Figure 98

The nose tip of the Crowned Face:

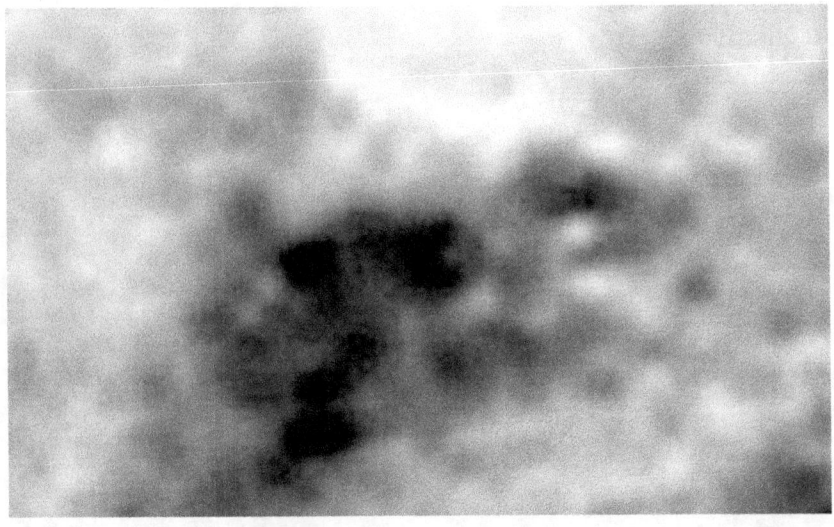

Figure 99

lines up almost perfectly with a formation on the Cydonia Face, which could have been a nose tip.

Figure 100

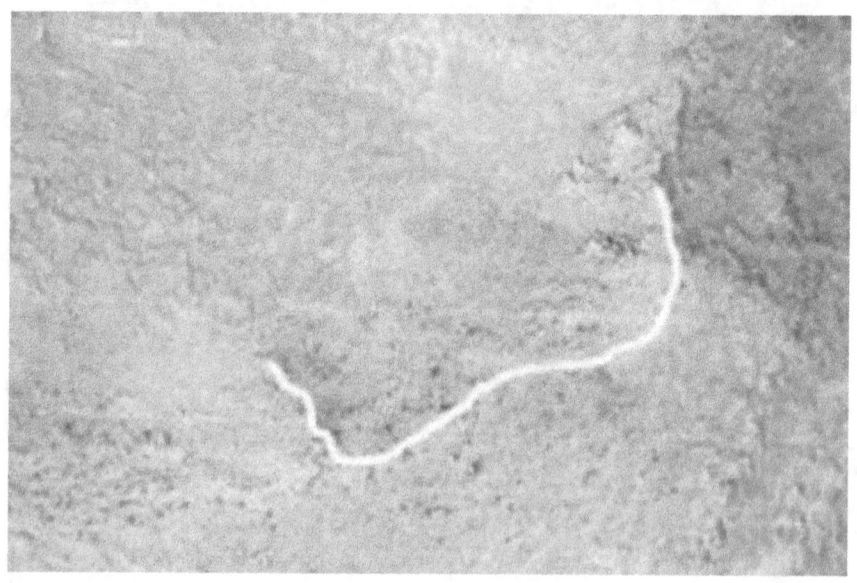

Figure 101

The nose tip shape outlined on the Cydonia Face.

On the right hand side of the Crowned Face crown, the shape shown below is similar on both Faces.

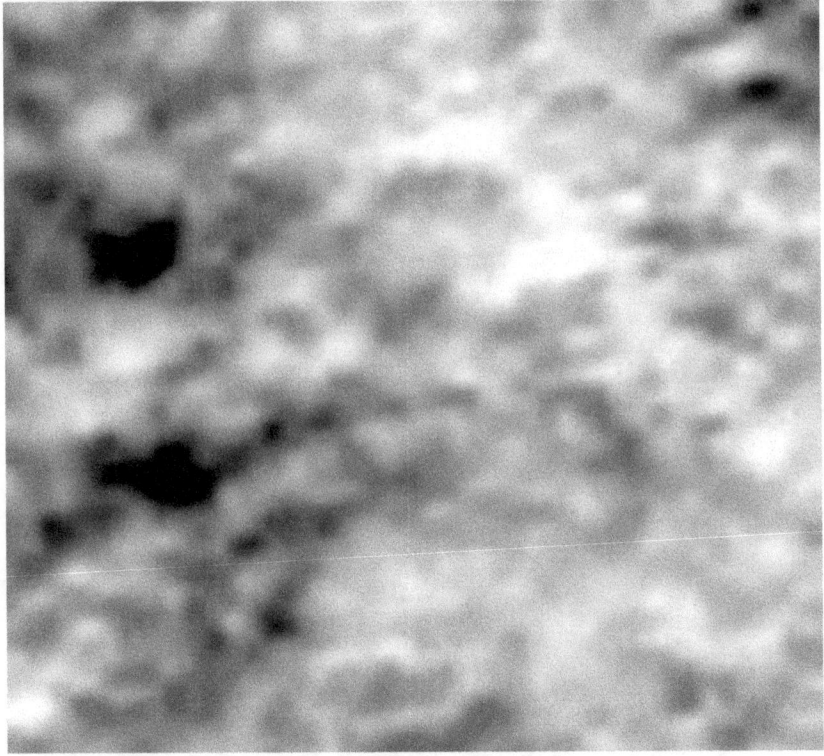

Figure 102

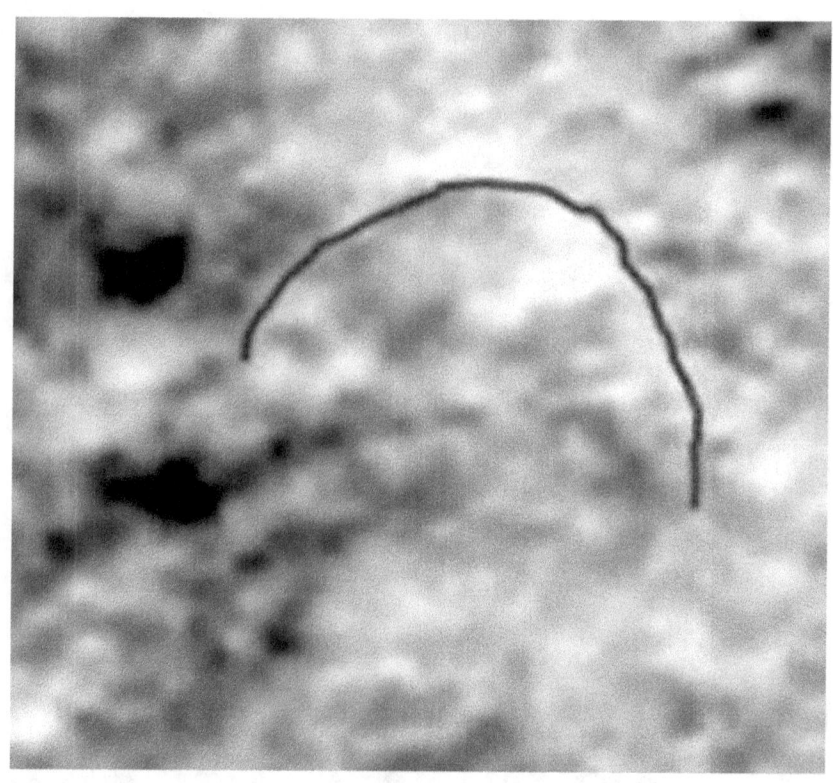

Figure 103

The shape on the Crowned Face is outlined above.

This is nearly the same as this shape on the Cydonia Face.

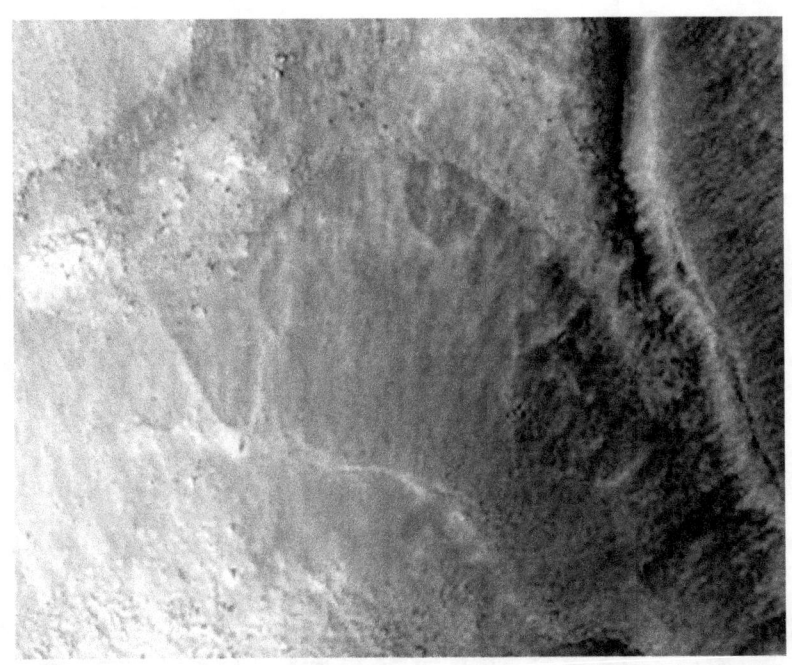

Figure 104

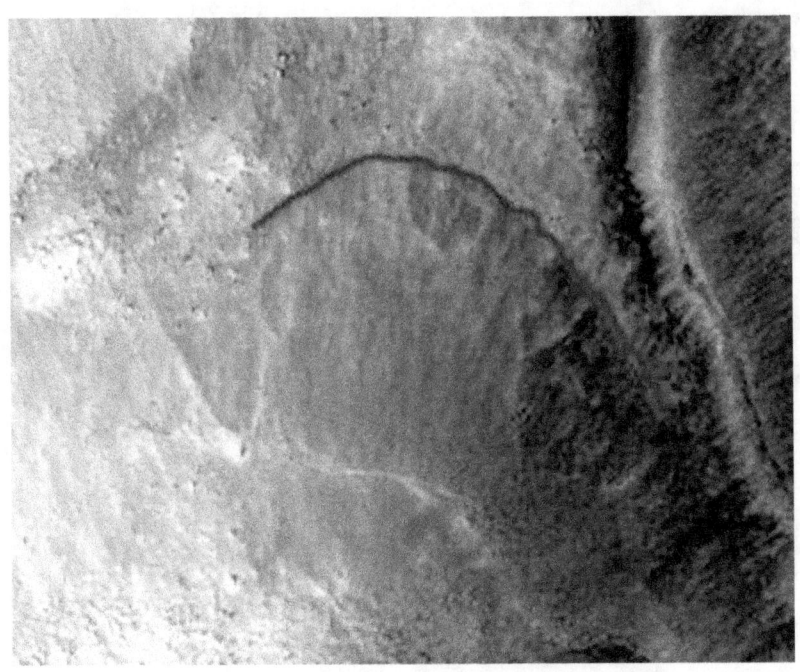

Figure 105

This outlines the shape on the Cydonia Face, similar to the shape on the Crowned Face.

To the right of the Cydonia Face left eye there is a shape.

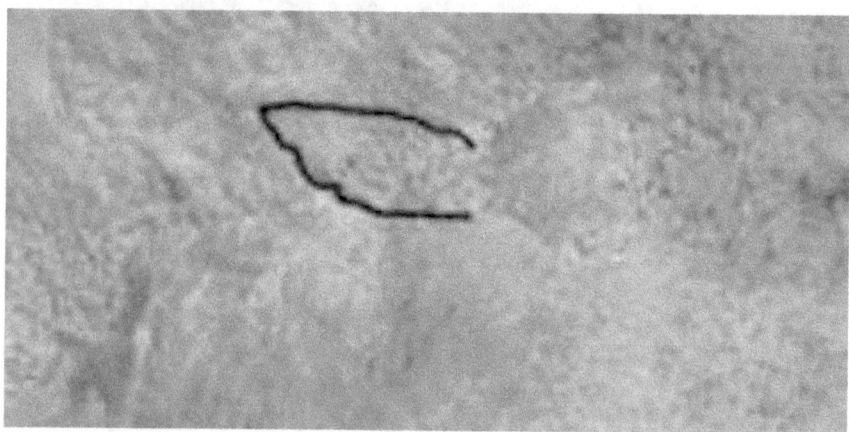

Figure 106

This is on the Cydonia Face.

This shape overlays nearly perfectly on another shape to the right of the Crowned Face left eye:

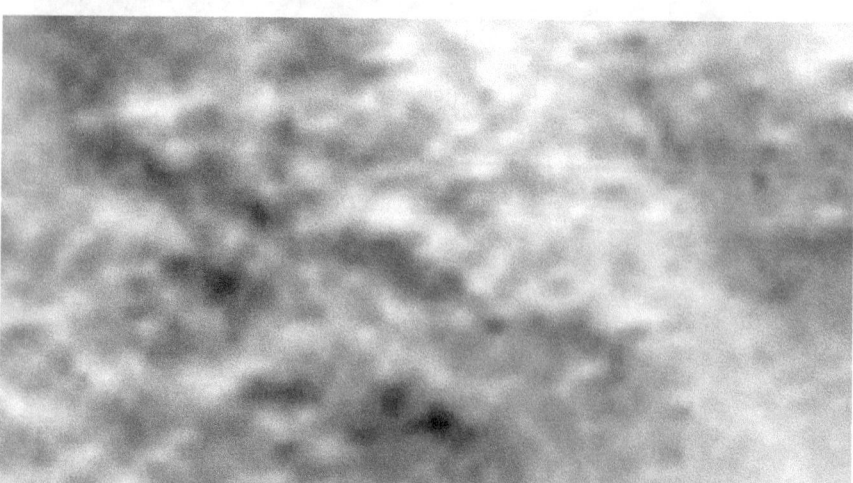

Figure 107

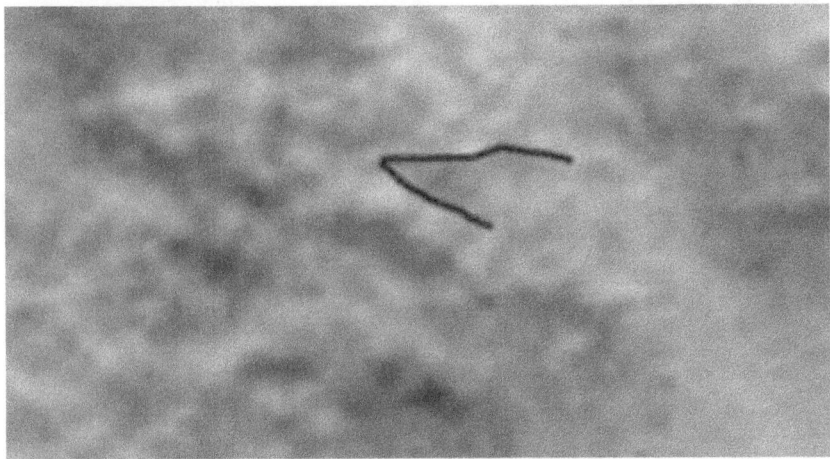

Figure 108

This is the shape above outlined.

There are at least 14 similarities between the two Faces, giving an odds against chance of between 10^{14} and 10^{84}. Because the first 10 similarities are excluded for the reasons explained earlier this gives a total of 10^{2} and 10^{12} to 1 against chance. This seems highly conservative that so many similarities can be only 100 to 1 against chance; if this was true then 1% of Mars would have to be covered in Cydonia Faces. It does emphasize that these odds are conservative in many ways. So far this gives between 10^{55} and 10^{330} to 1 against chance.

Chapter Sixteen

The Cydonia Face compared to the Meridiani Face

As before these two faces are overlaid onto each other and the similarities between them are pointed out.

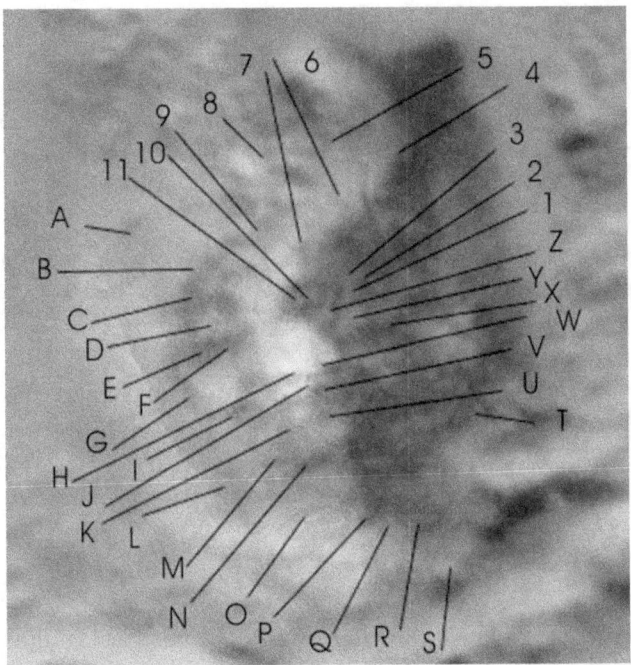

Figure 109

A is a dark patch on the edge of the Cydonia Face helmet.

B is the edge of the round shaped plateau above the left eye of the Cydonia Face. it also follows the line of the helmet, which goes in an arc over to the other eye in both images.

C Note the way the dark area curves right around the plateau.

D This is the raised plateau, which is also a similar feature on the Crowned Face.

E The eye is very similar on both faces with the same position for the iris. It is also the same position of the eye and iris on the Crowned Face.

F The eye terminates here in much the same way on all 3 faces.

G. The dark area follows a trough on the Cydonia Face, and also a jewelry like formation on the Crowned Face.

H This is a small ravine on the Cydonia Face and so might catch the light on its edge like here.

I This is a raised area on the Cydonia Face corresponding to a jewelry like shape on the Crowned Face and here there is a dark area at the end of the jewelry shape.

J This is the end of the nose on both the Crowned and Cydonia Faces as mentioned on a previous link. Here the light area ends on the Meridiani Face at the end of the noses on the other two images.

K shows a depression on the Cydonia Face like with U, both dark patches on the Meridiani Face.

L This dark area abuts the ridge of the helmet, like a shadow from it.

M There is a small ravine here on the Cydonia Face.

N The edge of this ravine is almost the same on the Cydonia Face and the Meridiani Face as well as the Crowned Face.

O The dark area here on the Meridiani Face matches closely the edge of the ravine, like the bottom edge of the Cydonia Face mouth.

P There is a small light spot here at the same position as the peak, and is dark to the right where there is a depression on the Cydonia Face.

Q The dark area here and just above it is similar in shape to the rocky outcrop on the Cydonia Face.

R The dark area ends here about the same place as the edge of this rocky outcrop.

S The end of the ridge on the Cydonia Face ends here where the light spot is.

388

T This dark shape is similar in shape to a rock outcrop on the Cydonia Face.

U This dark spot matches a depression on the Cydonia Face.

V Matches a dark area between two ridges, which might cast a shadow on the Cydonia Face.

W The shape of the bottom part of this light area is similar to the shape on the Cydonia Face.

X Perhaps a small depression here on the Cydonia Face.

Y A higher area on the Cydonia Face where this light area is.

Z A bottle like shape is similar on both faces.

1 The left edge of the depression like an eye socket fits in here.

2 The right hand edge of the eye socket. So the eye socket on the Cydonia Face is similar to an area here.

3 This area is like part of the helmet on the Crowned Face.

4 The light area ends on the edge of the Cydonia Face here.

5 Compare this area to the outer helmet edge of the Cydonia Face.

6 There is a lighter line here following a ravine on the Cydonia Face.

7 The upper part of this dark area abuts against the arc, as the front edge of the Cydonia Face helmet.

8 There is a dark line here following the outer edge of the Cydonia helmet.

9 Follows the arc of the Cydonia Face helmet edge.

10 There is a small hill in the Meridiani Face eye where the iris center would be in the Crowned Face eye, also matching an eye on the Cydonia Face.

11 The center of the eye is an arc shape pointed upwards on the Cydonia Face, similar to the arc shape on the Meridiani Face.

There are 37 points of similarity between the two Faces, giving an odds against chance of between 10^{37} and 10^{222} to 1. Reducing these by 10 gives odds against chance of between 10^{27} and 10^{162} to 1. This gives the total odds against chance so far of between 10^{82} and 10^{492} to 1.

As mentioned earlier there is an additional improbability of so many faces being found in one area, I will repeat this section. An approximation of the odds against chance can be worked out from this. There is about a 1 in 10,000 chance of selecting a group of 10 photos with a face on Mars in it. That is, about 10 photos contain virtually all the faces so far, so having 10,000 groups of photos each with 10 photos in it gives 100,000 photos in total. Any group except the one with the faces in it can have its 10 photos selected randomly before they are examined. Time after time the faces appear in this one group of 10 photos instead of in the other groups. So with 24 faces in these 10 photos (1 is too large to be in one photo) the odds against 24 faces over and over again appearing in these 10 images is $10,000^{24}$ which is 10,000 to the power of 24. This is 10^{72} to 1. Such a result cannot be random; there must be a nonrandom explanation for this. Even assuming half of the faces are not credible doesn't change the result significantly, this would still be 10^{36}. The point is not whether each face is obviously artificial or not, it is that faces are in a small set of photos instead of randomly distributed over Mars whether someone thinks the Faces are credible or not.

Multiplying this by the total odds against chance gives a total of between 10^{118} (82+36) and 10^{528} (492+36) to 1 against chance.

Chapter Seventeen

Comparing the three overlays

In this section I will examine all 3 faces for an overall view because these give us the best opportunity to see what the builders of these faces looked like. First there is one which is half Cydonia Face and half Crowned Face

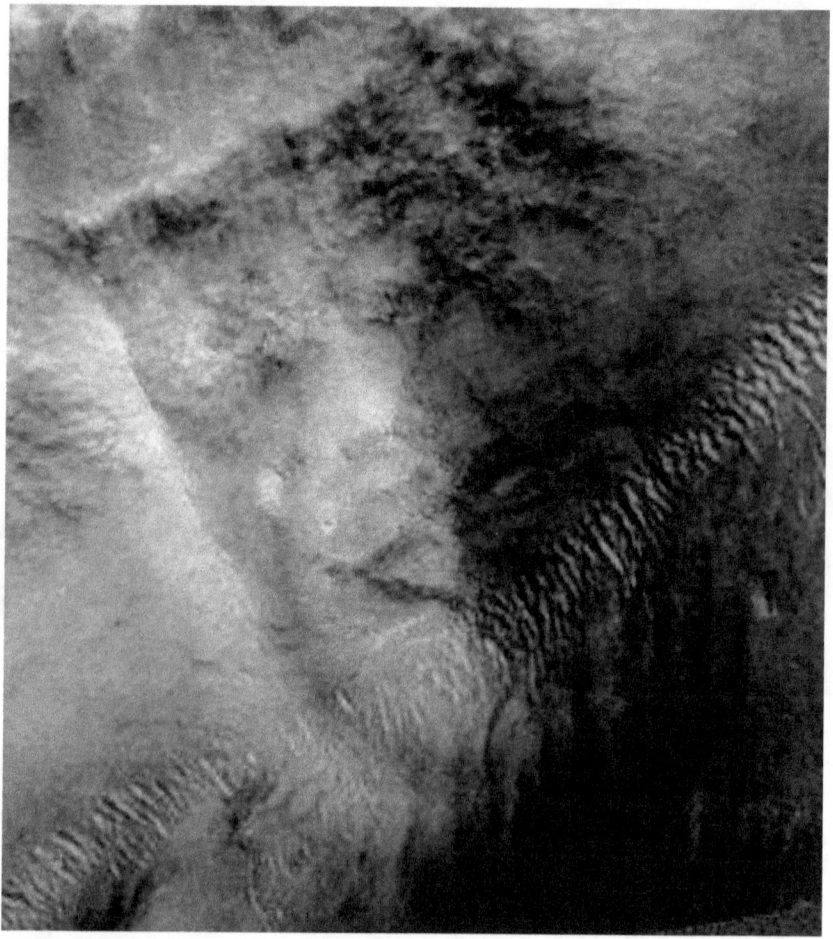

Figure 110

This is the Crowned Face and Cydonia Face. The overlay changes some of the features such as the left eye.

Next is the Crowned Face and Meridiani Face.

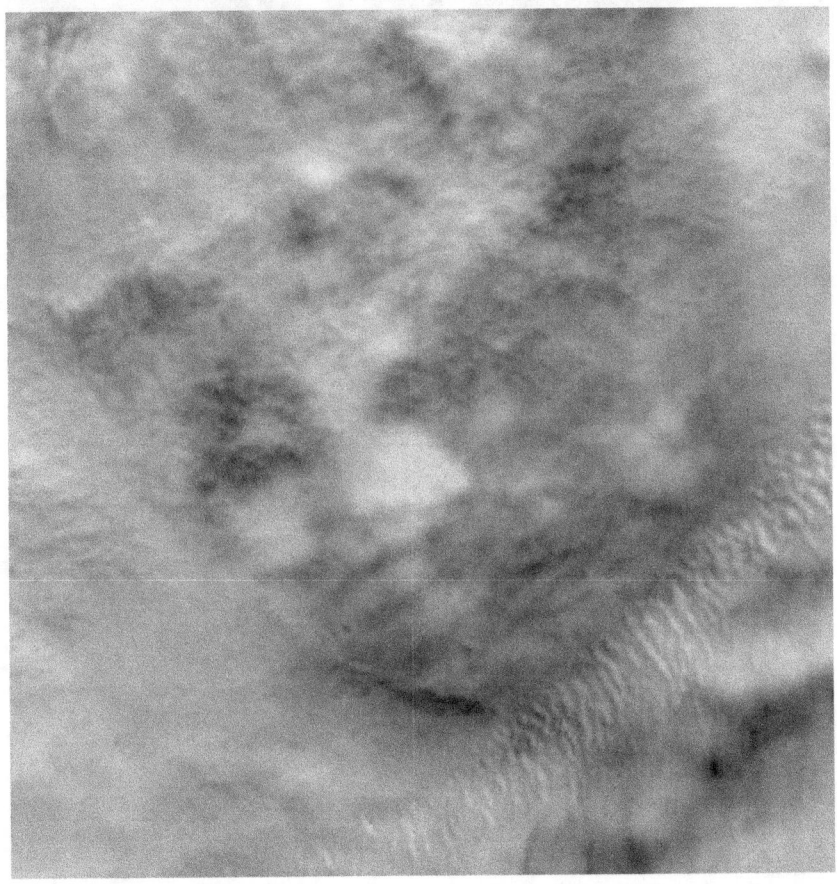

Figure 111

This gives a different emotional expression to the Face, perhaps more angry or annoyed.

Next is the Cydonia Face and Meridiani Face.

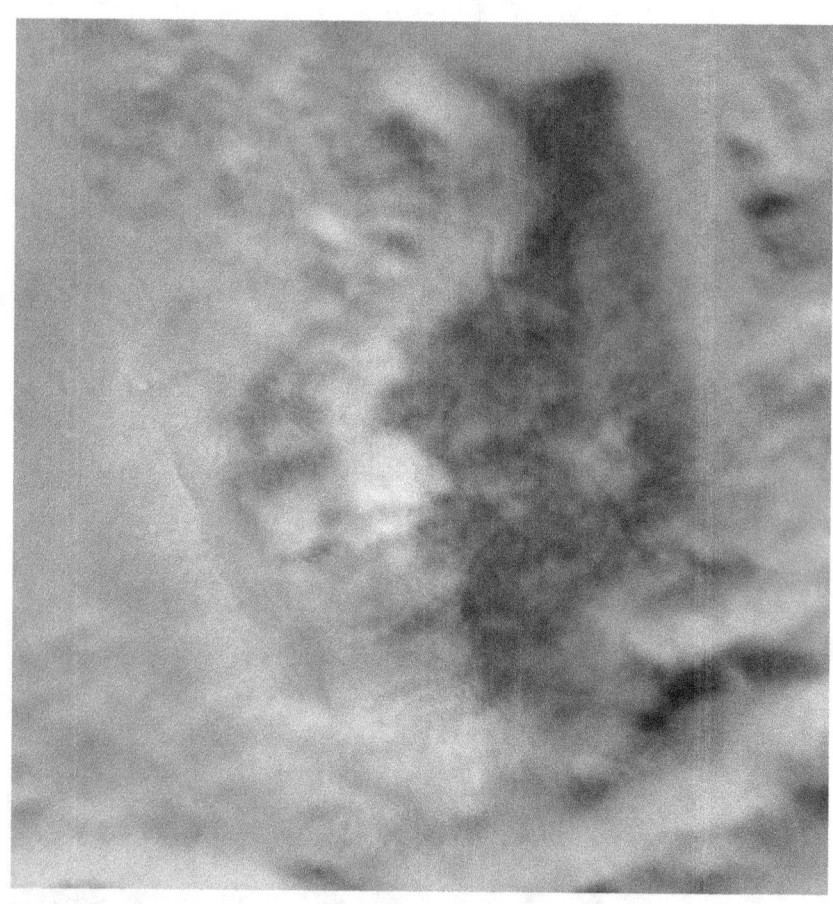

Figure 112

This may give an impression of wearing a space helmet, the border around the Cydonia Face may be intended to be this helmet. There is also a helmet shape around the head of the Spaceman formation which will be analyzed in an upcoming book.

Chapter Eighteen

Conclusions

This book is only a small part of the evidence of artificiality on Mars. I intend to catalogue all of this evidence in 4 books. There are 14 main areas of possible artifacts on Mars and many smaller formations which could also be artificial. Each requires extensive analysis to show all the evidence available.

"Truth is sought for its own sake. And those who are engaged upon the quest for anything for its own sake are not interested in other things. Finding the truth is difficult, and the road to it is rough." Ibn al-Haytham (Alhazen, 965–1039)

In this book I tried to analyze this evidence according to the scientific method.

1. Truth and belief. Confirmation bias is always a problem with scientific research, however in this case the evidence has almost invariably gone in a different direction to what I expected. Even today many researchers are as uncomfortable with the evidence as the skeptics because it has not been found according to either's expectations. No one on either side really expected so many faces; it goes against the notion of aliens coming to our solar system building solemn and impressive monuments. The idea of faces, giant birds, etc. sounds more humorous than we would like. This book tries to present the evidence the way it looks, not how we would like it to look. Even now the evidence is very uncertain in many ways, for example whether visiting aliens or indigenous Martians created these faces. Both have many problems associated with them, we believe it to be impossible to travel to other stars because of the cost and length of the journey. We also believe it was impossible for indigenous Martians to evolve on Mars because of how inhospitable it appears today. In an upcoming book I will give more details of the model that is evolving along with this evidence, this is difficult to do until enough evidence is presented and this requires several books to see it all in perspective first. However much of the outcome of this accumulating evidence is still uncertain, what does seem to be virtually certain however is that some of these formations were created artificially.

2. The four main aspects of the scientific method are explained well in the Wikipedia article referenced above. They are iterations,[35][36] recursions,[37] interleavings, or orderings of the following:

Characterizations (observations,[38] definitions, and measurements of the subject of inquiry)

Hypotheses[39][40] (theoretical, hypothetical explanations of observations and measurements of the subject)[41]

Predictions (reasoning including logical deduction[42] from the hypothesis or theory)

Experiments[43] (tests of all of the above)

The photos in this book represent observations made with NASA probes orbiting Mars; they are presented with as little distortion and alteration as possible. The hypotheses are simple at this stage, to prove that at least one part of one of these formations was not created by a natural process such as the wind, rain, erosion, etc. It is not the point of this book to try and prove who the builders were, or why they built these faces. Some ideas are offered at times but these are guides only, to prove these things we will probably have to mount a manned expedition to places like the King's Valley.

The predictions made have been implicit in this research for some time, that as more photos are taken more artificial looking structures will be found and that when these formations are reimaged at a higher resolution that more artificial looking structures will be found. The second implicit prediction is that this additional evidence will be similar to the evidence already found, i.e. that new faces found will resemble the older ones, that sites with possible artifacts will be near the edge of the former Northern Lowlands sea, that they will tend to be clustered around the old Martian equator, and that these sites will continue to line up with each other on a globe in degenerate triangles. The third implicit prediction is that when part of a face is missing there will be an apparent reason for it such as it appears broken off or hit with a meteor, it is buried under sand, etc.

The experiments will include more photos of Mars from HiRise and particularly of the areas with possible artifacts found so far. If this evidence becomes credible enough then it can be bolstered with a robotic or manned expedition to these areas.

Appendix

A history of Mars

Mars has a mysterious history. Many of the facts we know seem to be conflicting, for example it seems to have had a history of water flows which seems to contradict its current cold state. Valles Marineris is also hard to explain, a large rift valley with apparently no plates, which are needed on Earth to create them. There are many water flows from craters but there should be no water in those areas. Craters are unevenly distributed on Mars, covering only half the planet to a large degree and this coincides with a drop in height called the dichotomy boundary.

Here an attempted explanation is made based on large scale events, which may have driven most of the Martian changes and created the paradoxes we see today. It is a forced sequence of events so if the basic premises are accurate then it is quite likely something like this occurred, though perhaps in a different order.

The 4 largest outside influences we know about Mars are probably the last four big impact craters, Utopia, Isidis, Argyre, and Hellas. Since much of the confusion about Mars arises from the fact that we don't perceive its evolution to be understandable compared to our own, this theory attempts to explain it by outside events driving the changes.

If a planet was perfectly round a pole might tend to wander over time. Usually planets are a spheroid, which means they are like a slightly flattened ball, and the wider parts tend to be at the equator. This is because the extra weight tends to go to the equator where it wields more force because it spins faster there.

The converse of this is that an absence of mass such as an impact crater or large valley like Valles Marineris would tend to go toward a pole since they become the equivalent of the flattened part of a spheroid. We also know that on Mars large impacts typically form a Mons or mountain on the other side of the planet, though sometimes not exactly opposite the crater. Logically this Mons would take a long time to form and we know that the Tharsis Montes for example probably continued to grow and restart through much of Martian history.

So at first with an impact there is an immediate tendency for the crater to move to the pole, and then as the Mons grows on the other side this weight fights to move to the equator and eventually may overcome the lack of mass in the crater. Also the crater may partially fill up over time as Utopia Basin appears to have[1]. It may then be that at first the crater moves to the pole and then after time moves toward the equator as the Mons grows. By having 4 impact craters and their associated Mons doing this it can make the resultant climate on Mars vary wildly, and make it much harder to decipher.

Also as a crater moves toward a pole its gravitational influence diminishes, so it will likely stop near but not exactly on the centre of the pole. It may also fill partially with ice and this addition of mass in it can further stop it moving closer to the pole. If the climate then changes for example with the heat added to the atmosphere by the volcanism of the Mons then this amount of ice can further fluctuate moving the pole as well. This would be because the warmer climate might move ice from the pole to the equatorial regions, and perhaps sublimate CO_2 from the poles.

While a Mons can be initiated by an impact it can also be altered or even restarted by additional impacts if the shock waves are strong enough. For example in here I will show that Isidis may have weakened the Tharsis area, which was further affected by an oblique impact from Argyre, and then from Hellas. This would have contributed to these volcanoes restarting many times until they reached their enormous size.

Generally it is believed that a large impact like Hellas would naturally make a large Mons on the other side of Mars, but the record from large impacts is not so clear. Utopia Basin and Isidis Basin have no large Mons opposite them, Argyre has Elysium Mons, and Hellas is hard to line up with the Tharsis Montes and Olympus Mons though it lines up well with Alba Patera. There is however enough of a correlation to think they are related.

The main theory is that when the impact occurs the core acts like a lens focussing the shock waves onto the crust on the other side of the planet, either exactly opposite or somewhat offset if the impact is oblique. This theory has several problems however. One is that a spherical core is not really the right shape to focus shock waves like this; such a shape should tend to defocus the waves. Also we don't know the relative densities of the materials involved so we can't say for sure what angle of deflections would be made by such a lens. As Mars became progressively deformed impacts happened from different elevations and so it would seem unlikely all of these would just happen to give such a precise focus to create a Mons. The effect of shaping sound waves by a lens is well known, dolphins use it for example to direct their sonar.

A more likely explanation might be that as the shock wave spreads there is a small cone of the wave that goes directly through the centre of the core and other layers, and because over this area the surface of the core is relatively flat then this cone of the shock wave would tend to be not defocused. Outside of this cone the shock waves would tend to be defocused and perhaps distributed evenly over the opposite hemisphere to the impact crater.

People who need glasses can see a similar process. Squinting can improve eyesight because the light goes through a narrow aperture, which is similar to shock waves going through the centre of the core. The narrow beam that gets through the core without being defocused might by itself create only a very narrow circle of damage on the surface crating one small volcano, compared to a large area of devastation.

Other parts of the shock wave may even be reflected back so we may get bands of general volcanism as the force is distributed over a large area. For example the core and other layers should have an angle of reflection so that waves hitting at a shallow angle tend to bounce off it and waves at a steeper angle tend to refract into the core. Large enough impacts may even affect the magnetic field in the core and perhaps and stop the magnetic dynamo[2].

This general spreading of the shock waves could have induced a general increase in volcanism over the whole planet and perhaps in certain bands, so these may explain larger volcanic flows[3] [4]. As we will see later the two main areas of Noachian craters[5] may have been preserved by their having been under poles and thus have resisted volcanic resurfacing.

400

Some of the Martian names are shown on 2 maps[6] [7]. Also the polar wandering path of Sprenke et al[8] is referred to, these are adjusted to give a polar path more in line with impact craters and known deposits of ice found, which may correspond to old poles[9]. It is not possible to be exact with the positions of poles so long ago; also they may not have been circular. The current poles for example are not, which may be caused by the pole still moving.

The Utopia Impact

While previous impacts may have shaped Mars[10] [11] I begin here with the Utopia impact basin[12] [13]. Thomson and Head conclude this is an impact basin and also that it is very ancient, likely much older than Isidis, Argyre and Hellas since it is much shallower[14]. Isidis basin is on the edge of Utopia Basin, implying from its shape that it was formed later. There has long been controversy over whether this is an impact basin or an ancient sea. Thomson and Head argue the model of an impact basin as opposed to an ancient sea or ice sheet but the pole in the basin would perhaps make both indicators be seen together[15]. This area could have been a Northern ocean especially when the Utopia impact basin was still warm from the heat of the impact. I refer to the centre of Utopia Basin as **North Pole 1** and subsequent poles are in numerical order. The current pole is **Pole 5**.

Interestingly similar lacustrine and volcanic arguments are made[16] in Valles Marineris which we will see later may have formed from the Isidis and Argyre impacts. If the pole had moved to the area of Valles Marineris in combination with the volcanic activity from the impacts tuyas may have been formed as well.

Smith et al[17] believe Valles Marineris may have formed after Argyre and Hellas, but this may also indicate those impacts helped to increase the rifting after it began with the Isidis impact. Valles Marineris is a large negative gravity anomaly[18] and so should tend to move to a pole as well. With the Isidis impact it was already near a pole, so this tendency was already realised.

As time progresses it moves further from the pole and as we will see towards the equator where it largely remained through further polar wandering. This implies it did not get much larger, or that if it did the negative gravity was compensated by the increased mass nearby of Tharsis Montes, Olympus Mons and the area around Solis Planum. If its rifting had have happened separate from an impact then we should expect to see the advance of the poles towards these impacts to be reversed at some stage when Valles Marineris was formed. This implies then that either it was never a significant negative gravity anomaly which seems unlikely or it formed in circumstances where its lack of mass were already compensated for and this did not change with a widening unrelated to the forming of further Mons from impacts. It is likely then that Valles Marineris was formed around the same time as the Tharsis Montes.

Sprenke et al[19] show no such change in the polar path heading back towards Valles Marineris or even seemingly being affected by it. Smith et al[20] also show how the current gravity of the Valles Marineris area is dominated by Tharsis Montes and Olympus Mons, and also by the area around Solis Planum, which also indicates the Utopia and Isidis impacts could have made that area heavier, slowly negating the negative mass of the Utopia and Isidis impact basins and allowing the pole to move on and this area to move towards its equator at the time. This area is shown in **Figure 1**[21].

Isidis Basin also has rootless cones, which according to Martel[22] may be formed by lava flows over water or ice. This would also be consistent with a pole here, and avoids the need to assume large amounts of ice all over Mars. Moore et al[23] believe fluvial erosion occurred in the Noachian to Hesperian.

The centre of this basin is 45 degrees North 248 degrees west, so directly opposite this is 45 degrees south and 68 degrees west. This is shown as **A** in **Figure 2**[24].

Interestingly this area has no Mons at all, though normally there should be one or more from the shock waves. The ground is raised around Solis Planum however and all the way to the Tharsis Montes. It may be then that the rifts around Solis and Syria Planum may have been partially formed from the Utopia Impact, where the whole area was raised rather than a narrow Mons.

The Argyre impact basin is very close to **South Pole 1**, and it may be that this hit the Mons opposite Utopia Basin, and removed all traces of it. The other possibility is that the Mons was in the area of Syria Planum and was either damaged in the Argyre impact or had eroded to a smaller size. This might account for the generally raised area around Solis and Syria Planum.

The Isidis Impact

The next oldest impact may have been Isidis[25]. If it has happened before the Utopia impact then likely it would show the effects and perhaps have been buried. It seems apparent that water flowed from the Isidis Basin into the Utopia Basin[26] [27]. Argyre also appears to be much younger.

When the impact occurred it would have moved closer to the pole, and the pole opposite would have also moved. The centre of the Isidis Basin is 12.7N 272.6W[28], giving the other pole as near 12.7S 92.6W, shown as **B** in **Figure 2**[29]. This would place it in Sinai Planum. So from one impact to another the pole may have moved from 45S 68W from the Utopia impact to 12.7S 92.6W. We cannot say where exactly **South Pole 2** was, though it is likely near a line between **A** and **B**.

Isidis is also the landing place for Beagle 2[30] [31] [32] [33]. Being very flat it is also more likely to be older than Argyre and Hellas. It also has signs of fluvial activity[34], which could be from when it was at or near a pole. Toon[35] says that large craters and river valleys appear to be the same age. This can be from impacts melting water, but also from impact craters moving to a pole which attracts ice and perhaps water if the craters retain some heat.

The elevated areas between Solis Planum, Icaria Planum, and Aonia then may have been caused by these two impacts. It also turns out that even though water signs were not picked up here by ODYSSEY[36] [37] the area is thought to contain ancient water deposits[38] [39] [40].

Barlow et al suggest the area has contained ice since the Hesperian[41], which fits in well with an ancient pole having been there. Interestingly they suggest the water table may have tilted here with the formation of Tharsis[42], which agrees with the time lines suggested in this paper. Instead of or in addition to the water accumulating here from Tharsis it could have accumulated from a pole forming here.

This agrees well with the proposed poles by Sprenke et al, who begin their polar wander at 45S 90W and then move to approximately 30S. It is more difficult to see signs of the opposing **North Pole 2** around Utopia Basin though ice signs are seen[43].

Not only does the area around **South Pole 2** in Syria Planum not show ice signs from ODYSSEY but they show the opposite, of being some of the driest parts of Mars. This conflicts with what is seen in Solis Planum, where fluidised ejecta from craters is known[44]. It may be then that there was ice here at one stage but events drove the water out. Since this area is opposite the Utopia and Isidis impact basins it implies that heating from those impacts may have removed the ice. Some ice may still be there[45].

After the Utopia impact this area would have become saturated with ice and perhaps water as it became a pole, and we know from the ejecta ice was there.

The pole may have extended into Solis and Sinai Planum in the middle of the raised ridges and rifts, though only the ones to the south may have been there then. Since this is opposite the Isidis impact and roughly follows the path of polar wandering by Sprenke and Baker it seems a reasonable possibility.

While the impact of Isidis may explain the lack of water signs opposite Utopia Basin it cannot explain the lack of water signs opposite Isidis itself. An additional event may have happened to make the further rifts around this pole, I call **Pole 2** (**Pole 1** would have been formed by the Utopia impact). Valles Marineris may have been formed from this further event.

Syria Planum is said by Webb[46] to be surrounded by a raised annulus as well. Stresses then are likely to have created the rifts such as Valles Marineris and allowed these Planum to subside. Instead however of Tharsis forming this strain this may have been done by the combined impacts of Utopia and Isidis, and that forming a pole of ice here restricted the volcanic effects to their perimeter. Scott[47] argues that Syria, Sinai, and Solis Plana were formed with a mantle plume, which is consistent with the idea of heat from the Isidis impact. The subsidence may have been partially formed by the weight of the polar ice. Also Scott[48] says this upwelling may have been enough to form Valles Marineris[49]. This area may also have escaped resurfacing from Tharsis because of this higher elevation according to Smith[50].

Hartmann[51] says Solis Planum also has well preserved craters, with larger craters more preserved. This may be because of the pole being there for a long time, having buried craters in ice.

Arsia Mons[52] [53] is roughly at 12S 120W, the same latitude as **South Pole 2**. It is also the largest of the 3 Tharsis Montes as well as being closest to the opposite of the Isidis Impact. Head says evidence of glaciation[54] [55] has been found also on Pavonis Mons and Ascraeus Mons, which is suggested to be from the late Hesperian. This may also have been from the time the pole was nearby. Few signs of water are seen, which would be consistent with polar ice. Head and Marchant[56] also say that Arsia Mons had volcanic outflows into the Hesperian, which is consistent with later impacts restarting the flows. Sprenke et al[57] say that the Tharsis Montes may have been formed after the Martian global magnetic field ceased, which might imply the Hellas and Isidis impacts had affected this field. It may also mean the volcanoes were reactivated later in subsequent impacts. Vast amounts of ice may be there even today[58].

The Argyre Impact

Next the Argyre impact[59] may have occurred. Assuming **South Pole 2** was at Sinai Planum it would have been likely that this was an oblique impact. Argyre Planitia is shallower from the West to the North though it is deepest to the North West. Argyre is centred on approximately 50S 42W, and the Sinai Planum **South Pole 2** may have been at 12.7S 92.6W. Sprenke's[60] pole position would be 30S 90W so both give an oblique impact assuming the meteor came from within the ecliptic plane of the asteroid belt.

Looking at the higher elevations North West of Argyre Basin it seems these reach from Argyre all the way up to Valles Marineris and west to the northern edge of Icaria Planum, seen in **Figure 3**[61]. Illustrated in **Figure 4**[62] if you draw a line though Olympus Mons **A** and Pavonis Mons **B** then Arsia Mons and Ascraeus Mons **C** would be at close to right angles to this line as would the annulus **D** to the North West of Bosporos Planum. This also points at the centre of the Argyre Basin. The annulus west of Syria Planum and Solis Planum **E** would be approximately the same angle as Valles Marineris shown here as two lines **F** and **G**.

It may be then that while these features were partially formed by the Utopia and Isidis impacts the glancing blow of Argyre sent a shock wave in a shallow angle and created them. This would be the same mechanism as an impact on the other side of a planet except in this case the main shock wave comes from the side. The shock wave is always strongest in the path of the impact because the sound waves are most compressed along that line by the speed of the meteor and hence the frequency is most raised. So the strongest force in an oblique impact should be along the line of the impact trajectory.

405

This triangular shape can be approximated by shining a torch onto a circular bowl at an oblique angle, though the sides are more rounded. A shock wave would generally be cone shaped rather than cylindrical like a torch beam but the diffusion is unlikely to be great over this distance. As the beam goes through the pole at a shallow angle it would come out roughly in this triangular shape. The edges of the shock wave would tend to shear the ground creating an approximately triangular or egg shaped annulus and rifts. Rifting and faulting would tend to be in straighter lines making the shape more triangular.

The same thing may have created the triangular shape of Olympus Mons and the three Tharsis Montes, perhaps from a reflection from a subsurface layer. If a shock wave hits a denser layer below at an angle it may glance off like a light beam reflecting off a pane of glass at a shallow angle. The shape of this shock wave can be seen by shining a torch on a globe (representing the denser layer) at an oblique angle.

The shape is approximately the same, and faults would tend to be in straight lines to create the Tharsis Montes in a straight line. Here[63] Tharsis Montes and Olympus Mons are overlaid on this annulus shape together with Valles Marineris to show the similarity of the two shapes.

Another possibility is that the shock wave pointed directly at Tharsis and Olympus Mons, and a reflection made Valles Marineris and the rises. A geological layer would not reflect perfectly because of its rougher surface so some parts of the shock wave would go through and another part would reflect even when the angle of reflection favoured one or the other. A good analogy would be for light reflecting off frosted or roughened glass where some parts would reflect and others refract.

In this case then Valles Marineris would have been stressed but not rifted by the Isidis impact and upwelling and then sheared into a rift valley from the Argyre shock wave.

There are no signs of volcanic activity in any other direction from Argyre, which also implies a glancing impact. On the other side of Mars Elysium Mons would have been also formed from the impact. There are three Mons[64] opposite Argyre, Hecates Tholus, Elysium Mons, and Albor Tholus. This also connects the shape of Tharsis Montes plus Olympus Mons to the three opposite Argyre. The triangular shape of these three Mons may have occurred from a refraction of the somewhat triangular shock wave.

Hiesinger has out six scenarios for the evolution of the Argyre basin. Surious Valles has a delta shape, which may have been from water as **South Pole 2** melted. There is also evidence for a proglacial lake. Water signs are found there but too high for water to come in from the northern lowlands. A polar meltback[65] was proposed by Parker et al[66] though not in the **South Pole 2** position.

Elysium Mons is close to the Isidis basin which would have been the opposite **North Pole 2** at the time.

Such an oblique impact hitting the polar ice may have expelled large amounts of water and ice from Mars; indeed it is hard to imagine an event more likely to do so. Also if at this stage Mars had substantial amounts of frozen gases such as CO_2 at the pole this could also have been ejected leaving the atmosphere much thinner. Much material from the crater may have also been removed from mars which could have subsequently made its negative mass more influential relative to the coming shift to **Pole 3**. Some ejecta may have added to the annulus to the west of Bosporos Planum **D** and even to the raised area further North West.

The subsequent heat would have melted much of the pole and raised the overall Martian temperature. If gases were frozen this would have raised the air pressure and allowed more liquid water, also floods from the ice melting were likely. Signs of these are probably shown in Valles Marineris which has an elevation sufficient to carry water into Margaritifer Terra and Chryse Planitia[67] which are also shown in **Figure 4**. In Lunae Planum there is also evidence of large scale flooding, also on Xanthe Terra, but little south of Solis Planum[68] and perhaps some into Argyre Basin which would be too hot at this time for water to channel into it. The water may also have gone North West[69] if Tharsis was not large at this time.

South Pole 2 would try to reform polar ice and water but the heat would drive it away. This then may explain why this area had evidence of water from ejecta in craters but this seems to have disappeared according to Odyssey[70], also shown in **Figure 5**[71]. As shown in **Figure 6**[72] one red section appears to radiate out of Argyre from the west **A** to the north **B**, and the second one sits in Solis Planum **C** where **South Pole 2** may have been. This indicates the Argyre impact may have dried out these areas, which is why they showed fluidised ejecta in the past but now show little ice. It may even have dried out east of Ascraeus Mons.

Later water would drain from the Argyre Basin as it cooled and water ran into it from **South Pole 2**. The raised annulus west of Bosporos Planum may also have prevented water going to the basin, as well as south to Aonia.

Thaumasia[73] shows deformational features radiating from the Argyre Basin which may show the impact had influence this far. While also radial to Tharsis here they may be more related to Argyre.

The large negative mass of the Argyre basin would have begun a pole shift. Also as the impact helped to grow Tharsis Montes and Olympus Mons they would have tended to move to an equator which would have also helped to shift the pole. As we will see according to the polar wander path of Sprenke and Baker Tharsis and Olympus Mons move directly closer to an equator westward, which in turn moves the pole eastward.

Sprenke et al show a movement of the pole to approximately 0S and 30W and then to 330W. This would take the pole to North of Argyre Basin into Margaritifer Terra and then east to Meridiani Planum. Since it is unlikely the pole stopped in Margaritifer Sinus this is not given the name **South Pole 3**, that is for when the effects of the Argyre impact stabilise a new pole. So the pole wanders from Sinai Planum **A** through Margaritifer Sinus **B** to Meridiani Terra **C** heading eastward, shown in **Figure 7**[74].

According to Grant[75] Margaritifer Sinus contains high valley densities, which would be consistent with the pole moving and subsequently ice melting. Also the area was resurfaced several times[76], perhaps from the subsequent volcanism from the Argyre impact. While Grant[77] believes some precipitation occurred most would have been from ground water, which is consistent to a water table associated with a forming pole. This discharge[78] lasted a long time according to Grant[79]. The Parana Valles[80] drainage system is particularly extensive.

Hynek et al[81] believe this time of fluvial resurfacing lasted several hundred million years. A combination of rainfall and sapping[82] appear likely, which may have formed a lake for a time[83]. A moving pole then may link the two main theories of precipitation and sapping to explain the valley networks[84], and that according to Nelson a large build up of ice which periodically melted. Philips et al[85] examined Margaritifer Sinus and concluded much of the Tharsis bulge was already in place before the drainage channels were formed, which is consistent with the general rise in elevation in the area of Tharsis and Sinai Planum from the Isidis impact. Further growth of Tharsis could happen later, though at the late Noachian it was large enough to direct the channels northward. Large amounts of material from this area were removed along these channels probably from water erosion as the pole melted, and moved into Margaritifer Sinus.

Valles Marineris[86] [87] [88] is then likely to have formed from the stresses of the Isidis and then the Argyre impacts, making its origin harder to see[89] [90]. By this time water and ice would have accumulated in it as the pole melted and moved, which may explain the paleolakes[91]. Carr[92] suggested that ground water flowed into Valles Marineris and then[93] into Chryse Planitia, forming lakes. Rossi et al[94] believe there is good evidence of ice and glaciers which would be consistent with a polar area.

Lunae Planum, also shown in **Figure 7**[95] would also have received water from the moving and melting pole. Greeley and Kuzmin[96] show how Shalbatana Valles originates in the chaos on Lunae Planum. Interestingly it comes from a probable impact basin that formed a catastrophic outflow. This impact may have occurred before or at the same time as Argyre, though its shape (not mentioned by the authors) would likely be elliptical if from the time of **South Pole 2**. While it is suggested the impact breached an aquifer this would be unusual for Mars. It does link the area with large amounts of water and probably ice triggered from an impact, something not seen elsewhere. A pole here would supply the water, and once it carved a channel keep it going with more water. A nearby elliptical formation, Orcus Patera[97] may also have come from an impact, its shape implying a pole was near here.

Xanthe Terra also shows evidence of water flows. Nelson and Greeley[98] discuss 3 major fluvial events here. The first is a broad sheetwash from the Valles Marineris area perhaps coinciding with the Argyre impact. Then there was more extensive water forming Shalbatana, Ravi Simud, Tiu, and Areas Valles which might coincide with the pole moving to Margaritifer Sinus. Most of this water came from chaos areas[99] which would link to the Argyre impact. Subsequent flooding would be as the pole continued to move, and when further enough away the water would cease.

In the new Odyssey results of subsurface ice[100] there is a large deposit on the equator in Babaea Terra shown in **Figure 8**[101] and centred at 330 degrees west. Another one can be seen on the left edge of the map just below the equator shown in **Figure 9**[102]. This would correspond to the opposite pole. According to Sprenke et al the pole moves in a curve through this ice rich area to 0S 330W, almost the centre of the ice rich area. I call this area **South Pole 3**. Having icy areas opposite each other like this makes it likely they were poles.

This can be explained from the effects of the Argyre impact. As the Tharsis and Olympus Montes grow they accumulate more mass which seeks to go to the equator. The movement of the pole to this area allows these to get much closer to the equator. Tharsis and Olympus Montes are today on the equator at around 120W and the pole would have moved to 330W. This adds to 150 degrees so the Montes would have nearly reached the equator, which indicates their weight was dominating at this stage. The pole has assumed a position between the Argyre and Isidis impact basins as each would have had a tendency to be near the pole.

This would tend to be a stable configuration. The Montes have grown enough to get near the equator so little more can move them closer. Elysium Mons has probably also formed to some degree which reduces some of the negative mass tendency of the Argyre Basin. It is at 210W so there is more than 120 degrees between there and the pole. **Pole 3** then would be balanced with Olympus Mons and Tharsis Montes tending to go to the equator and Elysium Mons almost opposite them also near the equator, and Isidis, Utopia, and Argyre basins around **South Pole 3**.

This also implies the growth of these Montes is strongly linked to the Argyre impact as the pole moved from near the Argyre Basin to mid way between it and the Isidis basin. If the Argyre impact had not happened then, its happening now would scarcely move the pole. It would be unlikely an impact causing the Argyre basin happened then just at a time where it wouldn't move the pole. Also that Elysium Mons has moved towards an equator which it wouldn't have done if it hadn't formed yet. Logically then the Utopia, Isidis, and Argyre impacts had to happen in this order, forming their respective Mons in the order described for this polar wander path suggested by Sprenke.

Interestingly **South Pole 3** coincides with an area of heavy cratering[103] and the second cratered area corresponds well with the opposite **North Pole 3** in **Figure 9**[104]. One likely explanation is that the polar ice protected the craters[105] from erosion, and when they were exhumed from the ice they remained in more pristine condition. **Pole 3** probably lasted a long time to give this crater disparity. It also implies at this time that the surface was being altered severely and other craters were being buried or obliterated by lava flows.

This may be because of the shock wave effect of these impacts which may have sent shock waves over large parts of the surface initiating volcanism. This would explain how volcanoes have apparently restarted in Martian history and the surface being relatively young in parts. **Pole 3** likely remained here through this resurfacing. Since these crater areas are linked into what is termed the Noachian surface it may be that the time after the Argyre impact may be regarded as the Hesperian, obliterating much of the Noachian terrain except for these parts protected with polar ice. Some other areas with Noachian craters are also found around Margaritifer Sinus, implying the pole may have slowly moved and protected other areas for a time in its path.

In moving from **Pole 2** to **Pole 3**, the polar ice closely follows and may have formed the dichotomy boundary. The main boundary is seen between 180 degrees west and 90 degrees west, which is 270 degrees or ¾ of a total possible boundary. The rest is taken up by the land mass of Tharsis Montes, Syria Planum, etc. **South Pole 2** moved from 12.7S 92.6W eastward to approximately 0S 330W, which is approximately 122 degrees of longitudinal movement or approximately 1/3 of the total great circle. The opposite pole travels from 12.7N 272.6W to 0S 150W, which is where the dichotomy boundary ends against Olympus Mons, for a movement of 122 degrees. This makes then 244 degrees of movement over a dichotomy boundary of 270 degrees as a polar path. The rest can be explained by the width of **South Pole 3** at 330W, which makes it appear to extend further east. So of the total visible dichotomy boundary virtually all of it is on the same line as the movement of **Pole 2** to **Pole 3** which is unlikely to be a coincidence.

The pole then moves through Margaritifer Sinus and from here there is a green elevation path. This trail begins at east south east of **Pole 2** so the pole may have initially moved towards the Argyre crater, which is logical as its negative mass should move towards the pole. This implies the pole may have moved along this green area and lowered the terrain there.

The pole was probably moving on a slope, which may make the path easier to see than from **South Pole 3** to **South Pole 4** where the ground was not sloping. We already know the planet tends to slope towards the current North Pole, **North Pole 5**, and that this polar path is lower than the terrain south of it, and higher than the terrain north of it. This then implies the moving pole may have flattened part of a slope going into Acidalia Planitia and for the opposite Pole Elysium Planitia.

A pole moving on a slope like this would tend to have a runoff of water heading North through the journey. Depending on the temperatures and the air pressure at the time ice may have sublimated directly into water vapour and CO2[106] may also have been a primary erosional force. On the sloping ground water, mud and perhaps CO2 ice would tend to move like glaciers, with material in the ice moved north through avalanches and liquid CO2 as described by Hoffman. Dust that formed on the pole through dust storms then would be moved north and perhaps create a very smooth surface in Acidalia Planitia, Utopia Basin and Elysium Planitia.

As the pole moved new ice would tend to form on the ground ahead and melt on the ground behind it as the temperatures changed. The ice in front would tend to freeze into the soil and create a similar situation to the current **Pole 5** where approximately half or more of the soil is ice. When this eventually melted or sublimated the soil in the ice should have moved down the slope and spread out. If there was a high enough air pressure this should have created a seasonal water flow into Acidalia Planitia and created the smooth surface. CO2 might give the same movement at lower temperatures.

It is likely the temperatures of Mars were dropping from the Argyre impact, there are visible water channels in Lunae Planum, Xanthe Terra, and Margaritifer Sinus, but these are no longer seen as the pole moves eastwards. The edges of the green elevation may indicate the edges of the permanent ice cap.

This may mean then that the primary erosion was from ice and CO2. Some channels are found north of **South Pole 3** in Arabia Terra, but these may be from the Hellas impact later when the pole moves again. If so then this would again imply the temperatures and air pressure were too low after Margaritifer Sinus for water erosion. More investigation of this polar route should confirm whether channels existed.

The ice deposit at **South Pole 3** abuts a cliff to the north, which is an extension of the dichotomy boundary. This ice then implies that it is connected to the creation of this cliff and by extension created the cliff of the dichotomy boundary as the pole moved. As water ran down the slope at **South Pole 3** it would have eroded the ground, but where the ground was permanently frozen the ground would have been protected. This should then give a boundary to the north of the moving pole where the ground slopes more.

The speed of the polar wander should be according to how quickly the Tharsis Montes and Olympus Mons grew, with their tendency to go to the equator. Also as the pole moved away from Argyre it may have been held back because the negative mass of the Argyre impact basin would tend to be near the pole. As it came closer to the Utopia and Isidis impact basins it may have accelerated, releasing more water. These basins would counteract to some degree the negative mass of Argyre as they too would seek to be near a pole. It may also be that with the polar movement the channels would be regularly changing and so did not form as large as in Margaritifer Sinus.

The Hellas Impact

The Hellas basin is centred at approximately 40S and 290 W. This would have made it about 40 degrees from **South Pole 3** and so would have been an oblique impact, though not as much as Argyre. Almost exactly opposite Hellas is Alba Patera, again probably formed from the shock waves. The resultant shock waves may again have gone around Mars stimulating volcanism, perhaps restarting Tharsis Montes and Olympus Mons which are also close to opposite Hellas. It may also have stopped Mars' magnetic field. Sprenke and Baker[107] point out that the rotational poles closely follow the movement of **Pole 2** to **Pole 3**, but this does not appear to extend to **Pole 4**. This may be because the Hellas impact stopped the magnetic dynamo with the shock waves.

Anderson et al[108] analysed Syria Planum in comparison to Alba Patera. They concluded Syria Planum is Noachian to late Hesperian with intense activity that declined later. This would be consistent with its initial formation from the Isidis impact and later from the Argyre impact. They consider Alba Patera to be similar, which is plausible if it was formed by the Hellas impact. This is considered to be extending from the early Hesperian into the Amazonian and so is later than the Syria Planum volcanism. They believe[109] that Syria Planum had a greater impact on Tharsis than Alba Patera which is again consistent with the impact sequence.

The large negative mass of Hellas would have tended to move the pole towards it, and Sprenke found from elliptical craters that the pole probably moved to 45S 345W. This places it on the edge of the Hellas Basin in the direction of the Argyre Basin, probably with the two negative mass craters tending to both be near the pole. This is the same as in **South Pole 3** where Argyre and Isidis basins were both near the pole. It is closer to Hellas probably because Hellas is much larger than the Argyre

Basin, being younger. There are many large craters in this area as well; some may also have been preserved through burial under the polar ice. I call this **South Pole 4**.

In **Figure 10**[110] **A** is **South Pole 3** and **B** is **South Pole 4**. This is now relatively close to **Pole 1** at 45S 68W. There is approximately 83 degrees of longitude between **Pole 1** and **Pole 4**, and they are on the same latitude.

Also in this pole position Isidis is on the equator which means that **Pole 2,** Sinai Planum, and Tharsis Montes are also now on the equator. The pole then has moved to near Hellas while maintaining much of the weight of the Mons on the equator. This is consistent with the weight of Olympus Mons and Tharsis remaining on the equator, and the pole moving to the negative masses of impact craters.

Hellas is an oval shape approximately twice as long as it is wide[111], probably from the oblique impact. Since Hellas is approximately 45 degrees from **South Pole 3** this would imply an impact at 45N at the time, and the oval shape indicates an impact on the western side to give the oblique angle pointing mainly at right angles to **South Pole 3**. This is likely as to the west of Hellas there is a reddish area[112] with much less ice, shown in **Figure 11**[113]. The icy area of **South Pole 3** is elongated pointing along the path of the polar wander to **South Pole 4**. The section west of Hellas basin much drier is shown as from **C** to **D**. Since this is in a direct line with the longest part of Hellas it is likely this was formed from ejecta or shock waves making this area hotter for a long time, and eventually drying the area when **South Pole 4** moved. This is the same mechanism as might have happened with the Argyre impact drying west and north of it in **Figure 6**[114].

The icy blue area of the current South Pole reaches to Hellas and implies the pole may have wandered east into Hellas and then south to the current position, as shown in **Figure 13**[115]. The lack of ice on **South Pole 3** may also indicate the climate was warmer from the heat of the impact and stopped large amounts of ice forming.

The corresponding North Pole would then have moved to the east of Alba Patera in Tempe Terra. There is a possible platform there similar to that south west of Alba Patera. The pole may have moved as the gravitational influence of Argyre lessened, with the basin filling up and Elysium Mons growing larger. This would move **South Pole 4** to the east away from Argyre, there may be another platform on the eastern side of Hellas Basin. This would also be consistent with water gullies, which in other parts of Mars are near former poles. If the pole moved to eastern Hellas Basin this would explain gullies in Dao Vallis and Tempe Terra. It may also be that the polar ice extended eastward to there.

This would also be consistent with the shape of the current poles. Chasma Australe points to 270W and may have been formed by water melting as the pole moved, the pole perhaps still moving. Promethei Planum would also have been formed from the pole moving away from **South Pole 4**. A hot spot creating basal melting may have formed Chasma Australe but the moving pole could have also supplied the heat, the leading edge becoming colder, and the trailing edge warmer.

Hellas seems to have contained ice covered lakes which would be consistent with being near a pole. It can be seen that the moving pole may have made a lot of different areas appear to be ice rich and often to have fluvial flows, even glaciers and hydro volcanism. This could solve the mystery of why Mars has so many water signs and apparently not enough ice available to cover them all. It also can explain that even though the temperature has likely been too low for liquid water, channels are widely seen. The moving pole would have moved water and ice with it affecting each area in turn, looking as if there should be perhaps 10 times as much water on Mars.

Chemically Mars resembles a dry planet[116] so outside of these poles there may have been little ice or water, and CO2 erosion may even have predominated at times[117]. Each impact would have temporarily heated up the planet giving perhaps brief times of liquid water and perhaps higher air pressure through sublimated gases.

The various Mons might have been periodically restarted from shock waves from the impacts and the resulting heat kept the planet warmer for a time, until eventually all volcanic activity ceased and Mars reverted to the cold planet we see today.

Thomson and Head[118] believe glacial features, moraines, drumlins, and eskers are to be found in Hellas, consistent with being near a pole. According to them this could have been part of an ice sheet[119] and a proglacial lake[120], possibly middle Amazonian[121]. The lake they believe would have held enormous amounts of water that has disappeared, consistent with water from a pole that moved on.

Jakupova et al[122] have laid out a distribution of craters 10 km and over. This boundary of heavy cratering closely follows the movement from **Pole 2** to **Pole 3** and indicates when the resurfacing of the north may have occurred at this time.

This is consistent with the poles moving water and sediments northwards, and burying craters. There is an area above Margaritifer Sinus which is nearly devoid of craters; this is likely to have been resurfaced in the floodwaters in the movement of **South Pole 2** to **South Pole 3**. Isidis is comparatively devoid of craters, and this continues in a line with the movement of **North Pole 2** to **North Pole 3**. This would again be from transporting sediments and water north and removing craters. The area north of the equator and 60 degrees west extends high into the northern hemisphere with heavy cratering. This area was found by Odyssey to be drier and indicates that a lack of water is associated in this area with cratering.

Layers are Mars are thought to have been formed by dust alternating with CO_2 or water ice. As water and ice were moved north by the movement of **Pole 2** to **Pole 3** it would have deposited on these layers. With this pressure and with liquid water the tendency would be for the CO_2 and ice to melt and move upwards, which would make the layers collapse and the ground to lower in elevation. The movement of water northwards then could have created a lowering of the ground forming the Northern lowlands. This lowering in turn would enable a large sea, ice sheet or mud ocean to form and the collapsing of layers to become more and more widespread. If so then the southern hemisphere may have substantial amounts of CO_2 and ice still trapped in layers.

The northern hemisphere is seen as Amazonian and the Southern surface as Hesperian implying the southern hemisphere is older. This is however from crater counts and it is possible that the craters in the northern hemisphere may have been removed and buried in this process.

After the Argyre impact the formation of the volcanoes may have added a lot of ash into the atmosphere which would have tended to collect at the cold trap in the poles. This would have subsequently been moved northwards as the poles moved.

Permanent ice in the north[123] would tend to compress the ground leading to polygons when the ice was eventually removed. Dust and accretion from meteors would have built up on the northern ice as well as in the south. As meteors impacted in the north they would have fallen on ice and so not left a permanent mark. Head et al[124] show that the northern lowlands contain areas of polygons, craters with ejecta lobes, and potential coastlines.

Deviations in the Contact 2 coastline may be accounted for by changes in Tharsis and Elysium Mons which would be occurring as the pole moved, and later.

When eventually the planet became colder after the effects of Hellas and Tharsis wore off the air would have begun to freeze and go to the poles. Then the northern ice would also have sublimated, some may even still be buried as a frozen ocean. The material that had built up on the northern ice would have fallen down onto the craters that had been preserved under the ice, and buried some of them. This material would appear similar to that of the south, and likely not show signs of water as it came from the surface above the ice. Some of the northern ice sheet may have been liquid underneath around Tharsis Montes, Olympus Mons, Alba Patera and Elysium Mons because of their heat.

Some of these surfaces then may be smooth from having been sediments on an ocean floor in this way. Others areas may be smooth because as the ice sublimated the material fell smoothly.

In two areas ice seems to follow the dichotomy boundary, shown in **Figures 14**[125] and **15**[126]. This may be a residue from the movement of the polar ice from **Pole 2** to **Pole 3**.

Amazonis Planitia is thought to be flat from sedimentation or fluvial processes according to Head[127]. This is north of where **North Pole 3** stopped. Also the outflow channels may be partially from when **South Pole 2** in Syria Planum melted and began moving. Arsia Mons which is the most southern of the three Tharsis Mons may have been much smaller then. Some channels leading to Amazonis Planitia point north but to the east some point more to the North West.

Lucas Planum[128] is described by Cabrol et al[129] as an estuarine delta. If so then this may imply that the movement of **Pole 2** to **Pole 3** was accompanied by water flows as the pole melted. It may also have formed water locally from the heat of Apollinaris Patera[130], or from when the pole began to move north towards the future site of Alba Patera after the Hellas impact. Alba Patera has steep sides and may have formed in the polar ice of **North Pole 4**. Fuller et al[131] believe this area was resurfaced volcanically and with fluvial sediments. This could be for example from when the Hellas impact restarted some volcanism in Olympus Mons and started melting and moving **Pole 3**.

The Medusa Fossae Formation follows the path of **Pole 2** to **Pole 3**, and has formations similar to a pole according to Fuller et al[132].

McGill[133] refers to the younger material sitting on the older Noachian material, which is consistent with the dust layer settling. The buried materials are similar in age to the southern highland, which is consistent with the idea that this was buried under ice and then overlain with dust as the ice sublimated. In this case the air pressure would already be low so there would not be a liquid phase, hence no water to leave signs of the removal of the ice and chemical signs of water having been there.

Watters[134] shows lobate scarps are found south of the dichotomy boundary suggesting that compressional deformation was involved in the boundary's formation, which is consistent with the weight of polar ice. While he suggests that this occurred in the early Hesperian Anderson et al[135] believe Alba Patera was also formed in the early Hesperian to Amazonian, but the impact of Hellas may have defined the start of the Amazonian from the Hesperian. Since the Noachian, Hesperian, and Amazonian are calculated from crater records the impact of Hellas here may have changed these same crater records of the Hesperian to Amazonian at least around Alba Patera.

Head et al[136] believe much of the northern lowlands were resurfaced volcanically and in some areas as sublimation residue from frozen ponded bodies of water[137]. This may have occurred by the diffusion of shock waves from the Hellas impact over the northern hemisphere of **Pole 4**.

Tanaka et al[138] believe the northern lowlands were smoothed by glaciation. This would be consistent along the dichotomy boundary while it was forming, if there was a larger icy area north of the moving pole that was melting and reforming. If the air pressure was too low then this could have been from ice sublimating.

Hoffman et al[139] believe flood channels from Cerberus Rupes may be from CO_2. This area is to the north as the pole moved southwards from **North Pole 2** in Isidis Basin to the **North Pole 3** position. This then may give CO_2 through this area, in combination with flood water from the melting pole. The actual flooding depends on the temperatures but this area is equivalent to Margaritifer Sinus. **South Pole 2** had begun moving and released water into Lunae Planum, Xanthe Terra, and Margaritifer Sinus, so the opposite pole would likely be releasing water at the same time as it moved.

Since CO_2 typically sublimates on the current poles in summer this same mechanism would presumably be operating as the pole passed this area. It is not known at this stage whether CO_2 can account for these effects, but it is likely that it was available.

Burr et al[140] believe flood water originates to the north of the Elysium Basin and Marte Vallis[141]. A lake in Marte Vallis may have been fed from Medusae Fossae to the south. Ice from the pole moving to the **North Pole 3** position may have been heated by Elysium Mons which would be forming from the Argyre impact, and so may have provided heat to the area. They also conclude[142] precipitation was unlikely to form the channels because nearby areas show no erosion from rain. Groundwater is likely to form from a nearby pole, and the heat from Elysium Mons should turn this to water when the pole got close to the area.

Burr et al[143] see rootless cones though Athabasca Valles, which can from lava on wet ground. This can be from the volcanic activity from the Elysium Mons area caused by the Argyre impact, and water from the passing pole. The carrying of sediments[144] is consistent with the idea the moving pole may have carved out the dichotomy boundary, and floods moving the sediment north. Ice from north of Elysium Mons[145] may also have been melted by volcanism[146] giving flood water.

Elysium Mons[147] according to Bowling[148] had two periods of activity, which may correspond to the original formation from the Argyre impact, and being reactivated by the Hellas impact.

The ice on the edge of **South Pole 3**[149] cuts off on the dichotomy boundary as shown in **Figure 16**[150] which may indicate the pole helped form the boundary.

While the pole was here each summer, water or ice may have fallen down the slope of the dichotomy boundary off the northern edge of the pole, and the dichotomy boundary here could have been the edge of the permanent polar cap. The edge would form here because each summer water or perhaps ice or CO_2[151] avalanches would fall down the slope, eroding it away till it abutted the permanently frozen cap. In **figure 17** at **A** an ice trail, possibly from this water connects to higher ice areas. With higher ground to the south water may have tended to go north.

As the pole moved from **North Pole 3** to **North Pole 4** it moves close to Olympus Mons and the Tharsis Montes, which may have restarted from the shock waves of the Hellas impact. If so then the movement of the pole near such hot areas should sublimate frozen CO_2, so this time was probably one of higher air pressure. As Alba Patera formed from the Hellas impact at **Pole 4** the heat would also have kept CO_2 from freezing there and raised the air pressure. Since this would keep one pole from having as much frozen CO_2 the overall air pressure should have been substantially higher for a long time.

This places the **North Pole 4** just to the north and North West of Olympus Mons shown in **Figure 19**[152]. Milkovith et al[153] interpret this area to the North West of Olympus Mons to be glacial, but much large than terrestrial glaciers. This would be consistent with polar ice. Each of the Tharsis Montes according to Head[154] also shows glacial signs, perhaps from as the pole was passing. Since a pole should slow as it nears its resting point there may have been a pole on the Western edge of the Tharsis Montes for some time.

Pole Five

This is the current Martian pole. As time progressed Alba Patera became larger, perhaps cancelling out the negative mass of the Hellas basin. The current poles have the major Mons all near the equator just as they did at **Pole 3** and **Pole 4**. This indicates that from the time of **Pole 3** the polar wander had to move so as to keep these large masses on the equator. Since **Pole 3, 4,** and **5** are roughly in a straight line this would have been controlled by the Mons and indicates they are older than these three pole positions.

Pavonis Mons is on the equator and Arsia Mons the largest is at 10 degrees south. Olympus Mons is at 20 degrees North and Ascraeus Mons at 12 degrees north. Elysium Mons is at 25 degrees north.

Once the ice in the northern hemisphere started to sublimated and move to the poles this would have created a large negative mass that would have acted like a crater. The pole would have tended to move to the gravitational centre of this which would move it from Hellas Basin to the current position. Since most of the ice came from the north this may explain why[155] [156] the **North Pole 5** has more ice.

Since the shift to **Pole 5** temperatures may not have allowed this additional ice to sublimate and equalise the amount at both poles.

Chasma Boreale on the **North Pole 5** points to approximately the **North Pole 4**, and Chasma Australe on **South Pole 5** points approximately to **South Pole 4**. Both of these are the largest chasma on their respective poles. The poles may even still be moving which would explain why the South Pole is asymmetrical[157].

The North Pole seems similarly asymmetrical [158]. Both shapes seem to elongate at right angles to the previous pole positions. This would follow as the pole moved the forward edge would represent a line of temperature low enough to form a permanent CO_2 cap. Clearly the elongation could not point into the movement of the pole as this would be against the temperature gradient.

Byrne et al[159] found evidence of short term change on the current Martian South Pole, in the "Swiss cheese" formations. This is consistent with the idea that **Pole 5** is still moving, though it seems unlikely such short term changes would be associated with the pole moving. Changes may occur in spurts as areas collapse with the changed temperatures. One analogy might be the changes in glaciers on the Earth's pole which can change suddenly from the slower global warming.

These structures are found on the forward edge of **South Pole 5**, and the spider formations are found directly opposite this on the other side of the pole. It may be then that these "Swiss Cheese" formations may be older spider areas that have slowly been moved into colder areas and are now permanently frozen. In this way the similarity between the Swiss Cheese shapes and the spider bushes can be explained.

Some of the spider formations would then be left behind as the pole moved on its trailing edge, and we see this as spiders that merge into apparently inactive areas there.

Pathare et al[160] believe recent changes in the Polar Layered Deposits may have been caused by changing obliquity though these could also have been caused by the moving pole. Layering is seen along the path of **Pole 2** to **Pole 3**; implying layers may be formed as a pole moves.

Malin Space Science Systems[161] recently reported in Science more examples of changes on the South Pole. Hoffman[162] shows gullies on the current South Pole may be undergoing changes, again consistent with a moving pole. These pitted areas however are also significant in relation to **South Pole 4**, and the gullies may have been formed at that time. M1003736[163] mentioned by Hoffman is at 70.91S 358.7W, which is closest to **South Pole 4**. This would explain their pristine condition if they were moved into the polar area after the pole shift from **Pole 4** to **Pole 5**.

The Ages of Mars

The three ages of Mars, Noachian, Hesperian, and Amazonian are primarily based on craters counts. If the polar wander theory is correct then these time scales will be distorted by resurfacing after each of the four major impacts. The Argyre impact may have been so influential it might be said to have begun the Hesperian, forming Valles Marineris, Olympus Mons, the Tharsis Montes, the dichotomy boundary and Elysium Mons.

As an alternative guide the four impacts might themselves be defined as the start of an age. There would then be the ages of Utopia, Isidis, Argyre, and Hellas. This can be much easier to work out the ages of various formations as the beginning and end of each age is a fixed date. Ages could also be defined according to volcanoes, e.g. the Age of Tharsis, Olympus, Elysium, and Alba Patera.

An approximate age can be determined for each impact, and then a tree of cause and effect can be created following on from each impact. Then the age of each event that follows from an impact is determined and added to the tree. This in turn enables the age of each impact to be more precisely determined. Other events that were sufficiently independent could be portrayed as separate trees of cause and effect.

The age of Utopia may have begun to form some of the area around Solis and Bosporos Plana. There may also be areas to the north west of Elysium Mons that could be dated according to this impact. It may have formed a Mons that was destroyed by the Argyre impact, if so then signs around the Argyre Basin may be dated according to Utopia as a benchmark.

Many of the effects of each impact would happen relatively soon afterwards and there would be a long time between impacts dates. Many formations should then be able to be connected with an impact and more accurately determined. This is especially useful where each impact changes an area in turn. For example the area around Tharsis Montes may have been altered by all four impacts.

The age of Isidis could be initially estimated by comparing the relative age of the Isidis and utopia Basins. This in turn may date some of the changes to the Solis, Syria, and Sinai Plana, and perhaps earlier changes to the future Tharsis and Valles Marineris. Geologically it is easier to calculate the ages of these formations by showing how craters counts are changed by resurfacing.

The age of Argyre may have formed the Tharsis Montes and Olympus Mons. If so then craters on them may help date Argyre Basin. The beginnings of Valles Marineris, Candor, and Ophir Chasma could be dated to shortly after the Argyre impact. After this the water channels of Lunae Planum, Xanthe Terra and Margaritifer Terra could be estimated.

In turn this can be compared against the age of the dichotomy boundary which would be formed later. This may in turn allow the time of the formation of the northern lowlands to be determined, if much was formed after the Argyre impact.

The age of Hellas would move the pole from Lucus Planum northwards and begin the formation of Alba Patera. This may also date the restarting of Olympus Mons and Tharsis Montes from the shock waves of the impact. The combined heat from these volcanoes may have resurfaced the northern lowlands.

The current age may be dated from the time the existing poles were formed.

Narrow angle images

Previous poles have left many changes on the Martian surface. To examine smaller scale changes I have examined 730 MOC narrow angle images[164] out of a larger randomly acquired collection[165], separating them into various kinds of formations such as water signs, dunes, and layers. These were accumulated over several years, before the ideas in this paper were conceived so there is no relevant unconscious bias in their selection.

Fluid signs

The collection of fluid signs[166] was first examined in reference to **Pole 4**. This was done by converting the coordinates of each image to its latitude under **Pole 4**[167].

This gave a list of coordinates between 0, equal to 90 degrees old north and 180 which is 90 degrees old south.

Fluid signs gave an unusual distribution[168] with a large number clustering around the **Pole 4** equator. On further examination these were around 37 to 43S and 140 to 200W, and 39N 19W.

The first cluster is on the bottom edge of an ice ridge area identified by Odyssey[169], which abuts the old equator at these coordinates. This was previously identified as **Pole 3**. The second cluster abuts another ice rich area on the top, again on the old equator and also near the dichotomy boundary. This is at the opposite **Pole 3**.

The water could have several possible origins. Some ravines could have formed on this pole, but this is unlikely as it should be very ancient and the water flows are more recent[170]. The second possibility is these flows occurred on the equator of **Pole 4** because it was warm enough there for some water to melt from the remaining ice of **Pole 3**. In that case we may be seeing relics of this flow. The third possibility is we are seeing ice from **Pole 3** melting because the current latitudes are suitable for this. In all cases this implies the source of the water was ground water or ice in the soil from **Pole 3**.

Since the water only seems to have been flowing in small areas it seems likely that these areas started flowing on the equator in **Pole 4** in a restricted area close to the equator. Though the pole has shifted channels in the ground are still connected to this ice, and at certain mid latitudes this can still flow[171] [172]. As we will see the positions of gullies is also correlated with layering[173] relative to **Pole 4**. Hartmann et al[174] compare these gullies with Icelandic gullies, which is consistent with the idea of seepage from a former polar area.

Others seem to cluster around 35S 270W, and 29S 38W. The first cluster is on the north eastern edge of Hellas Basin and may represent an area on the eastern edge of the **South Pole 4** ice cap.

The second cluster is close to the south western edge of the **South Pole 3** ice cap and so is likely to have formed with the same mechanism. This then gives all 4 main clusters the same mechanism, of ice from an old pole now at mid latitudes suitable for a water flow.

Another cluster is at 37 to 53S 320 to 356W. This is close to **South Pole 4**. Again the mechanism seems to be a part of the ice from an old pole position which corresponds to the suitable latitude for water to flow now.

It would seem then the reason the craters with water signs are so rare is that old polar ice deposits and suitable current mid latitude temperatures only coincide at a few positions.

Schmidt[175] points out additional areas with craters, which are also examined.

Gorgonum Chaos[176] [177] is situated at[178] 37S 173W, which would be on the southern edge of **North Pole 3**. Wilson et al[179] believe these gullies are not so young and represent only a limited discharge. This would be consistent with the older position of **Pole 3**, which still had to move to **Pole 4** and on to the current **Pole 5**. One theory according to Moore et al[180] is that these knob fields were formed by water and near a lake, which would also be consistent with a previous **Pole 3** position.

Stewart et al[181] examine the idea that CO2 could have formed these channels, which is also consistent with a pole position. This would depend on the temperatures of **Pole 3** at the time. As seen earlier the pole may have cooled in its movement from **Pole 2**, as the number of channels may have diminished. The temperatures may have risen after the Hellas impact as the Pole moved northward to the **North Pole 4** position around then to be formed Alba Patera. Many of the objections to CO2 arise from not having a mechanism to maintain enough of it long enough to create these formations, however a pole may be able to supply this reservoir. There may be then in some of these areas a mixture of the two kinds of erosion[182], without knowing the temperature at the time it is probably not possible to determine. Other ravines may be more recent[183].

Newton Crater[184] is situated at 41.1S 159.8W, also on the edge of the **North Pole 3** area. Cabrol et al[185] refer to the large amounts of water released into Newton Crater, which represents the dilemma of abundant water on such a dry planet. Being on the edge of a pole however could supply this water while non polar areas could remain dry.

Tempe Terra[186] [187] is found at approximately 42N 73W. This area would again have been on the edge of an ice cap, in this case **North Pole 4**. Volcanism[188] in the area may have come from the Hellas impact, like Alba Patera. Hauber et al[189] point out chaos similar to Gorganum Chaos, which is also on the edge of a former pole. Syria Planum, associated with **South Pole 2** also has these formations. Even though these three formations are relatively close to each other, each could have been formed from a different pole. They also show signs of glaciation, which is consistent with the edge of a pole.

Nirgal Vallis[190] [191] is approximately at[192] 27.5S 317W, on the edge of **South Pole 4** near Hellas Basin. This may account for its appearance of having water more recently. Lee and Rice[193] compare Nirgal Vallis to Devon Island, and find indications of the decay and retreat of an ice cover, consistent with the edge of a Martian **South Pole 4**.

Hale Crater[194] [195] is located at 36S 37W. This may have experienced outflows from Argyre basin when **South Pole 2** was moving eastward. Alternatively it could have received water from **South Pole 1** and the movement to **South Pole 2**. This could have been a slow movement and generated a lot of water in this area.

Maunder Crater is on the south western edge of **South Pole 4**.

Rabe crater is situated at 44S 325W, which is around the area of **South Pole 4**.

Dao Vallis is located at 33S 266W, on the north eastern edge of **South Pole 4**. Arfstrom[196] interprets the area as glacial, which is consistent with being on the edge of a pole. He believes[197] this may be part of a larger ice flow.

All these areas[198] referred to by Malin et al are on previous pole positions. This makes it likely groundwater rather than snow is the cause of these gullies and viscous flow features[199], though snow[200] [201] may also be forming more easily in shapes made by flowing water. In some cases CO2[202] from previous poles may also have formed gullies.

Schorghofer et al[203] map locations of slope streaks, and one cluster is approximately on **South Pole 3**. This may imply they are indeed related to water. Other areas tend to follow the path of the opposite **North Pole 2** to **North Pole 3**, and then part of the movement to **North Pole 4** around Alba Patera. Residual ice in the soil may also make it conducive to landslides.

Palermo et al[204] believe these are more likely to be fluid flows. No actual gullies seem to be formed with streaks though a fluid reservoir implies they should flow over and over at least sometimes. The shapes probably represent the path to lower elevations water or dust would take.

Over time water should make some kind of mark or channel on the surface like we see in even in Russell Crater on sand. Moisture from underground ice deposits could cause landslides, and explain the link between old pole areas and the streaks. Sullivan et al[205] point out the streaks are similar to snow avalanches, which may be consistent with icy soil under a dusty surface layer. It may be fluid flows and dust avalanches both occur at times. Sullivan et al[206] refer to dust avalanches as the most likely explanation.

Another possibility is a small area of ice sublimates and the resulting vapour dislodges some dust and creates the landslide. If so then an example of this may have been imaged. MOC photo AB102003[207] shows a possible plume of vapour found by Spires, colloquially known as "Dan's smoker"[208]. There is a slope streak in this image pointing in the same direction, and other images in the area such as M2300332[209] contain slope streaks. The pale mark seems to go down one mound and then climb the other, which makes it unlikely to be a streak. If this was reimaged and the light streak was gone then it is more likely to have been vapour. E1103683[210] just misses this formation.

Hematite[211] has been found in the area of **South Pole 3**, which is consistent with the having water around a polar area. The area is believed to have been recently exhumed, by Lane et al[212] which is consistent with the pole moving and exposing this area. According to Hynek Aram Chaos and Valles Marineris[213] [214] also have hematite deposits, which is consistent with the path of the moving pole from **South Pole 2** to **South Pole 3** giving water to create hematite. Aram Chaos[215] is to the west of the **South Pole 3** position and is connected to the Areas Valles outflow channel. This would also be consistent with the pole depositing water as it moved.

Dunes

Images of dunes[216] were also compared to their **Pole 4** latitudes[217]. Dunes were generally found to be evenly spread around the current pole longitudinally and to be confined within 40 degrees north and south of the current equator. The distribution in regard to **Pole 4** was also even but generally from 50 degrees north to 50 degrees south. This is an usual correlation as **Pole 4** is at 45 degrees south, and so the distribution should be skewed. This implies that dunes formed in a band between 50 degrees old north and 50 degrees old south in relation to **Pole 4**. In turn this implies[218] a high enough air pressure for dunes to form at the time. After the pole shift the dunes would appear to be moving to the same kind of formation relative to **Pole 5**.

Dunes on the current **South Pole 5** were also examined[219]. These dunes were typically found between 75 to 85S and 100 westward to 350W. In relation to **South Pole 4** all of these dunes were found to be between 30 and 50 degrees old south (that is, south compared to **Pole 4**) which implies they were actually formed before the Pole shift of **Pole 4** to the current **Pole 5**. If so then these dunes may have been frozen in position since then. Malin et al[220] say there is no evidence dunes on Mars are presently mobile.

Layers

Layer images[221] were also examined[222] and mainly fell in the old southern hemisphere of **South Pole 4**, particularly from the equator to 50 degrees old south (old south is relative to **Pole 4**). This is surprising as it implies much of the layering shown was actually formed under **Pole 4** though earlier layers may be buried.

Layers found on the current South Pole, **Pole 5**, were typically at 85S and concentrated at 175 to 275W. This in relation to **South Pole 4** placed them between 30 and 50 degrees old south, also implying they were formed in the time of **Pole 4** rather than on the current pole. Malin et al[223] point out layers in Candor Chasma, which is near the equator of **Pole 4** seems relatively young compared to nearby Arsia Mons. The absence of craters may imply these layers were made in the time of **Pole 4**, or perhaps exhumed by processes in that time.

If so then much layering[224] has been a relatively recent occurrence associated with the Hellas impact, perhaps because older layering may have been buried. **Pole 4** may have existed for long enough for layers[225] under it to dominate.

Dunes and layers have a similar distribution in the southern hemisphere of **Pole 4**, while water signs seem to be related to different poles. It may be then that air rather than water is responsible for many of these layers. CO_2 may also have formed some layers, if the temperature dropped enough. The change from **Pole 4** to **Pole 5** may have occurred when the temperature dropped enough to freeze the air, so ice sublimated and went to the poles.

If the air pressure dropping was enough to make the pole move to **Pole 5** then that implies it had not dropped to that level earlier, or the pole shift would have already happened. The dropping of the temperature then may have represented a final phase change, or the temperatures may have periodically risen with changing obliquity. If so then the pole may have wandered[226] to some degree between the positions of **Pole 4** and **Pole 5** depending on the level of ice depositions in the northern lowlands.

In this time CO_2 may have been involved in the formation of layers[227]. Odyssey thermal data[228] implies layers were not formed by water. Layers are found in the area of **Pole 3**[229], which may have formed in that time just as some layers form on the poles today. Layers could have formed under all the previous poles; it may be that **Pole 4** dominates because it is the most recent.

Some layers may have been formed in the movement of **Pole 2** to **Pole 3**, and because this is on the dichotomy boundary this will appear in the equatorial region of **Pole 4**. Also since Tharsis Montes, Olympus Mons, Hecates Tholus, Elysium Mons and Alba Tholus all are near the **Pole 4** equator in the time ash may form layers also appearing to be caused by **Pole 4**. Some layers may also have been formed along the equatorial region of **Pole 4** by water if the temperature was high enough.

In terms of astrobiology the impacts certainly warmed Mars, and depending on their timing may have created a climate that could have supported life for long periods.

In an ideal scenario the Utopia impact would have heated the planet for a long time, as ice and CO_2 forming in the Utopia Basin would have been warmed. Since even now Mars is close to being warm enough to sublimate all the CO_2 from the poles it is reasonable to believe that this impact basin at the pole would have made it warm enough to raise the air pressure, especially with higher obliquity[230].

If the Isidis impact occurred before the planet cooled then this would have added heat, as **Pole 2** forming on Isidis Basin would have sublimated CO_2, as would the pole opposite it around Solis and Syria Planum.

Next the Argyre impact may have warmed the poles followed by the Hellas impact. If these were spaced at the right times then Mars may have had a long time of higher air pressure and occasional liquid water. These conditions would possibly have allowed microbial and even substantially more evolved life forms to survive, especially since the higher air pressure would have offered some protection from radiation. An impact may have decimated life much like impacts did on earth[231] but subsequently kept the planet warm enough for the survivors to adapt. If the Hellas impact destroyed the Martian magnetic field then this could have led to the extinction of some life. Without sufficient life to create oxygen this may have led to the air freezing more as CO_2 and ice sublimating to the poles.

Eventually the impacts and volcanoes ceased and life if it still or ever existed would have had to adapt to the much colder climates.

The more likely scenario is that life may have evolved somewhat during these times of higher temperatures only to die off or survive at a low level when the planet cooled and the air froze. With the next impact some life may have evolved again if the times were long enough, perhaps being reseeded with meteors from Earth.

Conclusions

Much of the Martian terrain is consistent with the idea of polar wander. This may have been controlled by 4 main impacts, Utopia, Isidis, Argyre, and Hellas.

The Isidis impact may have formed some of the deformations around Solis and Syria Planum.

The impact of Argyre may explain the formation of the Tharsis Montes, Olympus Mons, Elysium Mons, Hecates Tholes, and Alba Tholus.

The movement of **Pole 2** to **Pole 3** may explain the formation of the dichotomy boundary and all the fluid channels north of it. It may also have formed the Northern Lowlands.

The number of narrow angle images examined here is too small to give a definite conclusion, but they give a picture of **Pole 4** having a large influence on many current Martian features. This may be an artefact of the selection process but they were accumulated well before the ideas in this paper.

Fluid signs in craters and valleys are found on the edges of older poles, which are now at suitable latitudes for water ravines to form.

Some dunes are found to be evenly spread compared to **Pole 4**, even on the current **South Pole 5** and imply a time of higher air pressure, and perhaps that since then there has been insufficient air pressure to move them from this pattern.

Layering may have partially occurred in the time of **Pole 4**, and some layers on the current South Pole may be from before the pole shift to **Pole 5**.

Equations and Monte Carlo simulations on the distance between stars with habitable worlds

By Ness, Peter. MS (Ba App. Sc. Ap. Geol.), MBA, Mfin, PHD student*

Abstract

We derive formula to calculate the function (fH_p) for the number of stars with habitable worlds in a galaxy. The density Dh of stars with habitable worlds is calculated. An equation is then derived for calculating the distance (Rh) between stars with habitable worlds.

The function (fH_p) for the number of stars with habitable worlds in a galaxy includes the probability that a star has a planetary disk, the number of planetary disks upgraded via supernova, the likelihood that a world lies in the "goldilocks zone" with the correct orbit, and whether the world has the correct elements to support/sustain life.

Monte Carlo simulations are run for the above equations and also on Drake numbers. The number of different species we are likely to encounter within a 250 light year radius of earth is also discussed.

Introduction

The distance to the next star with earth-like habitable planets or moons (a star with Habitable worlds) may determine the speed at which we colonize space. This paper provides formulas and Monte Carlo simulations that allow us to place more definitive parameters on the likely separation of stars with habitable worlds. It provides some first-pass estimates on the proportions of the numbers of habitable worlds and various types of life forms likely to be found within 250 light years (ly) of earth.

Fermi's Paradox

In 1950, Enrico Fermi (Jones, 1985[cxiii]) posed the question as to why, if a multitude of advanced extraterrestrial (ET) civilizations exist in our galaxy there is no physical evidence (spacecraft, probes etc). This is the Fermi Paradox.

Given ET exists we should be receiving signals, not just from stars but also from other galaxies (Wesson, 1990)[cxiv]. If ET comes from outside our visible horizon, or beyond the local group of galaxies, they may be too far away to communicate and travel is likely impossible.

Thus, Fermi's Paradox is only a paradox if numerous ET exist in our galaxy, or in the local group of galaxies. Any messages received from galaxies outside the local group may be red-shifted and hard to find or decipher. By the time ET receive light from our sun and/or signs of our existence and/or attempt contact we may be like the dinosaurs - extinct.

The Drake Equation

In 1960, Dr Frank Drake developed an equation in an attempt to solve the problem with the Fermi Paradox. In the following year, Drake and his colleagues agreed that at any one time up to 10 advanced civilizations might co-exist in our galaxy (Sagan and Drake, 1972[cxv]).

Recent work by Cotta and Morales, (2009)[cxvi] also confirmed 10 ET as the most likely number, but some of their assumptions are not very realistic. For example, given large separations between stars and the short period in time that civilizations tend to last on Earth (e.g. typically <1,000 yrs), it is questionable whether ET would continue for millions of years. In contrast Drake used a figure of 10,000 years for the average time span of a civilization.

There are over 200 billion galaxies, many with over 200 billion stars, with the distribution of the "seeds of life" abundant throughout the universe (Wickramasinghe, 2009)[cxvii]. This is sufficient to convince many people that Drake number of D=10 is correct, even though there is no real scientific evidence to support it. In fact, continued negative results from over 30 years of SETI strongly imply that either we are looking for signals in the wrong place, or the Drake equation overstates the number of ET.

With such a large number of galaxies in the known universe, even if the Drake number (D) were as low as 1 in 200 billion that still does not preclude our existence. Given the number of UFO sightings, the conclusion that ET are few and far apart is difficult for many people to accept.

If the Drake number is 10 or greater then it may place severe restrictions on space exploration. We may well share the same space as another ET;

whom may not want us wandering around in territory that they have already claimed: They may see us as a potential threat which requires eliminating.

On the other hand, if the Drake number D <1 then there is no Fermi Paradox. We will be less concerned with upsetting a potentially (lethal) hostile neighbor and more concerned with finding the next habitable world so we can develop the technology to go there.

This paper shows that the Drake number is unlikely to be greater than 1. It provides equations both for the number of stars with habitable worlds and mean separations of those stars.

If ET is nearby then we will know before too much longer because we now have the technology to find earth-sized worlds, in the "goldilocks" zone. This is the distance from a star where the surface of a planet or moon would have comparable temperatures to earth and where liquid water should exist. If the vast majority of habitable worlds are unlikely to host any advanced life forms that is good news: provided ET is not too close we may be able to colonize some of these worlds.

Knowing the likely separation between stars with habitable worlds provides us clear objectives as to:

    a.   How many there are in our sector of space,
    b.   How long it will take to go there,
    c.   How easy it is to go there,
    d.   The speeds at which we need to propel spacecraft to make colonizing space a realistic possibility, and
    e.   The costs involved, which are likely astronomic.

Quite clearly, if there is some probability that a star with habitable world(s) lies within 4-6 light years from our sun (SOL) then there are only four targets. Two are G-type stars (Alpha Centauri A and B) and two are red giants (Proxima Centauri and Bernard's Star). So, we would know exactly where to look. We may even get as much funding as required. If a habitable planet (with water and an atmosphere we can breathe) is confirmed within 20-40 ly radii, the momentum to send a probe may eventually become unstoppable.

However, if the next habitable world is 200 to 500 ly away, or even 100 ly, then the budgets are as academic as getting there.

434

The aim of this paper is to calculate an equation to determine the average separation of stars with habitable planets in our (250 ly radius) sector of the galaxy.

Methodology

We need to know the number of stars with habitable worlds in order to calculate their average separation.

For the base case we assume that the number of stars in the Milky Way galaxy, $N^* = 200$ billion.

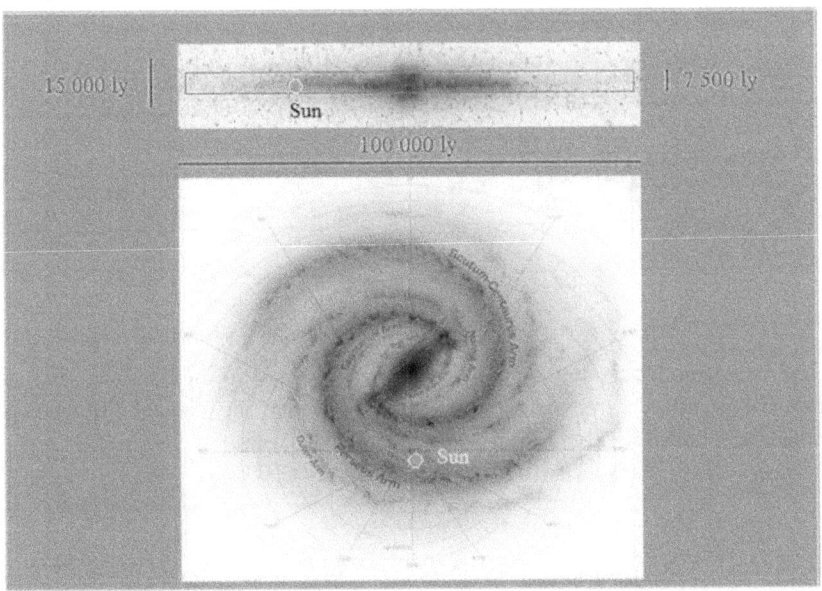

Figure 1 The radius of the galaxy r is 50,000 ly (21,118 pc). The average width w is 7,500 ly (3,178 pc). The volume is the area of the disk multiplied by disk width $= \pi \times r^2 \times w$, measured either in pc³, or ly³.

Volume and Density calculations

The next step is to calculate the volumes and density of the local space as compared to the galaxy as a whole. The distance between stars with habitable worlds will vary due to spatial density of stars at various places within the galaxy.

The density of stars in the Milky Way galaxy may be calculated as the number of stars divided by the volume (Fig. 1; Benjamin, 2008)[cxviii]. Stars in this spiral galaxy form a thin disk with diameter of 100,000 ly and with a central bulge of 15,000 ly. The radius of the galaxy r is 50,000 ly (21,118 pc). The average width w is 7,500 ly (3,178 pc). The volume is the area of the disk multiplied by disk width = $\pi \times r^2 \times w$, measured either in pc³, or ly³.

Dg = Density of Stars in the galaxy,

$\quad$ = Number of stars / Volume of space

$\quad$ = N* / (w $\times$ $\pi \times$ r²) $\qquad\qquad$ ....(1)

$\quad$ = 200,000,000,000 / (7500 $\times$ 3.14159 $\times$ (50,000)²)

$\quad$ = 0.003395 stars per ly

The density of local stars is calculated based on assuming a radial span of control in every direction. The local density for the closest 30 stars is calculated based on the volume of a localized sphere.

Ds = Density of local Stars

$\quad$ = number of stars / (4/3 $\times$ $\pi \times$ r³) $\qquad$ ....(2)

$\quad$ = 30/ (4/3 $\times$ $\pi \times$ 11.83³)

$\quad$ = 0.004326 stars/ ly

Including our sun, there are 30 stars within an 11.83 ly radius (sphere) of earth (Fig 2). The density of local stars is higher than the average star density in the galaxy, with almost 1/3 more stars in the same volume of space.

The density of stars within a radius of 250 ly = $260{,}000/(4/3 \times \pi \times 250^3)$ = 0.00397. This represents 8.8% fewer stars per volume of space than in our local area. This is an important concept and we will come back to it later on when we discuss the concept of *clustered habitats*.

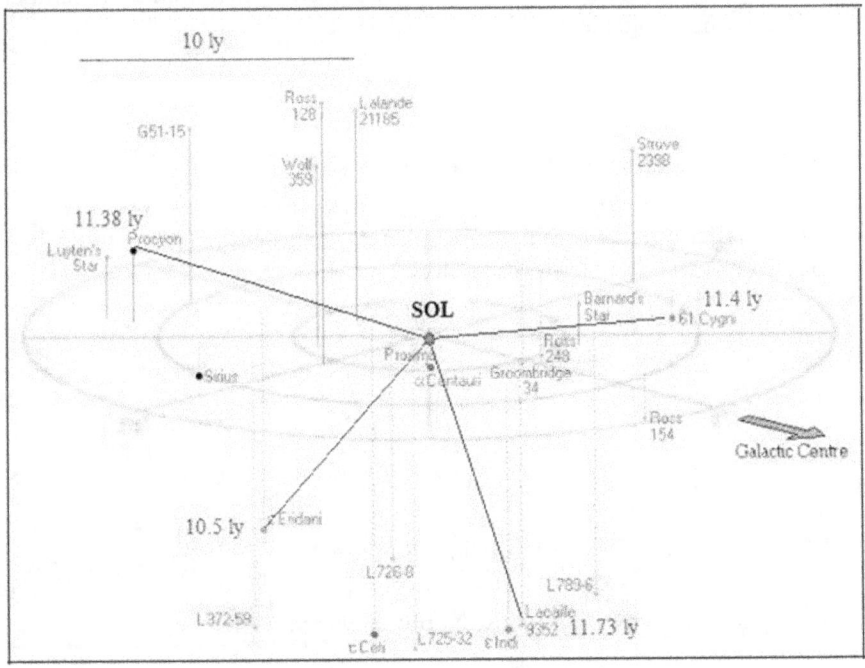

Figure 2 Including our sun, there are 30 stars within an 11.83 ly radius (sphere) of earth.

The Equations

There are two key equations. The first calculates the radial distance between stars with habitable worlds (Rh). This is given in equation (4), below.

Dh =Density of stars with habitable worlds
................(3)

437

$$Rh = [1 / [Dh \times 4/3 \times \pi]]^{1/3} \text{ ly} \qquad \qquad \ldots\ldots\ldots\ldots(4)$$

To calculate Rh we need to know the density of stars with habitable worlds (Dh). In order to calculate Dh, we need to determine the number of stars with habitable worlds in the galaxy, from equation (5) below.

Number of stars with habitable worlds

The number of stars with habitable planets in a galaxy fH_p, is a probability-based equation. It is derived as follows:

$$fH_p = \{ N^* \times fp \times fs \times fd \times fi \} \qquad \ldots\ldots\ldots\ldots(5)$$

This equals the number of stars in the galaxy $N^* = 200$ billion,

multiplied by each of the following probability variables:

a. The proportion of stars that have planetary disks $fp = 50\%$
b. The proportion of stars (planetary disks) with the correct chemistry to support life fs *(this is a supernova-related function)*
c. The probability fd that a planet/moon surrounding such a star is in the habitable zone i.e. the probability that a planet is in the correct place in a stable planetary system, with a circular orbit, and is able to support life
d. The proportion of those planets/moon in the habitable zone that have the correct elements to support life that then go on to develop life $fi < 100\%$ (certainly much lower)

Readers will observe that equation (5) is structured similar in form to the Drake equation but focuses on the astronomical requirements required to make habitability possible and for life to be sustained on a world.

438

The calculations and/or assumptions for each of the above parameters will be discussed in detail, but first we need to explain the role of supernovas in making a world habitable.

Upgrading by Supernovas

The supernova function fs = {[Density of Stars x $(4/3 \times \pi \times r^3)$] x ([(age of universe x $SN/100$)]) + $fs*$ ....(6)

For a planet or moon to be habitable it needs to have the right proportion of elements. These are derived from supernova explosions in a star nebula (nursery). This paper does not assume supernova seed life - there are arguments against it (Burton, 2009).[cxix] It only assumes that supernova provide the prerequisite chemistry required for life and to sustain it.

C-type asteroids are derived from supernova explosions. These may have occurred just prior to, or just after, the collapse of the dust cloud that created the sun and its proto-planetary disk.

C-chondrite meteorites are the equivalent of C-type asteroids that have impacted with our planet. The metals from C-chondrites are derived from comets that collide with the earth. Impacts upgrade the planet with the elements (Fe^{60}, water, heavy metals etc), which are essential for life as we know it. C-chondrite compositions match that of the pre-solar disk. Therefore, supernovas are a requirement for a habitable world to form.

Supernovas are often found in giant molecular clouds. Stellar evolution typically begins in gravitational collapse of a giant molecular cloud. GMCs are typically 100 light-years or more across with around 6,000,000 solar masses (Ungerechts, et. al., 2000[cxx]; Prialnik, 2000)[cxxi]. As stars form in the GMC some grow too fast become unstable and turn supernova.

Individual supernovas partly enrich many proto-planetary disks - and dust clouds that coalesce into stars. The rate that supernova occur, SN, depends on the supernova and host galaxy type according to the Hubble sequence.

The number of supernova events SN is a function of the number of stars in the galaxy, dispersion of star nurseries and its age. For younger spiral galaxies SN is smaller by a factor of 2 (Vanbeveren et. al., 1998).[cxxii] The outer spiral arms of M31 and other galaxies have a higher proportion of metal-rich stars than the Milky Way (Hammer et. Al., 2007)[cxxiii]. This may be a result of galactic collisions. Therefore, although we assume that supernova (*SN*) form at a constant rate, it is likely not constant.

The average supernova rate in the Milky Way is $SN =1.9$ each century (Diehl et. al., 2005)[cxxiv].

Stars that go supernova (*SN*) are typically very massive (75-100 M), found in clusters of over 10,000 members (Williams and Gaidos, 2007) [cxxv]. Large massive stars that go supernova are more likely to cause enrichment due to their shorter main-sequence lives.

A typical supernova would create on average 4 new solar masses and cover a 10 ly (Fig. 3) region of space (Stephenson and Green, 2005[cxxvi]; Hartmann et. al., 2002)[cxxvii]. For stars to have the same level of enrichment as our solar system they need to be within 0.4 pc (for a 100 M progenitor). 1 pc equals 3.26 light-years, so r=0.4pc, and 0.4 pc = 1.304 ly.

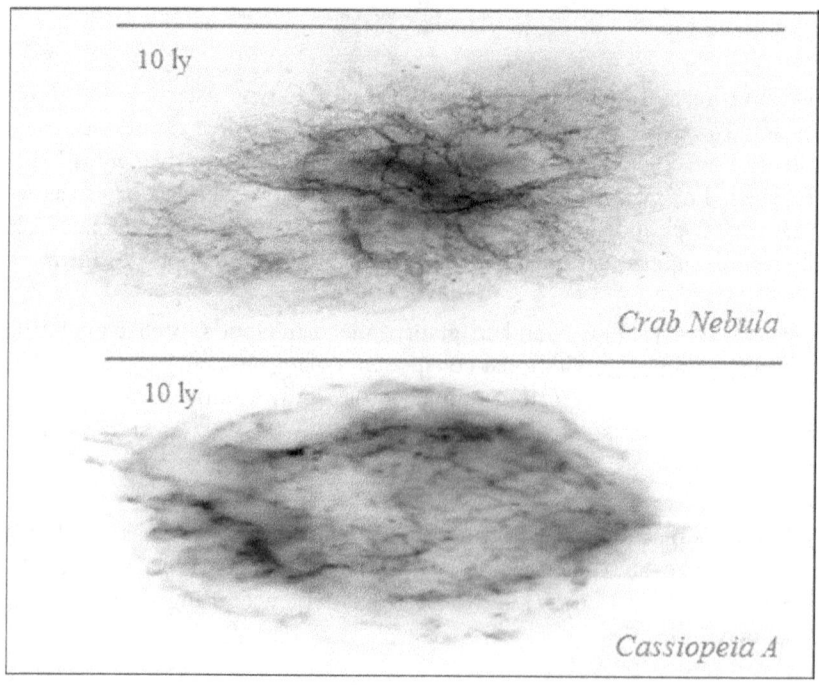

Figure 3 A typical supernova would create on average 4 new solar masses and cover a 10-ly region of space.

In practice, only a few percent of planetary disks are close enough to a supernova to be destroyed and some of the interstellar debris will eventually coalesce to form future stars. Even though supernova from some stars (O stars) can rapidly photo-evaporate the outer radii of proto-planetary disks (Storzer and Hollenbach 1999), sufficient mass to form planetary systems may remain bound to the star. Thus, a planetary system hit by a supernova would typically be enriched but not destroyed.

The calculations for each of the variables in the equation (5) will now be explained.

Calculations of parameters for Number of stars with habitable worlds

Equation (5) is as follows:

$$fH_p= \{ \ N^* \times \ fp \ \times fs \ \times fd \ \times fi \ \} \qquad \dots\dots\dots\dots(5)$$

Calculation of N*

We have already explained N*, which is the number of stars in the Milky Way galaxy. We have used a value of $N^* = 200$ billion as a base but have provided simulations for $N^*=100$, 200, 300 and 400 billion stars. While the number of habitable worlds changes significantly, there is not a great deal of difference in the separation between them by changing N^*.

Calculation of fp

At least 50 to 60% of all stars have planetary disks $fp = 50\%$. It is also possible to calculate fp by assigning a probability function: fp (Mn=0.1 , ML=0.6, Mx=1).

Calculation of fs:

There are roughly 6,000,000 solar masses in a GMC. From equation (2), the density, Dp, of proto-planetary stars in a typical GMC nursery is:

Dp =Density of proto-planetary Stars

$$= 6,000,000 \ / \ (4/3 \times \pi \times 100^3) \qquad \dots\dots\dots\dots(7)$$

$= 1.43239$ stars$/$ ly

This is an average sized GMC, so will give an "average" result. As a comparison, the density in NGC 6397 globular cluster is almost 1 million times greater than our local region (Cool and Nell, 2003)[cxxviii], so would give a biased result.

Our 13.6 billion year old Milky Way galaxy which produces $1.9 SN$ $/100$ years would produce a total of:

$SN* = 13,600,000,000$ yrs x $1.9/100$ yrs $= 258,400,000$ supernova events.

Assuming that the SN rate is constant (at $SN=1.9/100$ yrs) and the larger proportion of supernova occur in star nurseries such as GMC, then from equations (2) and (7) the number of directly enriched stars by supernova, fsa is:

fsa $=$ Density of local Stars x $(4/3 \times \pi \times r^3) \times SN*$................(8)

where $SN* =$ [age of universe x $SN/100$] ....(9)

fsa $=$ Density of Stars x $(4/3 \times \pi \times r^3) \times 258,400,000$

$= 1.43239 \times (4/3 \times \pi \times 1.304^3) \times 258,400,000$

$= 3,437,754,082$ stars initially enriched

This equates to 1.719% enrichment of stars in a 200 billion star galaxy.

The assumption is that all stars in the 1.304 ly volume of supernova space are upgraded. This calculation does not include any dust in that envelope that eventually coalesces into stars, or stars further away that may get partially upgraded. It excludes upgrading by supernova that are not in a GMC. Therefore, we need to adjust the equation to make it more realistic.

Supernova enrichment of proto-planetary disks affects only a small proportion of stars in the galaxy, but multiple supernovas may enrich many disks. If say another 1% of other stars in the galaxy $fs*$ are partly enriched by multiple supernova, or coalesce from enriched gas, then the total number of enriched stars that may produce planets that may support life is 5,437,754,082 stars or 2.719% of stars.

If we double, halve, or triple this value it changes separation of stars with Habitable worlds by <1 Stdev, so the figure is quite robust.

Equation (6) used here to calculate fs, the number of stars with correct elements to support life is:

$fs = fsa + fs*$

= {[Density of Stars x (4/3 x π x r^3)] x [age of universe x SN/100] + $fs*$

.............(6)

These equations do not differentiate by spectral class.

Now lets compare our value of fs with an approximation:

Kaler, (1999)[cxxix] estimates that 4% of Milky Way stars may be F-type, 9% G-type, and 14% may be K-type stars. If 10% of these stars have been upgraded by supernova or formed from upgraded dust coalescing then fs = [0.1 x (0.04 + 0.09+0.14)x 0.6 = 1.62%. Around 90% of stars are main sequence M-K type stars, of which roughly 74% are M-type (Delfosse, 2004)[cxxx]. If some M-type (1 in 100 = 0.74%) were able to support life, then 0.0164 + 0.0074 = 2.38% of stars would have the correct elements and be of the correct spectral type.

This gives an error of around 12% between the calculated value and the approximation, so either method could be used.

Calculation of fd:

The key variable widely used in the search for habitable planets is the probability whether a planet/moon is in the habitable zone, fd. This should account for the probability of a planet being in the correct location based on star size, stability, radial velocity, and eccentricity of orbit etc.

This value also needs to exclude planets or moons that are too small to hold on to their atmosphere, lack a magnetic field, ozone layer, surface water, or those affected with a run-way greenhouse (like Venus), or desertification (like Mars).

It is important to recognize that our own habitable zone could support two viable planets within the orbits of Venus and Mars, and a Jupiter sized planet in this same zone could support a number of habitable moons. This increases the probability of finding a habitable world in this "goldilocks zone". So far very few planets have been found in the habitable zone of nearby stars, so the probability is lower than expected.

It is difficult to accurately calculate the value for fd with our current knowledge, but based on the number of known planets the proportion of planets in the habitable zone it is in the range Mn = 0.01, ML = 0.1 and Mx = 0.40, so we can assign it a most likely value or estimate it by using a probability distribution.

Calculation of fl:

The Drake equation assumes that the proportion of those planets/moon in the habitable zone that have the correct elements to support life that then go on to develop life f_l = 100%. The scientists selecting values for the Drake equation were almost certainly in "Group think" mode: many of their choices for "average values" were not at all conservative. In this case: they chose the maximum value possible as the base number for f_l.

The value cannot be more than 100% and may be considerably less. There are numerous reasons why life may be precluded from forming on a potentially habitable world:

a. The star may expand into a red giant or go supernova,
b. Radiation levels may be too high,
c. Solar flares may wipe out life,
d. The elements to support life may be insufficient,
e. The gravitational attraction of the Jovian parent may destroy a moon,
f. A meteorite impact may turn the habitable world into a hot house etc.

The value may be less than 1, but again it may be assigned a probability function: fl (Mn=0.01 , ML=0.1, Mx=1).

Calculation of the number of Stars with Habitable worlds, fH_p

Equation (5) for the number of stars with habitable worlds in our galaxy was derived using the parameters explained in sections 1) to 5) above.

There should be around:

$$fH_p = \{ \; N^* \times \; fp \; \times fs \; \times fd \; \times fi \; \} \qquad \dots\dots\dots\dots(5)$$

$$= 200,000,000,000 \times 0.50 \times 0.02718863878 \times 1 \times 0.1$$

$$= 271,900,000 \text{ stars with habitable worlds}$$

The proportion of stars with planets that may be habitable in the galaxy is:

$$fH_p \; \% = fH_p / N^* \qquad\qquad \dots\dots\dots\dots(10)$$

$$= 271,900,000 / 200,000,000,000 = 0.0013595$$

This equates to 0.136% of stars, or 1 in 736 stars. Some stars may have multiple planets or moons in the habitable zone, so this figure may be conservative. Equation (5) counts only the number of stars – not the number of habitable worlds.

Calculation of the Separation of stars with habitable worlds

We need to use the density information from earlier equations to calculate the separation Rh of stars with habitable worlds given by equation (4).

To determine the density of habitable stars in the galaxy we use the values calculated from equation (1): the Density of all stars in the galaxy $Dg = 0.003395308$ stars / ly, and those from equation (10): the proportion of stars with planets that is habitable $fH_p \% = 0.0136\% = 0.0013595$.

The density of habitable stars in the galaxy is the proportion habitable multiplied by the star density of the galaxy:

$$fH_D = Dg \times fH_p \% \dots\dots\dots\dots(11)$$

$$= 0.003395308 \times 0.0013594$$

= 0.000004616 density habitable stars /ly

The density of stars in the galaxy Dg = 0.003395308 stars/ly and from equation (2) the density of local stars is Ds=0.00432 stars/ly. Thus, there should be M=1.272X as many habitable stars in our immediate vicinity as compared to the average in the galaxy. In our immediate sector of space 0.1730% of all stars should contain habitable worlds (1 in every 577). In a 250 ly radius of earth the figure drops off slightly to 1 in 629 stars or 0.159% (M=1.16926X the galaxy), still well above the average for the galaxy.

There are 260,000 stars in a 250-ly radius of SOL. Given the density of stars with habitable worlds we can calculate the radius in which we would find "X" stars with habitable worlds.

Equation (2) was:

Ds = Density of local Stars

$$= \text{number of stars} / (4/3 \times \pi \times r^3) \quad \text{................(2)}$$

Dh is the density of stars with habitable worlds. By re-arranging equation (2), there should be 260,000 stars with habitable worlds within in a radius (Rh):

$$Rh = 260{,}000 / [(\text{Dh} \times (4/3 \times \pi))^{1/3}$$

Therefore, using equation (4) we can calculate the radius in which we would find "X" habitable planets. If X=1, then:

$$Rh = [1 / [Dh \times 4/3 \times \pi]]^{1/3} \text{ ly.................(4)}$$

$$= [1 / [(0.000004616 \times 1.16926) \times 4/3 \times \pi]]^{1/3}$$

$$= 35.36 \text{ ly on average}$$

446

Therefore, in our galaxy of 200 billion stars there must be 1 habitable planet every:

$$Rh = [1 / [0.000004616 \times 4/3 \times \pi]]^{1/3}$$

$$= 37.25 \text{ ly on average}$$

Separation is log-normally distributed, so halving the density of stars with habitable worlds increases separation by 9.7 ly. Doubling density of stars with habitable worlds reduces the separation by 7.7 ly (=0.5 of a Stdev). Changing a single parameter by a factor of 10 times results in a change in separation of 1 Stdev, or 15.35 ly.

The separations are sensitive to large decreases in a single parameter value: >5 times decrease = 63.7 ly; >10 times decrease = 80.3 ly; >100 times decrease = 173 ly; and >1,000 times decrease = 373 ly.

Monte Carlo Simulations

Based on the above equations, Monte-Carlo simulations were carried out for 10,000 iterations and 1,000 trials (10,000,000 data points). The minimum and maximum bounds assigned to uncertain variables were:

a. fp is the fraction of stars that have planetary disks: fp (Mn=0.1 , ML=0.6, Mx=1)
b. fs is the proportion of planetary systems with elements for life: fs (Mn=0.0021719 , ML=0.02719, Mx=0.25)
c. fd is the probability that a star has planets in the habitable zone fd (Mn=0.01 , ML=0.2, Mx=0.4)
d. fl is the fraction of those planets that actually go on to develop life at some point: fl (Mn=0.01 , ML=0.1, Mx=1)

The sensitivity was confirmed by first setting all values at their minimum, then at their most-likely, and then at their maximum values. If all values are set to a minimum the extreme separation is 1,373 ly, which accounts for << 0.001% of the distribution tail in each simulation run.

Even in extreme cases, reducing fs by +100 times and removing all constraints, the 95% CL of distributions is <120 ly and the mean is always <50-60 ly.

Fig 4, shows the distributions of some of the parameters based on simulations.

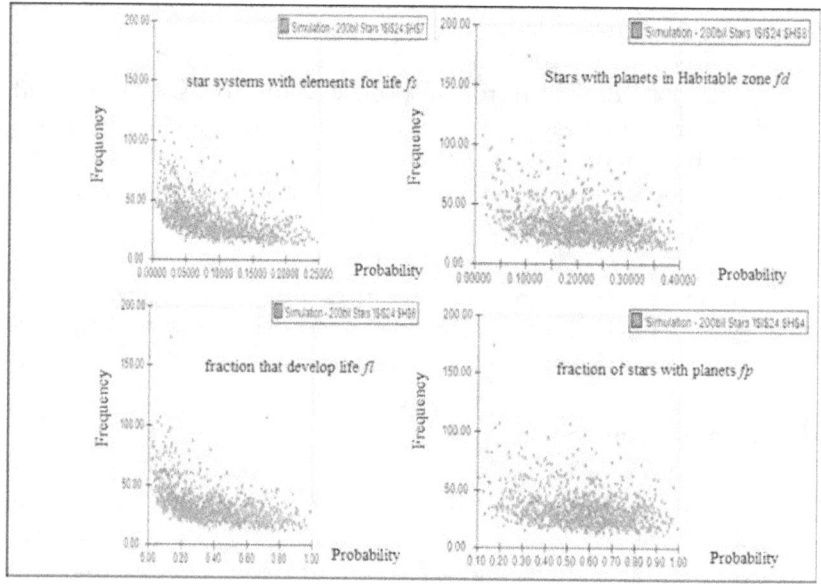

Figure 4 The distributions of some of the parameters based on simulations.

Simulations were run for 100-400 billion stars in a galaxy, with 200 billion stars being the base case, using the constraints given to set up simple distributions for each unknown variable.

The upper bounds for uncertain variables fp and fl were set at the numbers frequently used in the Drake equation. The lower bounds are much lower. The models were constrained so that negative values or those greater or less than the probability distributions were excluded. This has the effect of removing extreme values, such as those in the 0.001% CL tails.

Simulation Results

Statistics for a 200 billion star simulation are given in Table 1.

The mean separation of stars with habitable worlds is 34.6 ly (Stdev = 15.35 ly) for 200 billion stars, with maximum separation of 173.5 ly (Table 1, Fig. 5). The results are comparable to the hand calculations.

Value at risk (VAR) 95% is 62.2 ly, so 1 in 20 stars with habitable worlds would be at least 62.2 ly apart. The maximum separation of habitable worlds in 99% of cases is < 89 ly.

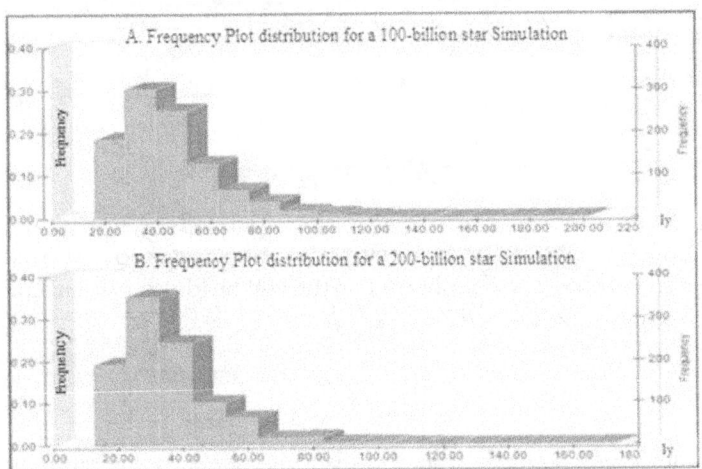

Figure 5 Mean separations for stars with habitable worlds for 100 and 200 million star galaxies. The main difference is the higher proportion being grouped to the middle of the distribution in the 200 billion star simulation model.

The separation for stars with habitable worlds for a 300 billion star galaxy (Fig. 6) is 30.3 ly with 99% of habitable worlds within 75 ly. For a 100 billion star galaxy (Fig. 5) the mean separation is 43.6 ly with 99% < 116ly.

For all the models, even totally unconstrained models give 95% of separations <120 ly. There should be six stars with habitable worlds within this distance of earth.

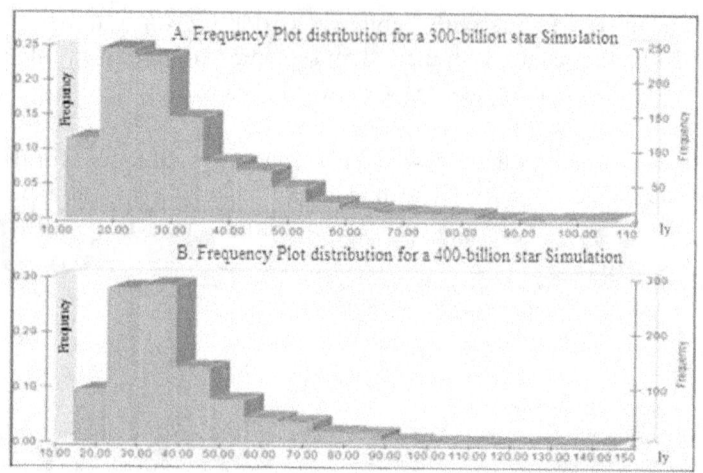

Figure 6 Mean separations for stars with habitable worlds for 300 and 400 million star galaxies. The distribution of the 400 billion star simulation is beginning to normalize around 25-45 ly.

Table 1 Statistics for a 200 billion star simulation. The mean separation of stars with habitable worlds is 34.56 ly, with an error of 15.35 ly.

450

Statistic	Value
Mean	34.56
Standard Deviation	15.35
Variance	235.4715455
Skewness	2.085666785
Kurtosis	8.927538874
Mode	28.67
Minimum	12.55
Maximum	173.5
Range	160.95
Mean Abs. Deviation	11.02
SemiVariance	67.34664266
SemiDeviation	8.21
Value at Risk 95%	62.23
Cond. Value at Risk 95%	32.21
Mean Confidence 95%	0.95
Std. Dev. Confidence 95%	1.57
Coefficient of Variation	0.443969262
Standard Error	0.49

Discussion

Problems and weakness with this method

Calculations of mean separation of stars with habitable worlds may be overstated or understated, but probably not by much. The equations are quite robust. Around 9% of stars in the galaxy are G-type stars, so if we simply plugged 60% of this value into the equation we would get separations of 29.6 ly. If we include other main sequence stars the separation decreases even further. If we exclude all stars less than 1 billion years old separations increase, but not greatly.

Life may take longer than a billion years to form. Alternatively, it may evolve much more rapidly than on earth. Life itself may even terra form a world and make it habitable, or it may flourish with limited elements plus a heat source. Not all life is carbon based and that should not be a pre-requisite.

Bacteria may be anaerobic and may not require earth-like atmospheres or habitable worlds. Sulfur-based forms also exist and they form around fumaroles of volcanoes in the ocean. These forms of life may represent the largest proportion in the universe and may go on to form advanced species. However, most would not pass the test of habitability. Terraforming these worlds may not be a viable option. We likely cannot live on, or colonize them. In addition, some worlds may not have been upgraded by supernova sufficiently for life to take hold or survive long-term.

A supernova is unlikely to upgrade the elements in a star more than a few percent, and most sun-like stars may form from gases of past supernova coalescing. The surface area of a debris disk has a diameter of 20 to 40 Au – much greater than it's parent star. Thus, a supernova is many times more likely to upgrade a planetary disk than a star. As such, the chemistry of a debris disk may be far more important to formation of life than the chemistry of the parent star.

Some stars may die or leave the galaxy, while new ones form as galaxies collide and the density and SN number vary. Planets may get re-set, or move in or out of the habitable zone. ET may terra-form planets or destroy them. Planetary systems may brush or collide, destroying or seeding new worlds. Panspermia or exogenesis may seed life on worlds where it may not normally occur – even in gas clouds in space (Weber and Greenberg, 1985[cxxxi]; Crick and Orgel, 1973[cxxxii]; Mautner, 1985)[cxxxiii].

Panspermia

Theoretically, life could start in one area and spread, most likely in the direction of higher star density. The transfer of life from one planet (or planetary system) to another via exogenesis (panspermia) requires microorganisms to survive the escape process from one planet, the journey through space as well as the re-entry/impact process in another planet (Henrique, 2009).[cxxxiv]

It is estimated that, on average, roughly 150 kg of 'hospitable' ejecta from earth rocks reach Mars each year (Mileikowsky et. al., 2000)[cxxxv]. Approximately 7% of hitchhiking microbes can be expected to survive the journey. The total mass of terrestrial fertile material delivered to nearby pre-stellar systems as the solar system moves through the galaxy varies from kilogrammes up to a tonne (Wallis and Wickramasinghe, 2003)[cxxxvi].

Microbes can form spores, become dormant and can adapt to surviving journeys through space (Nicholson et al. 2000[cxxxvii]; Horneck, 1993[cxxxviii]; Horneck et al. 2002[cxxxix]). The chances of survival and colonization would be enhanced if microbes were lucky enough to land in a warm, moist, spot (Ness and Orme, 2002).[cxl] This may increase the probability of life in our local volume of space. Therefore, as soon as one planet or moon has life, then contamination of other planets and moons in that planetary system (and those nearby) may be inevitable.

An organism detached from the Earth and pushed out into space by the radiation pressure of our sun would cross the orbit of Mars after twenty days, the Jupiter orbit after eighty days, Neptune's orbit after fourteen months, reaching Alpha Centauri (A & B), in nine thousand years (Arrhenius, 2009)[cxli]. Spores would reach most of the 29 stars in the 11.83 ly radius of SOL within 25,000 years and, all else being equal, it may only take 515,000 years to contaminate a small fraction of stars in a radius of 250 ly.

Realistically, over the life of our sun and assuming a major impact every 50 ma some 228,800 star disks may be randomly contaminated. However, the chance of a habitable planet or moon at large distances being contaminated is quite low, but even with a 1 in 10,000 chance up to 23 habitable worlds may be contaminated.

Furthermore, even with probabilities of less one-thousandth of 1%, many organisms (e.g. spores) may make it to Alpha Centauri A & B or to other planetary systems within 10-15 ly intact; so given the galaxy's current star density, some cross-contamination between proximal stars may occur.

If life itself terra forms worlds - as it appears to have terra formed earth - then the numbers of habitable worlds may be much larger in localized regions of space. This would imply that once we find one star with habitable worlds there might be others clustered nearby (Fig. 7).

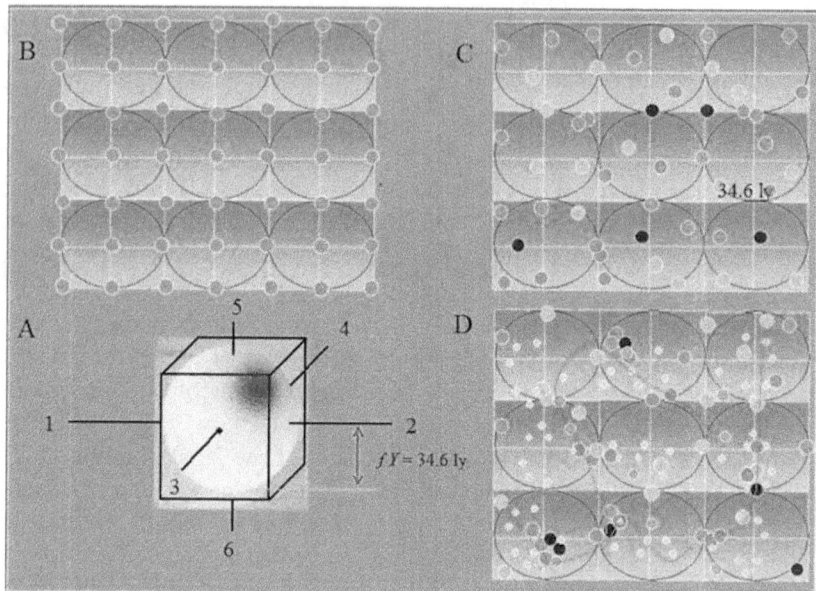

Figure 7 A. The calculations of Rh, the separation between stars with habitable worlds implies six stars with habitable worlds exist at the same radius. **B.** This is only true if all stars are equal distances apart. **C.** Stars will vary in their separation, as will stars with habitable worlds. D. Random walk theory suggests that clustered-habitats will form. Panspermia suggests that life (and ET) will permeate along the shortest route between stars.

Clustered-habitat Theory

Random walk theory suggests that stars with habitable worlds will not all be equally spaced. The closest habitable world will not be exactly 35 ly away. Some may be further apart and others closer together. If you find one then there may be a higher likelihood of finding two, or three or even more nearby – with a much larger gap to the next star with habitable worlds.

The density of stars in one area of space may be greater than the surrounding density of stars due to coincidence, because the stars have a shared history, due to higher proportion of binary or trinary stars, because these stars are in a safe-zone, or due to panspermia. In our local area of space the stars are clustered – and layered - due to a random walk (Fig. 8).

Once life develops on one world then it should soon spread via panspermia to surrounding planetary disks (Fig. 7) and along the shortest path. If we are part of the same clustered habitat the DNA from both worlds should match.

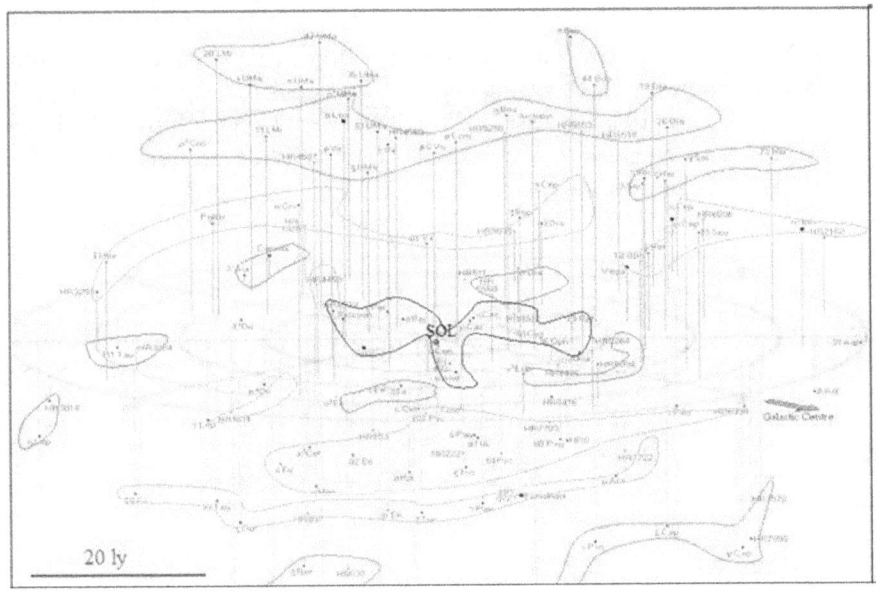

Figure 8 In our local area of space the stars are clustered – and layered - due to a random walk, so the average distance between stars with habitable worlds will vary.

Proportion of species required for each ET

The first evidence of life in the fossil record on Earth is in banded-iron formation, alternating magnetite and quartz dated at 4.28 g.a. in Canada (O'Neil et al. 2008)[cxlii]. The total number of species on Earth (Table 2) currently catalogued is 7,094,361 (IUCN, 2007;[cxliii] Jeffries, 1997)[cxliv].

Species growth follows an exponential growth curve. Our calculations suggest that it took at least > 11.29 million species to arrive at humans.

Around 1 in 629 stars (or 413 stars) should contain habitable worlds within 250 ly of earth. If the proportions of life on earth even remotely resemble those found on other habitable worlds, then roughly 43% will have virus and bacteria, 30% invertebrates (e.g. octopus, squid, jellyfish, worms etc), 23% lichen and simple land plants, and less than 5% (1 in 24) of habitable worlds may have complex land plants and simple land animals. Less than 1 in 238 habitable worlds would have vertebrates (complex land animals), but several of these worlds should exist in our 250-ly sector of space. Less than 1 in 11.29 million worlds should contain ET.

Table 2 The below table shows the numbers of species currently catalogued on Earth. By calculating the area under the exponential growth curve there had to be at least 11.29 million species on earth over the last 4 billion years. If earth is even remotely "normal" then we can expect 1 ET per 11.29 million species.

Current Species Diversity on Earth		
Vertebrate Animals		Sub-Totals
Mammals	5,416	
Birds	9,956	
Reptiles	8,240	
Amphibians	6,199	
Invertebrate Animals		29,811
Insects	950,000	
Molluscs	81,000	
Crustaceans	180,000	
Corals	2,175	
arachnid	780,000	
Others	130,200	
Plants		2,123,375
Flowering plants (angiosperms)	258,650	
Conifers (gymnosperms)	980	
Ferns and horsetails	13,025	
Mosses	15,000	
Red and green algae	9,671	
Others		297,326
Lichens	10,000	
Mushrooms	16,000	
Brown algae	2,849	
Nematode (worms)	15,000	
Fungi	1,550,000	
Bacteria etc		1,593,849
Mircrobial Eukaryotes (Protozoa	200,000	
Bacteria	1,050,000	
Prokaryota	1,000,000	
Viruses	400,000	
Nematodes	400,000	
Totals	7,094,361	3,050,000
	Totals	7,094,361

The reader will recall that the number of stars with habitable worlds was calculated as 271,900,000. One of these habitable worlds is earth. That means there are roughly 10 other habitable worlds in our galaxy with emergent (perhaps bipedal humanoid) races. It does not confirm that the Drake number is D=10. There were at least six failed species attempts prior to man on earth, suggesting that most of these will fail and the majority are far less advanced than we are.

The Drake Equation

The equations used in this text are similar to the probability functions used in the Drake equation, so they have similar biases. Drake based his calculation on a 100 billion star galaxy.

Drake's values give $N = 10 \times 0.5 \times 2 \times 1 \times 0.01 \times 0.01 \times 10,000 = 10$.

Doubling the stars doubles the Drake number. Lowering Drake numbers increases separation of stars with habitable worlds.

As a double check, we used the same parameter values with a Drake number of D=10 and 200 billion stars, and the first half of the Drake equation to calculate the number of habitable planets in this galaxy: $fH_p = 200,000,000,000 \times 0.5 \times 2 \times 1 = 200,000,000,000$. According to the Drake equation there should be 1 star with Habitable worlds every 4.13 ly for a 200 billion star galaxy. This represents an "outlier" in the simulations, but it could occur by chance due to random walk or panspermia.

4.13 ly also equals the average separation of stars in our part of the galaxy. Not only that, the Drake equation allows more habitable worlds in the galaxy than stars capable of having them, which is clearly wrong.

Some of the variables input into the Drake equation either represent the maximum value possible or are overly optimistic (e.g. $fl = 1$, $ne = 2$ etc). The fraction of civilizations that develop a technology that releases detectable signs of their existence into space, $f_c = 0.01$. We do. Yet, there is no guarantee we will even make it.

In other cases, the Drake equation simply uses one parameter to compensate for conservatism in another parameter. ne is the average number of worlds that can potentially support life per star. All $ne = 2$ does in the Drake equation is to compensate for the proportion of binary stars (that have no room for planets) in ns: the number of stars that are appropriate in terms of size, stability, and emissions.

Removing ne and adding the most likely values for fs and fd we can obtain a direct comparison to Drake's number, $N = 20 \times 0.5 \times 0.1 \times 0.0272 \times 1$

× 0.01 × 0.01 × 10,000 = 0.0544, or as few as 1 ET for 18.4 galaxies. This allows some chance of ET in our local group, and some chance that 1 or 2 of those other 10 bipedal humanoids species eventually making it to ET.

Next, we compared the simulation results (Fig. 10). In this case, the Drake number varies from a minimum of 1 ET per 1.15 million galaxies up to a maximum of 4 in our galaxy at any one time, with a most likely Drake number of 1 ET for every 76.62 galaxies. There is some, albeit slim, possibility that some of those other 10 bipedal humanoids are advanced species and some very slim chance that they are close by.

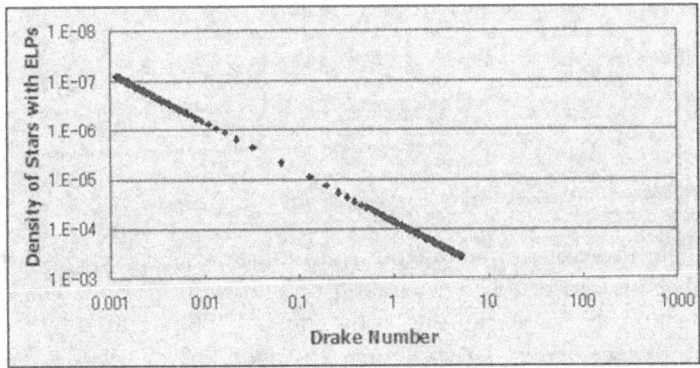

Figure 9 The Drake number varies with star density. In each case a D=10 was never reached, not even in the outliers.

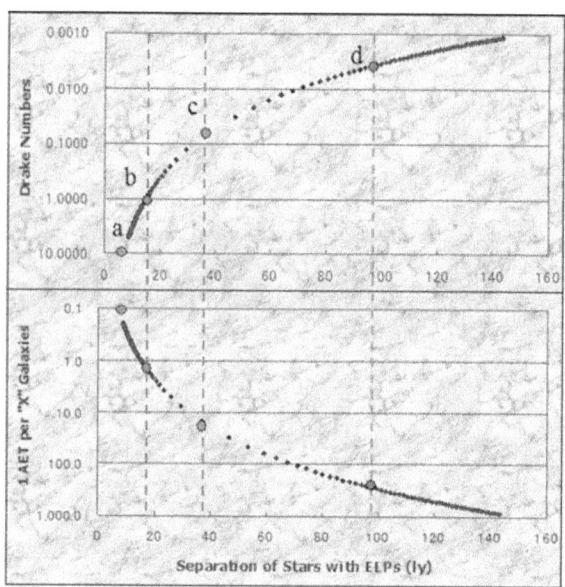

Figure 10 The top diagram shows the simulation data relationship for Drake numbers as compared to separation of stars with ELP (habitable worlds). There is a linear relationship at distances > 100 ly caused by the over-riding influence of star density. For a separation, Rh, of between 35-40 ly there is 1 ET per 10-18 galaxies. The Drake number is seldom > D=4. For that to happen average ELP separation must be < 10 ly. Point "a "represents the case where D=4, "b" where D=1, "c" the case approximating the simulation means and "d" the separation expected further out towards the edge of the galaxy disk.

In our simulations the Drake number varies with star density (Fig. 9) – and this is not at all implicit in the calculation itself.

When we place our most likely values into the Drake equation it reduces the Drake number to <1. If the Drake number is only 1, or is much less than 1, then there is no Fermi Paradox.

Conclusions

The average separation between stars with habitable worlds in our 200 billion star galaxy is most likely 34.6 ly with a standard deviation of 15.35 ly and a maximum separation of 173.5 ly. There should be six at that radius from earth within these limits.

The simulations for 100-400 billion star galaxies give separations that vary by less than 1 standard deviation. Our sector of space has a higher density of stars with potentially more habitable worlds than the average for the galaxy. There may be up to 577 stars with habitable worlds within 250 ly of space. This represents a very conservative value.

When our "most likely" parameters values used in the simulations are plugged into the Drake equation they give a Drake number of 1 every 18.4 galaxies. The simulations give much more conservative results, with a maximum value of D=4 and a minimum of 1 ET in as few as 1.15 million galaxies.

The Drake number is almost certainly less than 1 and most likely much lower. However, if earth data is even remotely representative there may be up a small number of habitable worlds in our 250-ly radius of space with land plants and complex organisms. There may even be up to 10 emerging bipedal humanoid species in the galaxy. The equations imply that less than 1 in 100 of these will ever make it.

If a similar number of emerging ET species exist in the local group of galaxies then there may be up to 290 emerging species. Also, if each ET also spawn on average 23 new home worlds (this is a minimum calculation from colonization simulations), and if only 1 in 10 of these survive long-term then the simulations Drake numbers would understate the real value by a factor of 2.3X. This would only improve the calculations to 1 ET for every 8 galaxies – but it does imply that there may be another one either in the Milky Way or its dwarf satellite galaxies. It also implies up to 3-4 may exist in the local group. If each of these have on average already spawned 10-12 worlds then the local group (of 30 galaxies – ignoring their size) may contain anything up to 43 species of ET, most of them genetically related.

Except where an ET has been around a very long time and has seeded planets, done wide-spread terra forming, and settled dozens of worlds Drake numbers of +4 are extremely unlikely, but not impossible.

Low numbers create a reverse Paradox: We may be so far apart that we are by definition alone.

Acknowledgements

The author wishes to express thanks the members of the Society for Planetary SETI Research (SPSR - http://www.spsr.utsi.edu) for correspondences on various aspects of this paper.

*Biographical Note:

Peter Ness has over ten years experience in the mining industry as a geologist and roughly nine years experience analyzing Mars geology. Peter has worked in finance for the last 8 years.

REFERENCES

The probability of extraterrestrial visitation (PEV) formula

By Ness Peter MS (Ba App. Sc. Ap. Geol.), MBA, Mfin, PHD student*

Abstract

This paper provides rigid formulas that allow one to calculate the probability that extraterrestrials (ET) have visited main sequence G,K,F and M-type stars, based on the assumption that older stars have a much higher probability of having been visited in the past. This is done for Drake numbers of D= 0.01, 0.1, 1, 10 and 100. The older the star, the more likely that ET have visited it, at least once.

There are three calculations: 1) The equations calculate the sum of the histories of probabilities based on a single visit, 2) The equations are then modified for the probability that ET will visit a habitable world, also based on a single visit, and 3) recalculated again based on the assumption that ET will visit multiple times if the planetary system is deemed habitable.

The results are rather intriguing. They suggest that it is highly likely that at least one or more species of ET have visited earth in the past, but the equations do not discern which species or when – they only provide the likelihood.

Introduction

Almost as many people believe in aliens as believe in religion. If ET is out there and resides in our backyard then we will soon have the capacity to find them. This may be the reason governments across the world are gradually releasing UFO files: to prepare for that eventuality. Yet, it is probably safer and less of a threat for ET to make contact with us now than wait for us to visit their home world(s).

One potentially habitable world, Gliese 581 g, has recently been identified. It is only 20.5 light years away. During the next 20-30 years scientists will likely find many more potentially habitable worlds. Some may be vacant lots, while others may already have occupants. During the next few decades we should also learn whether we are alone or not, and whether we really are being visited or not.

So, what is the probability that ET have visited us in the past? Ask a skeptic and they will laugh and say "never". Ask a believer and they will quote reels of supporting information, none of which is fact.

The only way to know for sure is to create formula to test the likelihood that: "we have been visited by ET" in the past.

Methodology

The chance that any ET will visit (or send a probe to visit) a planetary system depends on the age of the star and the type of star. We can calculate this using the Ness (2011) equation provided in this paper for the probability of extraterrestrial visitation (PEV).

The key assumption is that older stars have been around longer, so they have a much higher chance of having been visited by one or more species of ET.

The formula for ET visitation to visit main sequence (G, K, M and perhaps K) stars is:

$$PEV = [C \times \{ Sr \times p \times Z \}] \times D \times A \ldots\ldots\ldots\ldots(1)$$

    a. C is a constant $= 1,350,000$
    b. Sr is the ratio of the volume of control in space of ET vas a percentage of the galaxy
    c. p is the proportion of stars ET will visit in his sector of space
    d. Z is a factor that covers other unknown events
    e. D is the Drake Number
    f. A is the ratio of the age of a star As, versus the age of the universe Au

The equation can be modified for other stars if required. The proof for equation (1) and detailed explanations of each of the descriptions of various parameters are as follows:

A. Constant "C"

The value C in equation (1) is a constant. It depends on the average time span, Nt, that an average ET should continue their reign. Ng is the age of the universe or galaxy. Most scientists assume an average figure for Nt of 10,000 years. We have used a value for Ng of 13.5 billion years. The Time Span Reign Nt = 10,000 years, so the number of periods ET could exist in the universe is calculated as follows:

Periods Nr = Age of Universe or galaxy / Time span of ET =Ng/Nt
...............(2)

Nr = 13,500,000,000 / 10,000 = 1,350,000 periods

The constant C = Nr = 1,350,000

For C = 1.35 million, each civilization would survive 10,000 years. This is the widely accepted figure used to calculate the Drake Number (D). In other words, there would be 1.35 million periods (of i=1 to "k") civilizations, each spanning 10,000 years.

If there were only 1 ET in our galaxy at any one time (i.e. D=1), with no overlap at all between civilizations then 1.35 million ET may have existed in this galaxy over the last 13.5 billion years. The universe is of course much older than 13.5 billion years. The latest measurements place it as 13.75 billion years old, give or take 0.11 billion. An assumption is made that no advanced life forms were available in the first 250 million years of the life of the universe to visit other worlds.

B. Proportion of the galaxy "Sr" controlled by ET

Sr is the proportion or sector of a galaxy controlled by an ET during their entire time, survival, or reign as an advanced space faring civilization. The value for Sr in equation (3) below, depends on the "span of control" of ET and the size of the galaxy in cubic light years (ly^3).

Sr = Vs/Vg ly^3 ....(3)

Vs is the volume of space inhabited by ET. It is based on ET's span of control, measured as a radius from their home world in light years (ly^3). This is ET's sphere of influence. Rs is the radius of control and Pi = 3.14159.

Vs= Volume span of control of a ET civilization (= volume of ET's sphere of influence)

$$= 4/3 \times pi \times (Rs)^3 \quad ly^3$$
...............(4)

For example, if the radius of the span of control were 900 ly,

$$Vs = 4/3 \times 3.14159 \times 900^3$$

$$= 3,053,625,480 \ ly^3$$

In equations (1) and (3),Vg is the volume of a galaxy in ly^3. We live in a normal sized barred spiral galaxy. The volume of a spiral galaxy can be measured as a cylinder if the maximum and minimum thickness is known $(L = L_{max} + L_{min})/2$ and given that Rg is the radius of the major axis of the (flat part of the) disk itself.

Vg = Volume of galaxy

$$= pi \times (Rg)^2 \times L$$

$$= 3.14159 \times (100,000/2)^2 \ \times 7,500 \qquad ....(5)$$

$$= 58,904,812,500,000 \ ly^3$$

On average our galaxy is 100,000 light years across with an average thickness of 7,500 light years, being much wider near the core and narrower near the edge of the disk.

If we calculate Sr for a 900 ly span of control in any direction, we get Sr = 0.005184%. That is, ET's domain would cover roughly 0.005% of the total galaxy if it controlled a sphere of influence with 1800 ly diameter of space. A 900 ly radius was chosen based on colonization simulations that confirmed it as the typical span of control ET would reach after 10,000 years.

C. Proportion "p" visited per sector

ET is unlikely to visit all the stars in their sector of space, but they should visit some proportion, p, at least once during their entire civilization reign. They may physically visit or may prefer to send probes or satellites.

p = the proportion of stars in their sector that ET will visit at least once...............(6)

For argument sake we will assume that ET visit 1% of all stars in their (volume) span of control at least once. If so, p = 0.01 = 1%. The equations do not consider multiple visits.

D. The "Z-Factor"

There are many other unknowns. The value "Z" factors is all these other unknowns.

$$Z = 1 - \{Za + Zb + Zc.....+.... +Zn\} \quad ....(7)$$

The issues the Z-factor considers includes the following:

(a) ET may exist with quiescent periods, between periods of no emergent ET (Za),
(b) Some galaxies may not have the prerequisites to host life (Zb),
(c) Some stars have much less chance of being visited (Zc), and
(d) Other factors that may come into play (Zd to Zn).

Z acts to reduce the probability that ET may visit a star. Thus, Z is always less than 1. In reality Z is likely to be around Z = 0.1. We assumed the following   assumptions   for   Z:

> Za = 0.05 : That is, 5% of the time (675 million years), excluding the first 250 million years, there are no advanced beings in a galaxy. Some ET may be warlike and/or their interference may slow down species advancement or disrupt evolution on many worlds. *If we are in a quiescent period, then this would help explain the lack of signals SETI is receiving.*
> Zb = 0.05 : The physics laws may differ depending on when and where you are in the universe. Only 95% of galaxies may have

the right ingredients for advanced life to develop. Some may be too old, too energetic, too young, composed only of dark matter, and some may be too small.

Zc = 0 : We only consider main sequence stars, which make up +90% of stars.

Note: Over 90% of stars are main sequence M-K type stars, of which roughly 74% are M-type (Delfosse, 2004)[cxlv]. Kaler, (1999)[cxlvi] estimates that 4% of Milky Way stars may be F-type, 9% G-type, and 14% may be K-type stars.

E. Drake Number

D is the Drake Number, the number of ET coexisting in any galaxy at any one time. If D = 1 then a galaxy has only 1 ET. If D=10 then 10 ET may coexist at any one time.

F. Age of the Star

Equation (8) is the ratio of the age of any star as compared to the age of the universe or galaxy. It converts (reduces) the probability of PEV in equation (1) to the age of any sun.

$$A = As/Au = (Age\ of\ star/\ Age\ Universe) \dots\dots\dots(8)$$

Simplifying Equations

The initial assumption we made is that the chance that any ET will visit an individual planetary system depends on the age of the star. Older stars have the opportunity of being visited by multiple species of ET civilizations, multiple times.

Over time, the species on a world evolve and some of these species evolve to advanced life forms capable of interstellar travel. We assume that each ET control their volume of space for a long period in time, but for some reason they eventually either revert to pre-space technology or disappear (become extinct, evolve, or leave). Lets assume that there are k sets of D civilization periods in existence, {k = 1, 2, 3…..1.3 million, }. That is, we know that Nr = k = 1,350,000, so we can sum up the number of stars visited by all ET's in any given (say Nt = 10,000 year) period using equation (1).

If we combine these we end up with equation (9), below. There are 1.35 million sets of calculations to calculate as a sum of histories: we need to add up the probability that different periods (species) of ET civilizations will visit any given star in their sector of space during their (say Nt = 10,000 year) reign, plus the probability that any successive ET will also visit that star. This summation is given in equation (9):

$$PEV = D \text{ x } _{i=1}{}^{k} Z \text{ x } \{ (A_1) \text{ x } p_1 \text{ x } (Vs_1/Gg_1) + (A_2) \text{ x } p_2 \text{ x } (Vs_2/Gg_2) + (A_3) \text{ x } p_3 \text{ x } (Vs_3/Gg_3) ..+...+.... (A_{1.3m}) \text{ x } p_{1.35m} \text{ x } (Vs_{1.3m}/Gg_{1.35m}) \}$$
...............(9)

The purpose of equation (9) is to add up the likelihood(s) that any species of ET have visited a main sequence star at least once. The reader will notice that if the Drake number is large enough then the span of control of ET will overlap other existing ET. The span of control of an ET may also overlap pre-existing ET civilizations. In both cases this increases PEV.

By making the assumption that the parameter for each period is the same (i.e. Nt = 10,000 years), equation (9) can be simplified to equation (10), below:

$$PEV = A \text{ x } D \text{ x } [1,350,000 \text{ x } \{ Sr \text{ x } p \text{ x } Z \}]$$ ....(10)

Equation (10) allows us to calculate the likelihood of a star of any age, As, being visited by ET.

Please note that $A_1 = Au_1/Ag_1$ may differ to $A_2=Au_2/Ag_2$ and from $Au_{1.3m}/Ag_{1.3m}$ because as we go back in time both successive ages of the star and the age of the universe decrease by a constant value of Nt. Since Nt is constant (Nt=10,000 years) we have assumed that the ratio "A" remains constant. It does not remain constant, but it does remain "essentially" constant (to around 4 decimal places) provided the universe is more than 250 million years old and the age of the star is more than 100 hundred million years old.

If p = 0.01 and Z=0.1, the short equation is:

$$PEV = 135 (Sr) \text{ x } D \text{ x } A$$ ....(11)

In fact, if we only consider our galaxy and the same span of control 900ly) we can go one step further in equation (13):

$$PEV = (135 \times 0.00005184) \times D \times A$$

$$= 0.0069984 \ (D \times A) \ \dots\dots\dots\dots(12)$$

Discussion

If the volume of space Vs occupied by ET is based on a radius span of control of 900 ly, and the galaxy volume Vg represents a 100,000 ly wide spiral galaxy of average width $w = 7,500$ ly, then for D=1 in our galaxy: $PEV_{13g.a.} = 7\%$.

In other words, if only 1 ET exists per 10,000 year period (D=1) there is a 7%, or 1 in 14, chance that any 13 billion year old main sequence star has been visited by some sort of advanced life form (or probe) at least once. The odds are surprisingly high.

For a 4.57 billion year old star and for D=1, there is 1 ET in our galaxy at any point in time, there is a 2.4% chance (1 in 42 chance) it has been visited by ET. If only 1 ET exists in 10 galaxies at any point in time there is a 1 in 422 chance that we have been visited in the past. If D= 0.01 or there is 1 ET for 100 galaxies then the chance drops to only 1 in 4220. In other words, for low Drake numbers it is unlikely.

This calculation ignores the increased likelihood that ET will visit the star if they know or suspect that a habitable world may exist.

Varying the Drake Numbers

So far this discussion assumes a Drake number of D=1, that is at any point in time there is only 1 ET in our galaxy. However, if the generally accepted Drake Number of D=10 is correct then from Table 1 there is a 1 in 4 (i.e. 24%) chance we have been visited. If D=100 then it is a certainty we have been visited – and by at least two different species of ET.

These equations do not tell us how many times, they simply imply whether we have been visited.

The late Carl Sagan thought that there are as many as a million extraterrestrial civilizations in our galaxy alone[cxlvii]. This figure seems rather unlikely given the lack of signals received by SETI. Not withstanding, if we plug the numbers into the equations the results imply that thousands of species of ET, have visited us at least once in the past (close to 23,700 species)! This is hard to believe. According to SETI they have not received any "confirmed ET signals" in the last 2-3 decades. This implies that either:

a. Carl Sagan's assumptions are invalid (most likely),
b. SETI is looking at the wrong types of signals (more than likely),
c. ET is currently extinct in our backyard,
d. ET are deliberately avoiding us, or
e. SETI are not disclosing many confirmed signals to the public (hard to believe).

The most likely figure from Monte Carlo probability simulations for our sector of space is one star with habitable worlds every 35 light years in any direction. There should be 6 stars with habitable worlds at that distance from earth (like sides of cubes on a dice), some randomly closer and others randomly further apart based on a "random walk". The mean separation is most likely 34.6 ly (Stdev = 15.35) for a galaxy with 200 billion stars, with maximum separation of 173.5 ly expected. This is consistent with Gliese 581 g, only 20.5 ly away, representing a habitable world.

(e) **Table 3** PEV, probability a star has been visited by ET, For p = 0.1

Star Age Billion Years	Probability for D=0.01	Probability for D=0.1	Probability for D=1	Probability for D=10	Probability D=100
13	0.070%	0.70%	7.0%	70%	700%
12	0.062%	0.62%	6.2%	62%	622%
11	0.057%	0.57%	5.7%	57%	570%
10	0.052%	0.52%	5.2%	52%	518%
9	0.047%	0.47%	4.7%	47%	467%
8	0.041%	0.41%	4.1%	41%	415%
7	0.036%	0.36%	3.6%	36%	363%
6	0.031%	0.31%	3.1%	31%	311%
5	0.026%	0.26%	2.6%	26%	259%
4.57	0.024%	0.24%	2.4%	24%	237%
4	0.021%	0.21%	2.1%	21%	207%
3	0.016%	0.16%	1.6%	16%	156%
2	0.010%	0.10%	1.0%	10%	104%
1	0.005%	0.05%	0.5%	5%	52%

Yet, a habitable world does not equate to an advanced ET. For every world with bacteria only a few will have fish, even less will have land animals. Perhaps as few as only one in 11.3 million habitable worlds may develop advanced life forms. From Monte Carlo simulations, the Drake number may range from as high as 4 to as low as 1 in 1.35 million galaxies. It is probably $D < 0.1$, with the most likely value around 1 ET for every 18 galaxies.

Table 1 shows the likelihood an ET has visited a main sequence star, based on the stars age and on its Drake Number (D). Most stars (+74%) are M-type red dwarfs, so if ET comes from a M-type star and mainly visited red dwarfs it would not impact on the probability we have been visited. However, if ET were from a sun-like yellow or orange G-type (like our sun) or K-type star then the chances that ET would visit would be much higher. Likewise, if the star were close by then, regardless of the spectral type, the chances that ET may visit are higher (by some 1,000 to 10,000 X) compared to Table 1.

Goldilocks Planets

The reality is that many ET are likely to do exactly what we do, that is look for stars with planets in the "goldilocks" zone, the distance from a star where temperatures allow surface water to exist on a planet or moon. Any stars that have planets (or moons), even close to the goldilocks zone, would be visited. We can make this assumption because: a) it is exactly what we would do, and b) it is the logical thing to do.

Thus, for a star with planets or moons near the habitable zone p = 1, because if ET has the technology (and inclination) they should visit or send probes with 100% certainty. If they found a habitable world (or even think it may be habitable) then they would follow up with numerous probes, perhaps even culminating in manned flights. We know this would happen because that is exactly what we have done and will do with Mars.

(f) **Table 4**   PEV, probability a star has been visited by ET, given that ET preferentially visits all stars in its sector of space that contain planets or moons the habitable zone. P= 1.

Star Age Billion Years	Probability for D=0.01	Probability for D=0.1	Probability for D=1	Probability for D=10	Probability D=100
13	7.00%	70.0%	700%	6998%	69984%
12	6.22%	62.2%	622%	6221%	62208%
11	5.70%	57.0%	570%	5702%	57024%
10	5.18%	51.8%	518%	5184%	51840%
9	4.67%	46.7%	467%	4666%	46656%
8	4.15%	41.5%	415%	4147%	41472%
7	3.63%	36.3%	363%	3629%	36288%
6	3.11%	31.1%	311%	3110%	31104%
5	2.59%	25.9%	259%	2592%	25920%
4.57	2.37%	23.7%	237%	2369%	23691%
4	2.07%	20.7%	207%	2074%	20736%
3	1.56%	15.6%	156%	1555%	15552%
2	1.04%	10.4%	104%	1037%	10368%
1	0.52%	5.2%	52%	518%	5184%

From Table 2, assuming 1 ET for every 10 galaxies (D=0.1) there is a 23.7% chance that we have been visited by at least one alien species. Let us re-iterate that a one in four chance of a visit is exceptionally high. Most scientists would have been satisfied with a number 100 to 1000X lower.

Again from Table 2, for 1 ET currently (D=1), the probability our sun has been visited is 237%. In this case, we can be 100% confident that we have been visited and by more than 1 species of ET, and perhaps by 2 different species.

A Drake number of 10 is widely accepted as the most likely number of ET in our galaxy. If this number is correct, then the probability of ET visiting us is an astounding 2369%! It implies that we have been visited by up to 237 different species of ET sometime in the past.

It is important to re-iterate that these equations do not include individual species of ET visiting on more than one occasion, or different cultures of that species sending their own probes. The calculations do NOT consider multiple visits by multiple ET. They only count single visits by multiple ET. Humans have sent literally dozens of probes to other planets and moons in our own solar system, which implies that once ET find life on a habitable world the number of times they visit is likely to be in the tens or dozens (e.g. At least 20 probes have been sent to the moon, six missions were manned).

(g)

(h) Table 5 PEV, probability a star has been visited by ET, given that ET sends at an average of 5 probes to all stars in its sector of space that contain planets or moons the habitable zone. P= 1. X=5 probes.

Star Age Billion Years	Probability for D=0.01	Probability for D=0.1	Probability for D=1	Probability for D=10	Probability D=100
13	34.99%	349.92%	3499.2%	34992%	349920%
12	31.10%	311.04%	3110.4%	31104%	311040%
11	28.51%	285.12%	2851.2%	28512%	285120%
10	25.92%	259.20%	2592.0%	25920%	259200%
9	23.33%	233.28%	2332.8%	23328%	233280%
8	20.74%	207.36%	2073.6%	20736%	207360%
7	18.14%	181.44%	1814.4%	18144%	181440%
6	15.55%	155.52%	1555.2%	15552%	155520%
5	12.96%	129.60%	1296.0%	12960%	129600%
4.57	11.85%	118.45%	1184.5%	11845%	118454%
4	10.37%	103.68%	1036.8%	10368%	103680%
3	7.78%	77.76%	777.6%	7776%	77760%
2	5.18%	51.84%	518.4%	5184%	51840%
1	2.59%	25.92%	259.2%	2592%	25920%

If each ET sends on average only five probes (Table 3) the likelihood that we have been visited increases dramatically. If ET were even remotely similar to us then they would do what we do: they would send around 20-30 probes. This is 4-6 times the probabilities given in Table 3.

All values in Table 3 imply that there is a high chance that we have been visited in the past. Not only that, for only 1 ET per 10 galaxies all stars more than 4 billion years have likely been visited at least once. Even for 1 ET per 100 galaxies stars more than 4 billion years old have 10% to 33% chance that they have been visited.

These main problem with these equations is that we cannot say when we were visited or by what species of ET. Most of the visits may have been thousands, millions, even hundreds of millions of years ago. What we can conclude is that the results are overwhelmingly in favor of ET visitation.

3. Conclusion

ET are quite clearly much less likely to visit neutron stars, blue giants, stars about to go supernova, young stars in nebula, or stars that lack the compounds necessary for life. In fact, the likelihood that ET would visit those stars is very low; but the equations can be modified to allow for the calculation.

However, for main sequence stars and for stars with planets or moons near the habitable zone then frequency of visitations is surprisingly high. The PEV are sufficient to convince even the most skeptical mainstream scientists that we are not alone and have more than likely been visited in the past.

In defense of these equations we need to make the following points:

a. The equations are correct providing the underlying assumptions hold,
b. The equations suggest that there is a significantly higher chance of finding technologies of ancient races than actually meeting ET,
c. The equations imply that we have may have been visited, on more than one occasion by multiple (different) species of ET.

Most mainstream scientists should find these equations intriguing, if not compelling.

However, we need to make the following salient points: We cannot say when we were visited or by what species of ET, only that we most probably have been visited. Most of the visits may have been thousands, millions, even hundreds of millions of years ago. The equations do NOT prove that we are being visited by UFO's today, only that there is some chance that we are.

*Biographical Note:

Peter Ness has over ten years' experience in the mining industry as a geologist. He has 8 years' experience working in finance and roughly nine years' experience analyzing Mars geology. Peter is currently studying a Ph.D. in Planetary Geology with the Department of Complexity at Tokyo University in Japan. Inclusion of this information in this book does not imply that Peter agrees with other statements in this publication.

References

Colonization of space: It may take a lot longer than we think

By Ness Peter MS (Ba App. Sc. Ap. Geol.), MBA, Mfin, PHD student*

Abstract

This paper provides formulas to calculate the time it takes to colonize a habitable world. It also provides Monte Carlo simulations. The maximum span of control after 10,000 years is also calculated/simulated.

The history of colonization on earth is explained along with the reasons for failure and success. There are lessons to be gain from colonization failure that can be applied to speed up the space colonization process.

There really is a "one best way" to colonize which can and should be applied to space colonization.

The key to colonization is establishing "habitability" early on, setting strict goals and having the correct structure in place.

Introduction

Some authors (Savage, 1994)[cxlviii] believe that human expansion into space is our inevitable destiny and all it requires is that we follow a set pattern or approach. Others (Zubrin, 1996)[cxlix], believe that metals such as silver or deuterium may be mined on Mars and sold on Earth for profit. Space may be the next frontier but mining is unlikely to be an economically sustainable driver or provide the momentum to get us there, at least for the next +50 years. If the search for new mineral resources to replace depleted commodities controls colonization, then it is going to be a very

476

slow process without serious commitment to develop far quicker and cheaper modes of space travel.

Space tourism, and finding a habitable planet within say 30-50 ly from SOL are likely to be the key drivers for colonization both now and in the foreseeable future. If the aim is to colonize space, private enterprise has to be more involved. This will reduce costs, increase competition and speed up the process. It would make it more economic to mine near earth asteroids (NEA). Mining for local needs and for trade would then be inevitable.

This paper provides formulas that allow us to place definitive parameters on the likely colonization times to nearby stars with habitable worlds.

The Supply Chain is the Span of Control

History clearly shows that planning and the supply chain are critical factors in the success or failure of any mission. Napoleon, Hitler, the Roman Empire, the Egyptians and the Incas all learned this lesson the hard way. The maximum span of control (distance) was around 1,200km for Napoleon, roughly 1,500km for Hitler, 2,000 km for the Incas, 6,000 km for Columbus, and roughly 30,000 km for Magellan and Cook.

The maximum time you could be away from a port depended on the size of the cargo (food and/or fuel) you could effectively carry with you. The early seafarers had one huge advantage over space travel – they could obtain supplies from any port of call – and that dramatically increased their span of control.

A voyage may be self-sufficient in food and carry supplies of computer chips and instrument parts, but it is unlikely to have the capacity to manufacture complex replacement parts, heat shields, or rocket engines. Given an average life span of 100 years and a maximum speed of c, an advanced civilization is unlikely to spread too far in space (even if these are doubled or tripled).

By say 2025, NASA would like to have a Moon base and by 2035 a Mars base. However, even with multiple countries involved, we seem to be struggling to keep the ISS operating. A minor problem with toilets almost closed the ISS down. With current technology the supply chain could likely not cope with both the ISS and a Moon base. The span of control for manned missions by NASA is arguably 384,403 km, the distance from the Earth to the Moon. However, the current span of control for unmanned missions is the distance of Voyager and Pioneer spacecraft, now moving outward from the Sun beyond 40 – 60 Au into the Oort Cloud or Kuiper belt, (Cooper et.al., 2008).[cl]

The span of control is dictated by our technology and will increase, and perhaps even exponentially. A Moon base can be supplied from Earth, but a Mars base and any missions beyond Mars need to be self-sustaining. We may well overcome many of the most basic biological and psychological issues of humans living in space. Yet, even with dramatic improvements in technology and the speed of space travel, we may not have the resources to launch a manned trip to another star this century let alone have the supply chain set up to support such a trip. If a technical glitch occurs with a probe, rover or satellite on a Mars mission it can take weeks or months to get the communication channels working again. Unmanned probes to nearby stars will face diabolic challenges.

Even assuming we can send smart satellites, semi-android in nature (to say Alpha Centauri A & B), signals traveling at c take 4.4 years to get there. If a satellite sends a communication, then by the time we receive the signal and it receives new directions almost nine years has passed. That makes space exploration a very slow and painful process.

Sending dozens of satellites or probes in many directions requires teams of people to monitor them – and budgets that may continue for centuries. Technology tends to supersede itself rapidly. It is hard to imagine a probe sent 100 light years to another star with a project lasting 250 years. It is just not going to happen unless we radically adapt our current ways of thinking.

Distance, cost and time will limit the number of satellites/probes. They may only be sent to those star systems which we already know have life signs, or those where we can obtain very specific knowledge (e.g. a star expected to go supernova, or a young star with a planetary disk). Where we know a star system can sustain life from biochemical signatures obtained passively, satellites may be monitoring and sending us back continual one-way communications so that we can prepare for the manned mission some 30 to 80 years hence.

Once we establish that humans can live on a planet/moon and we are not going to be annihilated by some bacteria or other creature then some 10 to 100 years later the colony may arrive and establish a second supply chain.

Colonization of the Americas, Australia, and New Zealand

If we compare the discovery and subsequent colonization of North America, Australia and New Zealand (Fig. 1) we can see that discovery involved multiple trips by multiple races. In each case, it took between 10 and 81 years for private enterprise to get involved, and until they did very little happened and most colonies failed.

Technology and the distance to North America and Australia controlled the speed of colonization. Technology and the distance between habitable worlds will control the speed of colonization of space.

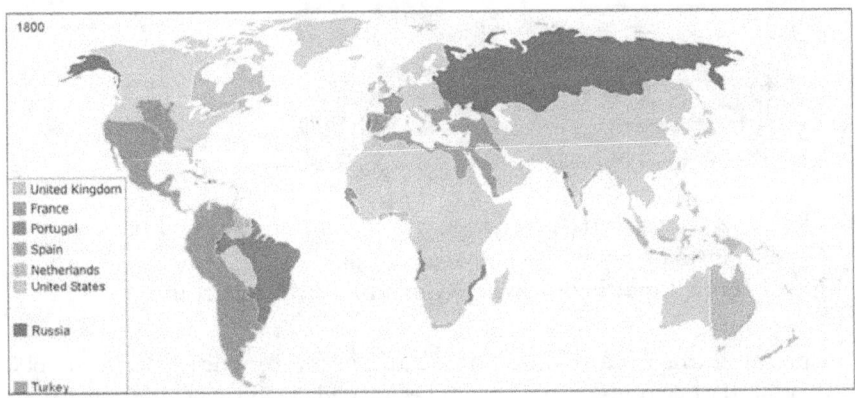

Figure 11 A map of 1800 showing colonization. Source: http://en.wikipedia.org/wiki/File:Colonisation_1800.png

Columbus discovered the America's in October 1492. Spain and Portugal established colonies in Central and South America. Spain had control as far south as Peru by 1532 (Fernandez-Armesto, 2006).[cli] They traded and pillaged for gold and spices for the next 300 years. Giovanni da Verrazzano mapped the area from Canada to Carolina in 1524 (Fig 2). There were small colonies in South Carolina as early as 1526: but, until the privately owned British, 'London Virginia Company' arrived at Jamestown in 1607 (81 yrs later) most North American colonies failed. After pilgrims colonized Massachusetts, in 1620, the North American colonies started to take hold (Andrews, 1933)[clii].

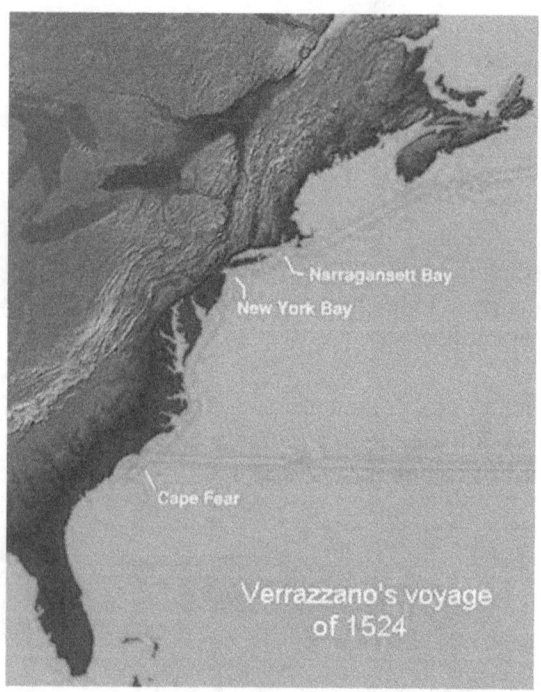

Figure 12 Giovanni da Verrazzano mapped the area from Canada to Carolina in 1524. Source:
http://en.wikipedia.org/wiki/Giovanni_da_Verrazzano

England learnt from colonization failures in the Americas. Between 1606 and 1770, a period of 165 yrs, an estimated 54 boats from a range of nations visited Australia (AOTM)[cliii]. However, it was the discovery of Botany Bay (Sydney), attributed to James Cook in 1770, which established "habitability" and directly led to colonization (Banks, 1997)[cliv]. In fact, the unstated mission statement was to "prove habitability".

Unlike the Americas, colonization of Australia was rapid and well orchestrated. This was essentially a military operation. The first fleet, comprising 11 ships and around 1,350 people, arrived at Botany Bay between 18 and 20 January 1788.

Sending even a tenth of this number of people to a nearby habitable world may be impossible without the involvement of private enterprise to adapt large commercial jetliners to operate in space. Military involvement helps control and speed up the colonization process.

From 1788 until 1823, New South Wales was a penal colony: the prisons were over-flowing and Britain needed somewhere to put the excess criminals. The pressure to colonize was immense. Free settlers began to arrive within five years of establishing the colony, in 1793. Colonies were established in Tasmania in 1803, and Western Australia in 1827. Free settlers poured into South Australia from 1836 onwards; and by 1850 there were >36,000 settlers in that state (i.e. a migration rate of 2,570/yr).

If we discover 2-4 habitable planets in a single binary star system, or a cluster of stars with habitable worlds close together, this process may be replicated in space.

It took 8 years after Cook's 1770 discovery of Australia to establish the first colony – the same time it took NASA to send men to the moon in the 1960's. Both programs were run by the military. The US government then placed NASA under civilian rule, letting researchers dictate the progress of space exploration. Nothing much has happened since. After 40 years of space exploration we still do not have a base on the moon or on Mars.

Unlike the Americas, the Australian colonization process was very rapid.

Colonization was rapid because the military controlled the initial stage and private enterprise was involved early on. The colony was making serious contributions to the British Empire by the 1899, and perhaps even by 1869 when Darwin was established. Australia did not become a federation until 1901, but it was self-sufficient by 1899, the date the new federal capital was relocated to NSW. Unlike the ad-hoc colonization of North America, government and military involvement ensured Australia's colonization was well planned and coordinated; the entire period from discovery of habitability to self-sufficiency took a mere 99 years.

So what about New Zealand?

Abel Tasman discovered New Zealand in December 1642, on a privately funded ship owned by the Dutch East India Company (Wilson, 2007)[clv]. James Cook arrived in New Zealand in 1755 and established "habitability" (King, 2003).[clvi] The colonization of New Zealand by Europeans was informal and unplanned; starting with American whalers in 1791 (36 years after Cook) followed by opportunists, religious groups, and private individuals in the 1830's (Ward, 2001)[clvii]. Migration was ad hoc. There

481

was no formal involvement by the British government early on. The colony only became viable after the 'New Zealand Company', formed in August 1838, began to actively promote migration - and much of this was via Australia. By 1854, New Zealand was declared a state and achieved self-government (Mein-Smith P., 2005 [clviii]; Sutton et. al, 2007[clix]).

From Cook's discovery of habitability to colonization and self-sufficiency took New Zealand some 123 years. New Zealand was certainly self-sufficient early on, but was not in a position until the late 1880's to make any substantial contribution to the British Empire. By World War 1 - a period of 136 yrs for Australia and 123 yrs for New Zealand - both countries were able to sustain prolonged overseas military action. However, without Australia as a springboard it would have taken much longer for New Zealand to realize that goal.

The time taken to discover each world varied. The success rate for North American colonies was < 10% and for New Zealand was < 20%. With multiple countries involved in separate campaigns South America fared even worse.

The British learnt from their experience with the US and Canada. They only had one attempt with Australia and had no choice – they had to succeed: the military and private enterprise partnership made it happen.

Colonization of Australia was very efficient. Both military and private enterprise were involved early on. Migration was planned, rapid, and well orchestrated, with strict controls on the speed and timing of private enterprise-led migration (e.g. Hindmarsh surveyed Adelaide prior to the arrival of the first settlers in 1836).

In comparison, North America and New Zealand started essentially as semi-viable religious colonies looking for a Utopia, which never happened. The process was inefficient until private enterprise got heavily involved. The British model for colonizing Australia works best. There is a one best way to colonize space – this is it.

People may argue that colonization processes on earth bear little relationship to space. Our technological level is much higher than in the 17th-19th centuries. It takes 40-60 years to bring a third-world country to first-world status, so colonization will be faster. If this were true, there would be manned bases on the moon or on Mars. The last time we looked, there were none.

Colonization in space depends on the speed one can travel, as it did in both the 17th - 19th centuries – our technology is exponentially advanced, but so is the separation between habitable worlds. Technology in the 19th century was arguably more advanced than in the 17th century but even so, the colonization process was not much faster. A self-sustainable economy is required before a colony can viably contribute to the expansion of an empire. The natural resources of a habitable planet are much greater than those of a single continent on earth, reducing (slowing) the rate of expansion.

In each case on earth, colonization took some 10-80 years, or more, to even proceed after the key discovery of habitability – which was much more important than initial discovery. It took roughly 120-150 years for each fledgling colony to advance its economy to a position where it could move outside its immediate sphere of influence.

The initial success of any colony is controlled by the degree of involvement of government, military and private enterprise, and on how well the process is planned. However, the focus of governments was, and will likely always be, on one colony at a time. The reasons for this are:

1) The supply chain stretches the resources to the limit, and
2) Diminishing returns from competing projects.

It is the same in space. It would be difficult for government-funded projects to focus on both the Moon and Mars colonies simultaneously – especially with an ISS - because this project will compete for allocation of rare resources. Large corporations with billion dollar projects analyze multiple projects simultaneously but typically only one at a time is selected for final approval.

It is seldom the best project (or company) that succeeds. Rather, it is the project with the fewest problems; the one that competes best on resources, price and timing, or the one on whose success the company depends on for its very survival. At the end of the day, the project team with the best marketing skills usually gets the money. We eat MacDonald's hamburgers, drink Starbucks coffee, and use Microsoft software. That does not necessarily imply the products are the best on the market – they may well be – but part of the reason each firm is successful is that they focus on the core of what they are good at, and they have superb marketing skills.

The ISS may have the shortest payback period, but may not be a long-term proposition. As a dependent outpost, a moon base is a high risk and high cost option. It may never be a viable long-term option for a colony, but as a quarantine station it may offer the best defense against life threatening species. Mars has the ability to become self-sustaining. It may have been habitable in the past, and the Mars base(s) option always beats these other projects on a long-term basis.

Although they have a long project pay-back period (e.g. +150 years), and are extremely expensive to get to and start-up, the value-added from technology improvements and increased GDP benefits alone makes colonizing habitable worlds an irresistible and extremely attractive proposition.

The Key Drivers in Space Colonization

Given the numerous perturbations of possible habitable worlds, Earth replicas or planets and moons on which we can live, with atmospheres we can breathe - are likely to be rare commodities and some considerable distance apart. If so, the colonization time FC_t , is likely dictated by the variables in Eq. 1:

$$FC_t = [f\$_t + fR_t + (n_P {}_x ((R_b / Lim_{0=>c} (Cv))))] \qquad ....(1)$$

Where:

a. $f\$_t$ is the economic cost of establishing the new colony stated as a function of time from discovery (e.g. of habitability). The Apollo landings took place some 40 years ago, on July 20, 1969. If it costs $104 billion and takes 30 years to establish a base on the Moon then the economic cost of a Moon base is $f\$_{t(moon)}$ = 40+30 = 70 years.
b. fR_t is the time to consume new resources and develop a colony to a stage where it is self-sustainable and has the ability to expand further without support (fR_t = +125 years);
c. n_P is the (weighted average) number of trips (or migration cycles) required to establish a colony, which has the propensity to dramatically increase/reduce the colonization time. It is a velocity dependent variable;
d. $(Lim_{0=>c} (Cv)$ is the maximum speed we can travel which is controlled by technology and under current laws of physics, c the speed of light; and,

e. Rh is the distance between stars with habitable worlds. $Rh = [1 / [Dh \times 4/3 \times \pi]]^{1/3}$ ly, where Dh =Density of stars with habitable worlds.

The above variables are likely correlated to, and vary, with both incremental and major technological change, with development of a dominant design controlling success (Utterback, 1994)[clx]. The key variable in FC_t is the distance to the next habitable world Rh in any given sector of a galaxy, rather than the total number of them.

The critical variables in the expansion of humans into space are:

a. Distance, or separation, between habitable worlds Rh;
b. The propensity for a planetary system to have the correct elements required to host life fs, which may depend on supernova providing the correct elements essential for life (Boss et. al, 2008)[clxi]; and,
c. The probability that a planet lies in the habitable zone fd.

Knowing the likely separation between stars with habitable worlds provides us clear objectives as to the speed at which we need to propel spacecraft to make colonizing space possible.

Quite clearly, if a star with habitable worlds lies within 4-6 light years from SOL or even within 40 ly radii, and if it is known to pass the "habitability" test the momentum to colonize will be unstoppable. However, if the next habitable world is over 100 ly away, then it may be centuries before we take the first steps.

The test for habitability

The "habitability" test used by Cook was a climate with weather (temperature and rainfall) the same as his mother England. The Spanish and Portuguese were more concerned with trade and conquering new lands than habitability – which turns out to be a major design flaw. Their test was simple: if people lived there then it was habitable. Many Central and South American countries were livable but not self-sustaining for some +300-500 years, so they may have failed the test of habitability.

The strategy of the British, Spanish and Portuguese on indigenous populations was to attempt to negotiate trade. If the locals were non-hostile they would occupy them. If the indigenous populations were hostile they would conquer them, kill them, or leave. This may not prove very effective in space. It is not the most effective long-term solution for any colony.

The test for habitability in space may differ depending on the location, separation between habitable worlds, purpose for the occupation, the orbital eccentricity, planetary dynamics (weather and geologic processes)), and availability of life forms (= food). The current method to test for habitability relies on the position of the world in respect to the spectral class of host star, with liquid water and composition of the atmosphere being the key ingredients. This alone does not prove that humans can survive long-term on the world. Other things are just as important, for example, whether:

a. The world lies in the "goldilocks" habitable zone, with similar diurnal temperature fluctuations as earth

b. The world has liquid water. Lakes, rivers and an ocean would be a bonus.

c. The world contains a non-toxic, oxygen-nitrogen rich atmosphere of the range of say 730 – 1,360 mbar and roughly 20% (200 mbar) oxygen component. These are safe limits given more than a 73% change in pressure causes decompression issues (Gerth and Vann, 1995) [clxii]. We can live on worlds with lower/higher atmospheric pressures with some difficulty.

d. An ability for nitrogen fixation in soils (see below)

e. The world is in a "safe haven": i.e. It is away from proximal stars likely to go supernova, has a stable orbit around a stable star, or has satellites in stable orbits, radiation, gravity, magnetic fields of stars or proximal planets/moons and asteroid belts that are not likely to be a severe survival issue

f. No severe (or reciprocated) life threat from/towards any life forms occupying the world exists

g. Any indigenous ET or pre-ET population exists. Common sense suggests this should remove a world from the list of potential worlds to be colonized.

The type of DNA or life form present may be a controlling factor. It may inhibit colonization.

The ability for nitrogen fixation in soils is critical and is independent of the nitrogen in the atmosphere. All organisms found on earth use the ammonia (NH3) form of nitrogen to manufacture amino acids, proteins, nucleic acids and other nitrogen-containing components necessary for life (Lindemann and Glover, 2003)[clxiii]. Biological nitrogen fixation is the process that changes inert N2 to biologically useful NH3. This process is mediated in nature only by bacteria. Other plants benefit from nitrogen-fixing bacteria when the bacteria die and release nitrogen to the environment or when the bacteria live in close association with the plant. If the world has trees and plants or life forms that can be harvested for food or biotechnology that would be a plus.

Many planets or moons such as Mars or Europa may meet the test for partial habitability. They both have water ice. Temperatures of Mars make habitation possible and terra forming may improve chances of long-term survival. Europa has an ocean below a crust of ice, but in its present location it may be impossible to terra form. Many habitable worlds may fit this basic list but the atmosphere may be too thin/thick to breathe. The time to get there is the larger issue.

Time to Colonize a New World

Speed and direction of colonization are likely to be controlled by the maximum separation between stars with habitable worlds. The mean separation is most likely 34.6 ly (Stdev = 15.35) for 200 billion stars, with maximum separation of 173.5 ly expected.

If we can travel at $c = 0.3$ with a payload of 50 persons per trip (until 1,500 were on the habitable world) and up to a maximum of say ten flights proceeding at any one time, from Eq. 1:

$$FC_t = [(3) \times ((35/0.3)) + 125 + 70] \qquad = 545 \text{ years per colony}$$

Span of control = 10,000 yrs/ 545 yrs $\times$ (2 $\times$ 35 ly) = 1,284 ly

It may take anything up to 545 years to establish each colony, with a mere 92 colonies (*f nt* = 5 $\times$ (10,000/ FC_t)) established over a 10,000-year empire. This assumes equal spacing and "one-way" flights. It ignores *clustered-habitat theory*, which may reduce separation and colonization times. Small outposts and half-way-house refueling colonies are ignored. Sublight speeds are assumed. The equations count multiple habitable worlds in single or binary stars, colonized in parallel, as single colonies. Reducing separation between trips (migration cycles) or increasing the number of people per flight reduces the time.

Definition of a Clustered habitat: *A clustered habitat occurs where stars with habitable planets are randomly clustered close together. Random walk theory suggests that once we find one habitable world there may be others close by, separated by vast distances to the next habitable world.*

Simulations

Based on the above equations, Monte-Carlo simulations were carried out for 10,000 iterations and 1,000 trials (10,000,000 data points). The simulations were based on the following parameter bounds:

Separation between habitable worlds (Mn=4.4, ML=35, Max=120 ly)

A colony of (Mn=1,500, ML=36,000, Mx=100,000)

Ship velocity of (Mn=0.01, ML=0.3, Mx=0.9)

Ship capacity of only (Mn=10, ML=20, Mx=50)

Time to consume sufficient resources to be able to contribute to the empire (Mn=90, ML=125, Mx=300)

Economic cost of colony in time (Mn=10, ML=40, Mx=100)

Table 1 shows an example from the input data used in the simulation.

Table 6 Control parameters for the simulations. The over-riding control variable is the separation of the stars with habitable worlds. The "Time to get there" 99.22 yrs value is a (likely pessimistic) result of 1 simulation run to colonize and bring a habitable world 20.6 ly away (e.g. Gliese 581 g) to a position it could be self-sustaining in 301.91 yrs. Private enterprise & military involvement should be much faster.

		Unknown Variable	Minimum	Most Likely	Maximum
Ship Capacity	s	25	10	20	50
Number of People per period	p	612.47	1500	36000	100000
Number of ships Required	n	24.50	150	1800	2000.00
Number of ships sent per year	sy	19	1	10	20
Number of ships sent per period		5.19			
Number actually sent			51.92	103.84	259.60
Base Colony Required			1500	36000	100000
Time to get Base colony to New World	np	1.45	28.89	346.69	385.21
Time to get there		99.22			
Velocity of Travel	LimO=>c (Cv)	0.21	0.01	0.3	0.9
Habitable Star Separation	fHd	20.60	4.4	35	120
Time to Colony established		145.46			
Resource Consumption Time	fRt	126.02	90	125	300
Economic Cost of Colony	f $t	30.43	10	40	100
Time to Sufficiency	FCt	301.91			

Fig. 3 shows plots of the individual data points used for each parameter in the simulations. The time to get to the colony is liner as it depends on the distance to the habitable world. The time to use resources is partly linear and exponential as it tends to be controlled by population growth , which is fed by colonization.

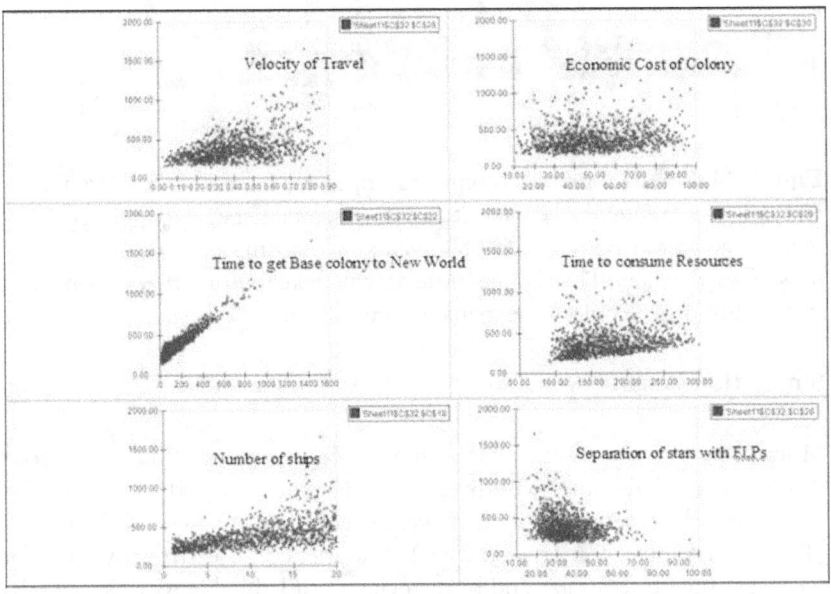

Figure 13 Plots of the individual data points used for each parameter in the simulations. The time to get to the colony is liner as it depends on the distance to the habitable world. An ELP = a habitable world.

489

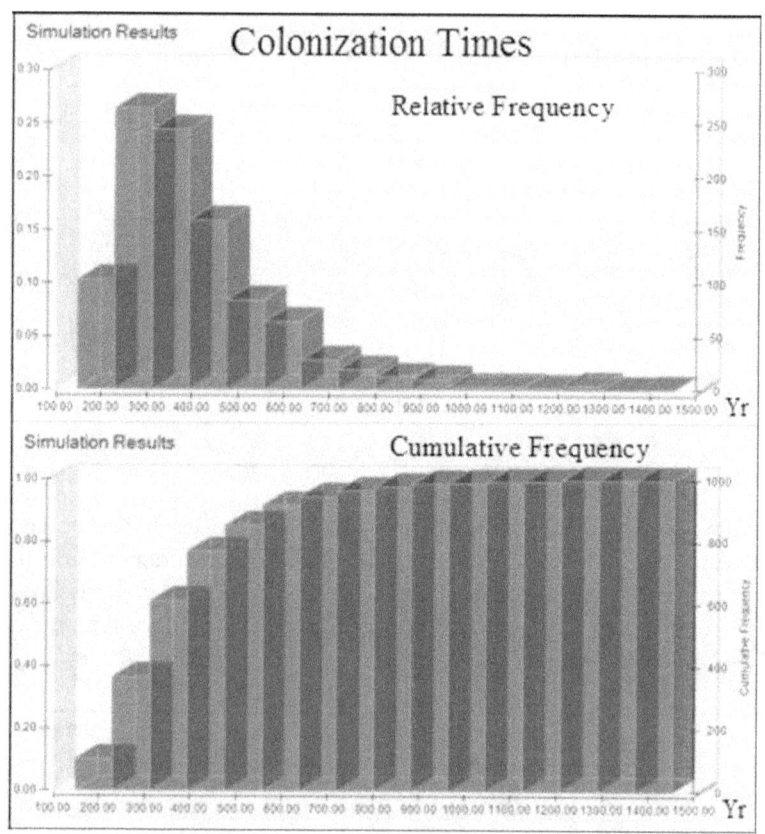

Figure 14 Simulations for colonization rates in space, shows as frequency and relative frequency. Colonization times are distance dependent, but tend to cluster at between 250-500 years as the time to develop a colony to self-sufficiency. This is the time it will take before that colony can viably support the colonies expansion much further.

Simulation Results

Monte Carlo simulations give a mean 394 years for the time from discovery of habitable worlds to colonization (Fig 4). Due to the separations between stars with habitable worlds the standard deviations (Table 2) are huge (Stdev = 162 yrs). Experience on earth proves that it takes less time to colonize and bring a cluster of habitable worlds to a self-sufficient economy as it does for a group of widely separated new worlds. Therefore, we will naturally migrate towards areas of higher star density in our area of space.

490

The distribution is skewed to the left with a minimum value of 146 years (Fig. 4, Table 2). Value at Risk 95% is 709, so 1 in 20 colonies may be separated from the empire and/or take 700 or more years to develop.

Table 7 Statistical output from the simulation runs give a mean 394 years to colonize and bring a habitable world to the stage it is self-sustaining. Due to the separations between stars with habitable worlds the standard deviations are huge (162 yrs).

Statistic	Value	Percentile
Mean	393.69	1%
Standard Deviation	162.73	2%
Variance	26480	3%
Skewness	1.54	4%
Kurtosis	3.69	5%
Mode	292.66	6%
Minimum	146.63	7%
Maximum	1427.69	8%
Range	1281.06	9%
Mean Abs. Deviation	122.76	10%
SemiVariance	8223	11%
SemiDeviation	90.68	12%
Value at Risk 95%	708.88	13%
Cond. Value at Risk 95	369.45	14%
Mean Confidence 95%	10.09	15%
Std. Dev. Confidence 9	12.03	16%
Coefficient of Variation	0.413	17%
Standard Error	5.14	18%

Discussion

Problems and Limitations

This simulation equation assumes that most habitable worlds are suitable for habitation. Many may not be. It assumes that colonization times on earth can be replicated to some extent in space. It ignores the number of habitable worlds per star, assuming that it takes roughly the same time to colonize a single world as for a clustered-habitat of habitable worlds. Other than assuming we will create the technology to colonize stars with habitable worlds, the method totally ignores technological change and the processes used. The time to colonize may speed up due to technology and the learning curve, so these figures may be overstated.

Due to the distribution pattern (Fig. 3,4), the simulations imply that as few as 127 planetary systems may be colonized over a 10,000 year empire. The number of clustered habitable worlds and terra-formed worlds colonized may increase this considerably. The average radius of control for colonies would be roughly 900 ly, so the span (diameter) of control diameter is 1,800 ly, encompassing a volume of 10 million stars. The radius of control for military and exploration purposes is likely to be much larger. The radius of control of the "core" protected volume of space under control is likely closer to 600-ly radius.

The colonization process is technology and velocity dependent. This provides a clear understanding of the time frames and likely span of control that we can expect to attain.

If we were able to colonize only 127 worlds then we would be cherry picking the best habitable worlds, and/or focusing on clustered-habitats. Other worlds may be too dangerous or too difficult to colonize. The alternatives are our span of control would be much less than 1,800 ly, or that technology and major gains in physics will allow us to travel at speeds greater than c and not conflict with the laws of physics. In that case, these colonization times are far too conservative.

Impact of Clustered-Habitats

If stars with habitable worlds form in clustered habitats, or more than a single habitable world exists in a planetary system the time to colonize all worlds would likely proceed in the same way as the colonization of Australia and New Zealand. As a result, the time to colonize all the worlds in the enclave may not be much longer than the time to colonize a single isolated world.

Random walk theory suggests that panspermia, ET, and humans will tend to look for and migrate in the direction of clustered-habitats along the shortest route between stars. This is also a prediction plot for expansion of any civilization. They will always seek to minimize the supply channel. Migration will be in the direction of higher star density where enclaves of clustered habitable worlds are likely to be higher in number.

This would reduce the time to colonize. It may make it easier to communicate, exchange technology, travel and maintain control over worlds. As such, it makes much more sense to colonize clustered habitats in preference to individual habitable worlds.

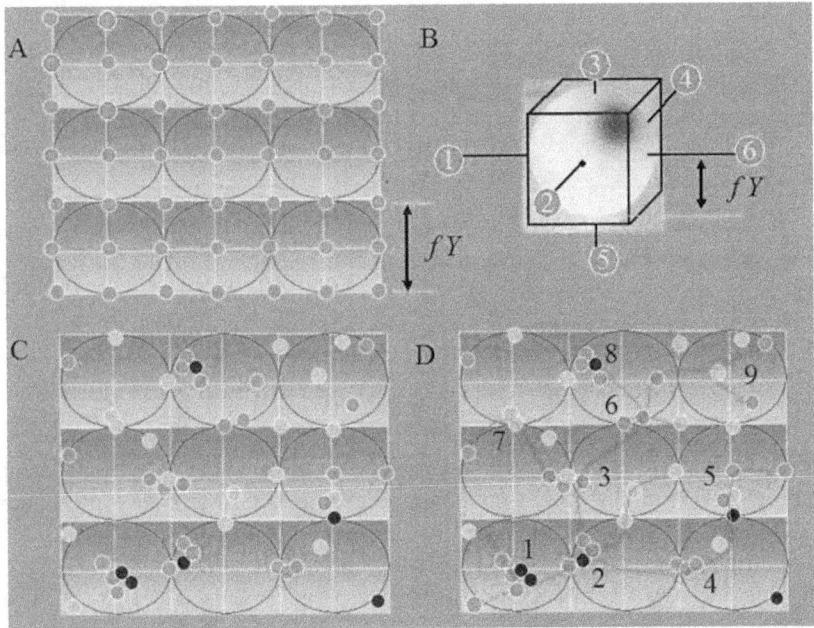

Figure 15 A. Grid if all stars are equal distances apart. They are not. **B.** The separation between stars with habitable worlds implies six stars with habitable worlds exist at roughly the same radius. **C.** Stars will vary in their separation. **D.** Random walk theory suggests that panspermia, ET, and humans will tend to look for and migrate in the direction of clustered-habitats along the shortest route between stars. This is also a prediction plot for expansion of any civilization. They will always seek to minimize the supply channel.

Conclusion

There is one best way to colonize and that is with well-planned programs controlled by the military, with private enterprise involvement early on.

Monte Carlo simulations give a mean 394 years with a Stdev of 162 yrs) for the time from discovery to colonization of each habitable world. The minimum time required to bring a colony to self-sufficiency would be 146 years with 1 in 20 colonies taking more than 700 to develop.

The simulations imply that we would be struggling to colonize more than 127 planetary systems over a 10,000-year time span. However, if we find clustered habitable worlds the number may increase dramatically.

The average radius of control for colonies would be roughly 900 ly, giving a span of control of 1,800 ly. This volume of space encompasses a volume of 10 million stars, which means we would be cherry picking the habitable worlds that are either closer to home or those that represent the lower risk opportunities.

Acknowledgements

The author wishes to express thanks to the members of the Society for Planetary SETI Research (SPSR - http://www.spsr.utsi.edu) for correspondences on various aspects of this paper.

*Biographical Note:

Peter Ness has over ten years experience in the mining industry as a geologist and roughly nine years experience analyzing Mars geology. Peter has worked in finance for the last 8 years.

REFERENCES

ABOUT THE AUTHOR

Greg Orme has been studying Martian images since 1993; this is his first book on the subject. He has several popular web sites on Mars.Ultor.org is where the latest information on possible Martian artifacts can be found. Harmakhis.org contains interesting photos of Martian geology. Martianspiders.com contains images of the enigmatic spider formation. He lives in Brisbane Australia.

[1] Bradley J. Thomson and James W. Head III "Utopia Basin, Mars: Characterization of topography and morphology

and assessment of the origin and evolution of basin internal structure" JOURNAL OF GEOPHYSICAL RESEARCH, VOL. 106, NO. 0, PAGES   1–22,   MONTH   2001 http://www.planetary.brown.edu/planetary/documents/2396.pdf

See Figure 1, comparing Utopia Basin with Hellas Basin.

[2]NASA, Jack Connerney, Mario Acuna, Carol Ladd "Martian Magnetic Map"

http://www.solarviews.com/cap/mgs/magmap.htm

[3] Nanjing University Syllabus "The Surface of Mars available online at

http://astronomy.nju.edu.cn/astron/AT3/AT31004.HTM

[4] Mike Caplinger" Determining the age of surfaces on Mars" Malin Space Science Systems February 1994 available online at http://www.msss.com/http/ps/age2.html [16/8/03]

[6] Jarmo Korteniemi "Main albedo features and full nomenclature" available online at http://www.student.oulu.fi/~jkorteni/space/mars/maps/ [16/8/03]

[7] USGS Astrogeology Research program Regional MOLA map available online at

http://planetarynames.wr.usgs.gov/images/mola_regional.pdf

[8] K. F. Sprenke and L. L. Baker "POLAR WANDERING ON MARS?" Lunar and Planetary Science XXXI 1930.pdf

http://www.lpi.usra.edu/meetings/lpsc2000/pdf/1930.pdf

[9] Aviation Now "Water Find Will Shape Mars Exploration Plan" available online at http://www.aviationnow.com/content/publication/awst/20020603/aw3 2.htm

[10] David E. Smith et al "The Gravity Field of Mars:

Results from Mars Global Surveyor" available online at

http://ltpwww.gsfc.nasa.gov/tharsis/smith.mgs.grav.pdf

[11] FREY, Herbert V "LARGE BURIED AND VISIBLE BASINS ON MARS: TOTAL POPULATION AGES OF THE HIGHLANDS AND LOWLANDS"

496

http://gsa.confex.com/gsa/2002AM/finalprogram/abstract_44774.htm

[12] http://www.harmakhis.org/history/1.jpg

[13] JOURNAL OF GEOPHYSICAL RESEARCH, VOL. 106, NO. 0, PAGES 1–22, MONTH 2001

http://www.planetary.brown.edu/planetary/documents/2396.pdf

[14] David E. Smith et al "The Gravity Field of Mars: Results from Mars Global Surveyor" 1 OCTOBER 1999 VOL 286 SCIENCE www.sciencemag.org available online at

http://ltpwww.gsfc.nasa.gov/tharsis/smith.mgs.grav.pdf

[15] [Duane et al., 1972;

Swift et al., 1972, 1973; Stubblefield et al., 1975; Swift, 1975;

Rice, 1994]

8.6. Bradley J. Thomson and James W. Head III

"Utopia Basin, Mars: Characterization of topography and morphology

and assessment of the origin and evolution of basin internal structure" JOURNAL OF GEOPHYSICAL RESEARCH, VOL. 106, NO. 0, PAGES 1–22, MONTH 2001

http://www.planetary.brown.edu/planetary/documents/2396.pdf

[16] Page 15 Volcano/Ice Interaction Workshop August 13-15, 2000 Reykjavík, Iceland available online at

http://astrogeology.usgs.gov/Projects/VolcanoIceWorkshop/abstract_volume_rev5.pdf

[17] David E. Smith et al "The Gravity Field of Mars: Results from Mars Global Surveyor" available online at

http://ltpwww.gsfc.nasa.gov/tharsis/smith.mgs.grav.pdf

[18] Fig 6 David E. Smith et al "The Gravity Field of Mars:Results from Mars Global Surveyor" available online at http://ltpwww.gsfc.nasa.gov/tharsis/smith.mgs.grav.pdf

[19] K. F. Sprenke and L. L. Baker "POLAR WANDERING ON MARS?" Lunar and Planetary Science XXXI 1930.pdf

http://www.lpi.usra.edu/meetings/lpsc2000/pdf/1930.pdf

[20] Fig 1a David E. Smith et al "The Gravity Field of Mars: Results from Mars Global Surveyor" available online at

http://ltpwww.gsfc.nasa.gov/tharsis/smith.mgs.grav.pdf

[21]http://www.harmakhis.org/history/1.jpg

[22] Linda M.V. Martel "If Lava Mingled with Ground Ice on Mars" PSR Discoveries [6/26/01] available online at http://www.psrd.hawaii.edu/June01/lavaIceMars.html

[23] DPS 2001 meeting, November 2001 Session 48. Mars Surface available online athttp://www.aas.org/publications/baas/v33n3/dps2001/221.htm

[24] http://www.harmakhis.org/history/2.jpg

[25] H. V. Frey, S. E. H. Sakimoto, and J. H. Roark "MOLA TOPOGRAPHY AND THE ISIDIS BASIN: CONSTRAINTS ON BASIN CENTER AND RING DIAMETERS" LPSC98 available online at

http://mars.jpl.nasa.gov/mgs/sci/lpsc98/1631.pdf

[26] Figure 1a

[27] J. M. Moore (NASA Ames), A. D. Howard (U. Va.), P. M. Schenk (LPI) "[43.04] The Topography and Basin Deposits of the Equatorial Highlands: A MGS–Viking Synergistic Study" 31st Annual Meeting of the DPS, October 1999 Session 43. Mars Surface: Structure available online at

http://www.aas.org/publications/baas/v31n4/dps99/98.htm

[28] J.A.Iluhina and J.F.Rodionova "AUTOMATED MAKING THE MAP OF ISIDIS'S BASIN" Microsymposium 36, MS037, 2002 available online at

http://www.planetary.brown.edu/planetary/documents/Micro_36/Abstr acts/037_Iluhina_Rodionova.pdf

[29] http://www.harmakhis.org/history/2.jpg

[30]"Isidis is generally very flat with low average slope values calculated from MOLA data." The Natural History Museum "Missions to Mars" available   online   at
http://www.nhm.ac.uk/mineralogy/mars/Marshtml/2missions.html

[31] http://www.space4case.com/mars/mars7/mars143.html

[32] http://www.space4case.com/mars/mars7/mars142.html

[33] Page 4 J. D. Farmer "Exploring for Martian Life: Recent Results and Future Opportunities" Astrobiology Volume 1 Number 3 [2001] available online   at
http://216.239.57.104/search?q=cache:XenIThQGrU8J:cips.berkeley.edu /events/discussion_group_2003_spring/farmer_astrobiology_space_miss ions.pdf+farmer_astrobiology_space_missions.pdf&hl=en&ie=UTF-8

[34]First   image   Ames   Research   Centre   available   online   at
http://amesnews.arc.nasa.gov/releases/2002/02images/mars/mars.html

[35] "When the river valleys on Mars were confirmed in the 1970s, many scientists believed there once was an Earth-like period with warmth, rivers and oceans," said Owen Toon, a professor at the University of Colorado and a coauthor of the Science paper. "What sparked our interest was that the large craters and river valleys appeared to be about the same age." Another piece of evidence arguing against a condition warm, wet period on Mars are images of river channels without any sign of tributaries flowing into the main channel. "We definitely see river valleys but not tributaries, indicating the rivers were not as mature as those on Earth," said Toon. Jeff Foust "New research explores past, present water on Mars" Spaceflight Now December 5th 2002 available online at http://spaceflightnow.com/news/n0212/05mars/

[36] Nancy Ambrosiano "Los Alamos releases new maps of Mars water" Los Alamos National Laboratory [2003] available online at

http://www.lanl.gov/orgs/pa/News/cover_epi.jpg

[37] M. L. Litvak "DISTRIBUTION OF CHEMICALLY BOUND WATER IN SURFACE LAYER OF MARS BASED ON DATA ACQUIRED BY HIGH ENERGY NEUTRON SPECTROMETER, MARS ODYSSEY" Microsymposium 36, MS062, 2002

http://www.planetary.brown.edu/planetary/documents/Micro_36/Abstr acts/062_Litvak_etal.pdf

[38] Nadine Barlow "SOLIS PLANUM, MARS: THE "OASIS HYPOTHESIS" REVISITED "2001-2002 Colloquium Series NAU Liberal Arts (Bldg 18, Rm 135), Thursday, 24 January 2002 available online at http://www.phy.nau.edu/EVENTS/colloquium/speakers0102/barlow.ht ml

[39] N. G. Barlow "SUBSURFACE VOLATILE RESERVOIRS: CLUES FROM MARTIAN IMPACT CRATER MORPHOLOGIES" Fifth Conference 1999 http://mars.jpl.nasa.gov/mgs/sci/fifthconf99/6082.pdf

[40] Ames Research Centre "Mars Watering Hole Found, Scientists Say" available online at

http://astrobiology.arc.nasa.gov/news/expandnews.cfm?id=1009

[41] N. G. Barlow, C. B. Perez (U. Central Fl.), J. Koroshetz (U. Fl.) "[39.06] A Volatile-Rich Reservoir South of Valles Marineris, Mars" 31st Annual Meeting of the DPS, October 1999 Session 39. Mars Surface: Evidence of Change available online at http://www.aas.org/publications/baas/v31n4/dps99/40.htm

[42] Ibid.

[43] Bradley J. Thomson and James W. Head III "Utopia Basin, Mars: Characterization of topography and morphology and assessment of the origin and evolution of basin internal structure" JOURNAL OF GEOPHYSICAL RESEARCH, VOL. 106, NO. 0, PAGES 1–22, MONTH 2001 available online at

http://www.planetary.brown.edu/planetary/documents/2396.pdf

[44] N. G. Barlow, C. B. Perez (U. Central Fl.), J. Koroshetz (U. Fl.) "[39.06] A Volatile-Rich Reservoir South of Valles Marineris, Mars" 31st Annual Meeting of the DPS, October 1999 Session 39. Mars Surface: Evidence of Change available online at http://www.aas.org/publications/baas/v31n4/dps99/40.htm

[45] Leonard David "Mars Watering Hole Found, Scientists Say" Space.com posted: 03:00 pm ET 13 August 2001

http://space.com/scienceastronomy/solarsystem/mars_ice_010813.html

[46] WEBB, Benjamin M. "NOACHIAN TECTONICS OF SYRIA PLANUM AND THE THAUMASIA PLATEAU" Paper No. 132-0 [2001] available online at http://gsa.confex.com/gsa/2001AM/finalprogram/abstract_28019.htm

[47] Evelyn D. Scott "SUB-LITHOSPHERIC 'SUBDUCTION' ON MARS: CONVECTIVE REMOVAL OF A LITHOSPHERIC ROOT. III: SYRIA PLANUM REGION" Lunar and Planetary Science XXXI 1331.pdf http://www.lpi.usra.edu/meetings/lpsc2000/pdf/1331.pdf

[48] Ibid.

[49] http://www.space4case.com/mars/mars6/mars125.html

[50] David E. Smith "The Global Topography of Mars and Implications for Surface Evolution" www.sciencemag.org SCIENCE VOL 284 28 MAY 1999 available online at

http://www.ciw.edu/library/solomon/sci_284_1495.pdf

[51] William K. Hartmann "MARTIAN CRATER POPULATIONS AND OBLITERATION RATES: FIRST RESULTS FROM MARS GLOBAL SURVEYOR" 1998, LUNAR AND PLANETARY SCIENCE CONFERENCE 29 (HOUSTON) available online at

http://www.psi.edu/projects/mgs/lpsc.html

[52] MSSS "Wide Angle View of Arsia Mons Volcano" MGS MOC Release No.   MOC2-179,   27   September   1999 http://mars.jpl.nasa.gov/mgs/msss/camera/images/9_27_99_arsia/

[53] Calvin J. Hamilton "Arsia Mons" 1997 available online at

http://www.star.ucl.ac.uk/~apod/solarsys/cap/mars/arsia.htm

[54] J. W. Head , D. R. Marchant "COLD-BASED MOUNTAIN GLACIERS ON MARS: WESTERN ARSIA MONS" Geophysical Research Abstracts, Vol. 5, 02770, 2003c European Geophysical Society 2003 available online at

http://www.cosis.net/abstracts/EAE03/02770/EAE03-J-02770.pdf

[55] James W. Head and David R. Marchant "MOUNTAIN GLACIERS ON MARS?: WESTERN ARSIA MONS FAN-SHAPED DEPOSIT SMOOTH FACIES AS ROCK GLACIERS:" Microsymposium 36, MS103, 2002

http://www.planetary.brown.edu/planetary/documents/Micro_36/Abstr acts/103_Head_Marchant.pdf

[56] James W. Head and David R. Marchant "MOUNTAIN GLACIERS ON MARS? CHARACTERIZATION OF WESTERN ARSIA MONS FANSHAPED DEPOSITS USING MGS DATA:" Microsymposium 36, MS105, 2002

http://www.planetary.brown.edu/planetary/documents/Micro_36/Abstr acts/105_Head_Marchant.pdf

[57] And see Figure 1 K.F. Sprenke and L.L. Baker "Magnetization of Arsia Mons, Mars" Lunar and Planetary Science XXXIII (2002) 1070.pdf

http://www.lpi.usra.edu/meetings/lpsc2002/pdf/1070.pdf

[58] Johann Helgason "Does Mars Hide Vast Water Deposits" MARSDAILY.COM SPECIAL REPORT Reykjavik - June 10, 2000 available online athttp://www.spacedaily.com/news/mars-water-00b.html

[59]"The   Argyre   Basin" http://ltpwww.gsfc.nasa.gov/tharsis/argyre_insight.html
502

[60] See Figure 1 K. F. Sprenke and L. L. Baker "POLAR WANDERING ON MARS?" Lunar and Planetary Science XXXI 1930.pdf

http://www.lpi.usra.edu/meetings/lpsc2000/pdf/1930.pdf

[61] http://www.harmakhis.org/history/3.jpg

[62] http://www.harmakhis.org/history/4.jpg

[63] http://www.harmakhis.org/history/shockwave.jpg

[64] http://www.space4case.com/mars/mars5/mars110.html

[65] H. Hiesinger, J.W. Head III

"GEOLOGY OF THE ARGYRE BASIN, MARS: NEW INSIGHTS FROM MOLA AND MOC" Lunar and Planetary Science XXXII (2001) 1799.pdf

http://www.lpi.usra.edu/meetings/lpsc2001/pdf/1799.pdf

[66] http://www.lpi.usra.edu/meetings/lpsc2000/pdf/2033.pdf

Parker et al., [2000], LPSC XXXI, 2033.pdf

[67] R.A. De Hon "MARTIAN SEDIMENTARY BASINS AND REGIONAL WATERSHEDS" Lunar and Planetary Science XXXIII (2002) 1915.pdf

http://www.lpi.usra.edu/meetings/lpsc2002/pdf/1915.pdf

[68] See Figure 1 Lionel Wilson and James W. Head III "Tharsis-radial graben systems as the surface manifestation of plume-related dike intrusion complexes: Models and implications" JOURNAL OF GEOPHYSICAL RESEARCH, VOL. 107, NO. E8, 10.1029/2001JE001593, 2002

http://www.planetary.brown.edu/planetary/documents/2584.pdf

[69] Lori Stiles "Scientists Find Largest Flood Channels in the Solar System" University of Arizona uanews.org [2001] available online at

http://uanews.opi.arizona.edu/cgi-bin/WebObjects/UANews.woa/wa/SRStoryDetails?ArticleID=3995

[70] http://www.lanl.gov/orgs/pa/News/cover_epi.jpg

[71] http://www.harmakhis.org/history/5.jpg

[72] http://www.harmakhis.org/history/6.jpg

[73] BY THE VIKING ORBITER IMAGING TEAM "Deformational features" NASA SP-441

http://history.nasa.gov/SP-441/ch6.htm

[74] http://www.harmakhis.org/history/7.jpg

[75] J. A. Grant "Valley Evolution in Margaritifer Sinus, Mars" Available online at

http://www.nasm.si.edu/ceps/research/grant/grant_marg2.pdf

[76] Ibid. available online at

http://www.nasm.si.edu/ceps/research/grant/grant_marg2.pdf

[77] Ibid. Available online at

http://www.nasm.si.edu/ceps/research/grant/grant_marg2.pdf

[78] Goddard Space Flight Centre "View inside Mars reveals rapid cooling and buried channels" March 9 2000 available online at http://www.gsfc.nasa.gov/topstory/20000309mars.html

[79] John A. Grant "DRAINAGE EVOLUTION IN MARGARITIFER SINUS, MARS" Paper No. 132-0 GSA Annual Meeting, November 5-8, 2001 Boston, Massachusetts http://gsa.confex.com/gsa/2001AM/finalprogram/abstract_27669.htm

[80] Dave Williams "Parana Valles drainage system in Margaritifer Sinus, Mars" available online at

http://nssdc.gsfc.nasa.gov/imgcat/html/object_page/vol_084a47.html

[81] Brian M. Hynek and Roger J. Phillips "Evidence for extensive denudation of the Martian highlands" available online at

http://ltpwww.gsfc.nasa.gov/tharsis/hynek.erosion.pdf

[82] Ibid. available online at

http://ltpwww.gsfc.nasa.gov/tharsis/hynek.erosion.pdf

[83]   Ibid.   Available   online   at
http://ltpwww.gsfc.nasa.gov/tharsis/hynek.erosion.pdf

[84] Bruce Moomaw "Mars: A World of Varied Catastrophes" MARSDAILY May 1, 2001 available online athttp://www.spacedaily.com/news/lunarplanet-2001-01a2.html

[85] Roger J. Phillips et al "Ancient Geodynamics and Global-Scale Hydrology on Mars" www.sciencemag.org SCIENCE VOL 291 30 MARCH 2001 available online at

http://ltpwww.gsfc.nasa.gov/tharsis/phillips.tharsis.pdf

[86] Goddard Space Flight Center Educational Programs [2002] image of Valles Marineris available online at http://education.gsfc.nasa.gov/experimental/all98invProject.Site/Pages/Vallis.Marineris.html

[87] Image of Valles Marineris

http://www.astronomija.co.yu/suncsist/planete/Mars/marindetalj.htm

[88] http://www.mmedia.is/~bjj/planet_rend/mars_vallesm.jpg

[89] Herbert Frey Geodynamics 2001 The Year in Review available online at

http://denali.gsfc.nasa.gov/annual2001/mgg6

[90] M.E. Purucker et al "Interpretation of a magnetic map of the Valles Marineris region, Mars" available online at

http://denali.gsfc.nasa.gov/terr_mag/abstract_mars.pdf

[91] Bruce Murray [1999] "PALEOLAKE DEPOSITS IN CENTRAL VALLES MARINERIS: A UNIQUE OPPORTUNITY FOR 2001" Second Mars Surveyor landing site Workshop available online at

http://web99.arc.nasa.gov/~vgulick/MSLS99_Wkshp/Murray_Paleolakes_VM_abs.pdf

[92] Ibid. available online at

http://web99.arc.nasa.gov/~vgulick/MSLS99_Wkshp/Murray_Paleolakes_VM_abs.pdf

[93]   Valles   Marineris   Outflow   Site   (MER-A) http://marsoweb.nas.nasa.gov/landingsites/mer2003/topsites/VMout/zoom.html

[94] A.P. Rossi, G. Komatsu, and J.S. Kargel "[46.03] Flow-like features in Valles Marineris, Mars: Possible ice-driven creep processes" 31st Annual Meeting   of   the   DPS,   October   1999 Session 46. Mars Surface: Evidence of Change Posters available online at http://www.aas.org/publications/baas/v31n4/dps99/158.htm

[95] http://www.harmakhis.org/history/7.jpg

[96] Ronald Greeley and Ruslan Kuzmin "SHALBATANA VALLIS: A POTENTIAL SITE FOR ANCIENT GROUND WATER" MSL99 Workshop available online at

http://web99.arc.nasa.gov/~vgulick/MSLS99_Wkshp/Greeley_Kuzmin_Shalbat_abs%20.pdf

[97] D.A. van der Kolk et al "ORCUS PATERA, MARS: IMPACT CRATER OR VOLCANIC CALDERA?" Lunar and Planetary Science XXXII (2001) 1085.pdf

http://www.geology.pomona.edu/Mars2000/1085.pdf

506

[98] D.M. Nelson, R. Greeley "XANTHE TERRA OUTFLOW CHANNEL GEOLOGY AT THE MARS PATHFINDER LANDING SITE" Lunar and Planetary Science XXIX 1158.pdf

http://mars.jpl.nasa.gov/MPF/science/lpsc98/1158.pdf

[99] Ibid. http://mars.jpl.nasa.gov/MPF/science/lpsc98/1158.pdf

[100] http://www.lanl.gov/orgs/pa/News/cover_epi.jpg

[101] http://www.harmakhis.org/history/8.jpg

[102] http://www.harmakhis.org/history/9.jpg

[103] Mike Caplinger February 1994 "Determining the age of surfaces on Mars" available online at

http://www.msss.com/http/ps/age2.html

[104] http://www.harmakhis.org/history/9.jpg

[105] A.D. Howard "Features of Martian Cratered Terrain" GEOMORPHOLOGY HOME PAGE University of Virginia available online at

http://erode.evsc.virginia.edu/marscrat.htm

[106] N. Hoffman et al "EMPLACEMENT OF A DEBRIS OCEAN ON MARS BY REGIONAL-SCALE COLLAPSE AND FLOW AT THE CRUSTAL DICHOTOMY" Lunar and Planetary Science XXXII (2001) 1584.pdf

http://www.lpi.usra.edu/meetings/lpsc2001/pdf/1584.pdf

[107] K. F. Sprenke and L. L. Baker "POLAR WANDERING ON MARS?" Lunar and Planetary Science XXXI 1930.pdf

http://www.lpi.usra.edu/meetings/lpsc2000/pdf/1930.pdf

[108] http://www.lpi.usra.edu/meetings/lpsc2002/pdf/1811.pdf

[109] R. C. Anderson et al "COMPARATIVE INVESTIGATION OF THE GEOLOGICAL HISTORIES AMONG ALBA PATERA AND SYRIA PLANUM, MARS" Lunar and Planetary Science XXXIII (2002) 1811.pdf

http://www.lpi.usra.edu/meetings/lpsc2002/pdf/1811.pdf

[110] http://www.harmakhis.org/history/10.jpg

[111] J. Richardon "Isostacy in the Hellas Basin on Mars" available online at http://www.lpl.arizona.edu/~jrich/work/hellasslides.pdf

[112] http://www.lanl.gov/orgs/pa/News/cover_epi.jpg

[113] http://www.harmakhis.org/history/11.jpg

[114] http://www.harmakhis.org/history/6.jpg

[115] http://www.harmakhis.org/history/13.jpg

[116] R. N. Clark and T. M. Hoefen "New Evidence Suggests Mars Has Been Cold and Dry "Red Planet" Abundant with Green Minerals" available online at http://speclab.cr.usgs.gov/mars.press.release.10.2000.html

[117] N. Hoffman "Water on mars? Who are they trying to kid?" School of Earth Sciences University of Melbourne available online at

http://www.earthsci.unimelb.edu.au/mars/

[118] B. J. Thomson and J. W. Head "THE ROLE OF WATER/ICE IN THE RESURFACING HISTORY OF HELLAS BASIN" Fifth International Conference on Mars 6200.pdf http://www.lpi.usra.edu/meetings/5thMars99/pdf/6200.pdf

[119] Ibid. available online at http://www.lpi.usra.edu/meetings/5thMars99/pdf/6200.pdf

[120] Ibid. available online at http://mars.jpl.nasa.gov/mgs/sci/fifthconf99/6200.pdf

[121] Bradley J. Thomson and James W. Head "HELLAS BASIN, MARS: EXAMINATION OF A GLACIAL HYPOTHEISIS WITH MOLA TOPOGRAPHY" 32-th Vernadsky-Brown Microsymposium / Abstracts available online at http://www.geokhi.ru/~planetology/Abstracts/Thompson%20et%20al.pdf

[122] Jakupova A. E. et al "PREPARATION OF AN ATLAS OF THE CRATERING OF MARS" 38th Vernadsky/Brown Microsymposium on Comparative Planetology available online at

http://www.geokhi.ru/~planetology/Abstracts/Jakupova%20et%20al.pdf

[123] NASA Mars Global Surveyor Project; MOLA Team "Possible configuration of ancient oceans on Mars: Topographic portrayal of the surface of Mars derived from Mars Orbiter Laser Altimeter (MOLA) data" available online at

http://www.brown.edu/Administration/News_Bureau/1999-00/99-060g.html

[124] See Figure 2 for shorelines and Figures 3 A to F for polygons and craters with ejecta lobes.

James W. Head III et al. "Possible Ancient Oceans on Mars: Evidence from Mars Orbiter Laser Altimeter Data" 10 DECEMBER 1999 VOL 286 SCIENCE www.sciencemag.org available online at

http://ltpwww.gsfc.nasa.gov/tharsis/mola.ocean.pdf

[125] http://www.harmakhis.org/history/14.jpg

[126] http://www.harmakhis.org/history/15.jpg

[127] Oded Aharonson, Maria T. Zuber and Gregory A. Neumann "Mars: Northern hemisphere slopes

and slope distributions" GEOPHYSICAL RESEARCH LETTERS, VOL. 25, NO. 24, PAGES 4413-4416, DECEMBER 15, 1998 available online at http://ltpwww.gsfc.nasa.gov/tharsis/grl98_slopes.pdf

[128] NASA/JPL/Arizona State University [2003] "Mars Odyssey THEMIS Image: Lucus Planum" http://www.marstoday.com/viewsr.html?pid=8867

[129] Cabrol et al "Duration of the Ma'adim Vallis/Gusev crater hydrogeologic system, Mars" *Icarus* 133, 98-108 [1998].

[130] http://www.space4case.com/mars/mars5/mars112.html

[131] Elizabeth R. Fuller and James W. Head III "Amazonis Planitia: The role of geologically recent volcanism and sedimentation in the formation of the smoothest plains on Mars" JOURNAL OF GEOPHYSICAL RESEARCH, VOL. 107, NO. E10, 5081, doi:10.1029/2002JE001842, 2002 available online at

http://www.planetary.brown.edu/planetary/documents/2682.pdf

[132] E. R. Fuller and J. W. Head, III "PROPOSING A HIGH VOLATILE CONTENT IN THE EQUATORIAL LAYERED DEPOSITS INCLUDING THE MEDUSAE FOSSAE FORMATION, MARS" Microsymposium 36, MS022, 2002 Available online at

http://www.planetary.brown.edu/planetary/documents/Micro_36/Abstracts/022_Fuller_Head.pdf

[133] George E. McGill "Geologic Map Transecting the Highland/Lowland Boundary Zone, Arabia Terra, Mars: Quadrangles 30332, 35332, 40332, AND 45332" available online at http://geopubs.wr.usgs.gov/i-map/i2746/

[134] T. R. Watters "THE TECTONICS AND TOPOGRAPHY OF THE DICHOTOMY BOUNDARY IN THE EASTERN HEMISPHERE OF MARS" Lunar and Planetary Science XXXIII (2002) 1692.pdf

http://www.lpi.usra.edu/meetings/lpsc2002/pdf/1692.pdf

[135] R. C. Anderson et al. "COMPARATIVE INVESTIGATION OF THE GEOLOGICAL HISTORIES AMONG ALBA PATERA AND SYRIA PLANUM, MARS" Lunar and Planetary Science XXXIII (2002) 1811.pdf

http://www.lpi.usra.edu/meetings/lpsc2002/pdf/1811.pdf
510

[136] James W. Head III et al. "Northern lowlands of Mars: Evidence for widespread volcanic flooding and tectonic deformation in the Hesperian Period" JOURNAL OF GEOPHYSICAL RESEARCH, VOL. 107, NO. 0, 10.1029/2000JE001445, 2002 available online at

http://www.planetary.brown.edu/planetary/documents/2575.pdf

[137] Mikhail A. Kreslavsky and James W. Head "Fate of outflow channel effluents in the northern lowlands of Mars: The Vastitas Borealis Formation as a sublimation residue from frozen ponded bodies of water" JOURNAL OF GEOPHYSICAL RESEARCH, VOL. 107, NO. E12, 5121, doi:10.1029/2001JE001831, 2002 available online at

http://www.planetary.brown.edu/planetary/documents/2686.pdf

[138] K. L. Tanaka et al "RESURFACING OF THE NORTHERN PLAINS OF MARS BY SHALLOW SUBSURFACE, VOLATILEDRIVEN ACTIVITY" Lunar and Planetary Science XXXIII (2002) 1406.pdf

http://www.lpi.usra.edu/meetings/lpsc2002/pdf/1406.pdf

[139] Nick Hoffman and Ken Tanaka "CO-EXISTING "FLOOD" AND "VOLCANIC" MORPHOLOGIES IN ELYSIUM AS EVIDENCE FOR

COLD CO2 OR WARM H2O OUTBURSTS" Lunar and Planetary Science XXXIII (2002) 1505.pdf

http://www.earthsci.unimelb.edu.au/mars/LPSC_2002_1505_Athabasca.pdf

[140] D. M. Burr et al "RECENT FLUVIAL ACTIVITY IN AND NEAR MARTE VALLIS, MARS" Lunar and Planetary Science XXXI 1951.pdf

http://www.lpi.usra.edu/meetings/lpsc2000/pdf/1951.pdf

[141] See images in http://webgis.wr.usgs.gov/mer/March_2002_presentations/Burr/Burr-Landingsite3.pdf

[142] D. M. Burr "RECENT FLUVIAL ACTIVITY IN AND NEAR MARTE VALLIS, MARS" Lunar and Planetary Science XXXI 1951.pdf

http://www.lpi.usra.edu/meetings/lpsc2000/pdf/1951.pdf

[143] D. M. Burr et al "EXTENSIVE AQUEOUS FLOODING FROM THE CERBERUS FOSSAE, MARS, AND ITS IMPLICATIONS FOR THE MARTIAN HYDROSPHERE" Lunar and Planetary Science XXXIII (2002) 1047.pdf

http://www.lpi.usra.edu/meetings/lpsc2002/pdf/1047.pdf

[144] D. M. Burr "TEMPORARY PONDING OF FLOODWATER IN ATHABASCA VALLIS, MARS" Lunar and Planetary Science XXXIV (2003)   1066.pdf
http://www.lpi.usra.edu/meetings/lpsc2003/pdf/1066.pdf

[145] Burr, D. M "4. Source of the Flood Water" GEOPHYSICAL RESEARCH LETTERS, VOL. 29, NO. 1, 10.1029/2001GL013345, 2002
http://www.agu.org/pubs/sample_articles/sp/2001GL013345/4.shtml

[146] S. De Silva "Elysium Mons" University of South Dakota available online at

http://volcano.und.nodak.edu/vwdocs/planet_volcano/mars/Shields/elysium.html

[147] Mars Odyssey and MGS Mars Orbital Laser Altimeter (MOLA) Science teams "Mars Odyssey Orbiter Watches a Frosty Mars" [2003] available online at

http://mars.jpl.nasa.gov/odyssey/newsroom/pressreleases/20030626a.html

[148] Lunar and Planetary Science XXX 1185.pdf

http://www.lpi.usra.edu/meetings/LPSC99/pdf/1185.pdf

[149] M. G. Chapman "2001 SITE IN NORTH TERRA MERIDIANI: THE TES CONCENTRATION AREA" MSL99 Workshop available online at

512

http://web99.arc.nasa.gov/~vgulick/MSLS99_Wkshp/Chapman_Hem_abs_pg1.pdf

[150] http://www.harmakhis.org/history/16.jpg

[151] N. Hoffman "Outburst floods as cold and dry avalanches" available online at

http://www.earthsci.unimelb.edu.au/mars/Outburst.html

[152] http://www.harmakhis.org/history/19.jpg

[153] S. M. Milkovich and J. W. Head, III "OLYMPUS MONS FAN SHAPED DEPOSIT MORPHOLOGY: EVIDENCE FOR DEBRIS GLACIERS" Sixth International Conference on Mars (2003) 3149.pdf

http://www.lpi.usra.edu/meetings/sixthmars2003/pdf/3149.pdf

[154] See Figure 1, J. W. Head "MOUNTAIN GLACIERS ON MARS? CHARACTERIZATION OF WESTERN THARSIS MONTES FAN SHAPED DEPOSITS USING MGS DATA" Mars atmosphere modeling and observations workshop [2002] available online at

http://www-mars.lmd.jussieu.fr/granada2003/abstract/head.pdf

[155] "NASA's Revealing Odyssey" MEDIA RELATIONS OFFICE JET PROPULSION   LABORATORY
http://mars.jpl.nasa.gov/odyssey/newsroom/pressreleases/20021207a.html

[156] Mars Odyssey and MGS Mars Orbital Laser Altimeter (MOLA) Science teams "Mars Odyssey Orbiter Watches a Frosty Mars" available online   at
http://mars.jpl.nasa.gov/odyssey/newsroom/pressreleases/20030626a.html

[157] T. N. Titus (Oak Ridge Associated Universities), H. H. Kieffer, K. F. Mullins (U.S. Geological Survey) "TES Observations of the South Pole" available online at http://www.mars-ice.org/crocus.html

[158] Timothy N. Titus, Hugh H. Kieffer, Kevin F. Mullins, Phillip Christensen "Slab Ice and Snow Flurries in the Mars Northern Polar Night" available online at

http://www.mars-ice.org/cold.html

[159] Lunar and Planetary Science XXXIV (2003) 3112.pdf

http://www.lpi.usra.edu/meetings/sixthmars2003/pdf/3112.pdf

[160] Asmin V. Pathare and David A. Paige "Enhanced Ice Flow at High Martian Obliquity: A Rheological Model of the Polar Layered Deposits" Lunar and Planetary Science XXXI 1571.pdf

http://www.lpi.usra.edu/meetings/lpsc2000/pdf/1571.pdf

[161] Malin Space Science Systems Team "Evidence for Recent Climate Change on Mars" available online at

http://mars.jpl.nasa.gov/mgs/msss/camera/images/CO2_Science_rel/malin_etal.html

[162] N. Hoffman LPSC [2001] available online at

http://www.earthsci.unimelb.edu.au/mars/LPSC2001_Hoffman_Polar.pdf

[163]
http://www.msss.com/moc_gallery/m07_m12/images/M10/M1003736.html

[164] http://www.harmakhis.org/paper/

[165] They can all be seen online at

http://www.harmakhis.org/

[166] http://www.harmakhis.org/paper/water/

[167]

[168] http://www.harmakhis.org/paper/water/craterchanneldata.htm

[169] http://www.lanl.gov/orgs/pa/News/cover_epi.jpg

specifically

http://www.harmakhis.org/history/9.jpg

[170] MGS MOC Releases MOC2-234 to MOC2-245, 22 June 2000 available online at http://mars.jpl.nasa.gov/mgs/msss/camera/images/june2000/

[171] K. S. Edgett et al "POLAR- AND MIDDLE-LATITUDE MARTIAN GULLIES: A VIEW FROM MGS MOC AFTER 2 MARS YEARS IN THE MAPPING ORBIT" Lunar and Planetary Science XXXIV (2003) 1038.pdf

http://www.lpi.usra.edu/meetings/lpsc2003/pdf/1038.pdf

[172] N. Mangold et al "FORMATION OF GULLIES ON MARS: WHAT DO WE LEARN FROM EARTH?" Sixth International Conference on Mars (2003) 3048.pdf

http://www.lpi.usra.edu/meetings/sixthmars2003/pdf/3048.pdf

[173] K.S. Edgett et al "POLAR- AND MIDDLE-LATITUDE MARTIAN GULLIES: A VIEW FROM MGS MOC AFTER 2 MARS

YEARS IN THE MAPPING ORBIT" Lunar and Planetary Science XXXIV (2003) 1038.pdf

http://www.lpi.usra.edu/meetings/lpsc2003/pdf/1038.pdf

[174] William K. Hartmann et al "COMPARISON OF ICELANDIC AND MARTIAN HILLSIDE GULLIES" Lunar and Planetary Science XXXIII (2002) 1904.pdf

http://www.lpi.usra.edu/meetings/lpsc2002/pdf/1904.pdf

[175] "Global distribution of observed gully landforms" diagram in "True Colors of Mars by Schmidt [2001] available online at

http://silver.neep.wisc.edu/~neep533/FALL2001/lecture19.pdf

http://www.harmakhis.org/history/gullies.jpg

[176] "Evidence for Recent Liquid Water on Mars: Gullies in Gorgonum Chaos" MGS MOC Release No. MOC2-236, 22 June 2000 available online at http://www.msss.com/mars_images/moc/june2000/gorgonum/

[177] "Mars Odyssey THEMIS Image: Gullies of Gorgonus Chaos" Mars Odyssey THEMIS Tuesday, June 11, 2002 available online at http://www.marstoday.com/viewsr.html?pid=5736

[178] A. D. Howard and J. M. Moore "THE CURIOUS SHORELINES OF GORGONUM CHAOS" Sixth International Conference on Mars (2003) 3190.pdf http://www.lpi.usra.edu/meetings/sixthmars2003/pdf/3190.pdf

[179] G. Leone "GORGONUM CHAOS: ARE THE SEEPAGE-RUNOFF FEATURES REALLY RECENT?" Lunar and Planetary Science XXXII (2001) 1649.pdf http://www.lpi.usra.edu/meetings/lpsc2001/pdf/1649.pdf

[180] J. M. Moore and A. D. Howard "ARIADNES-GORGONUM KNOB FIELDS OF NORTH-WESTERN TERRA SIRENUM, MARS" Lunar and Planetary Science XXXIV (2003) 1402.pdf

http://www.lpi.usra.edu/meetings/lpsc2003/pdf/1402.pdf

[181] Sarah T. Stewart and Francis Nimmo "Surface runoff features on Mars:Testing the carbon dioxide formation hypothesis" JOURNAL OF GEOPHYSICAL RESEARCH, VOL. 107, NO. E9, 5069, doi:10.1029/2000JE001465, 2002

http://bullard.esc.cam.ac.uk/~nimmo/paper16.pdf

[182] Bruce Moomaw "The Case For Outgassing" July 5, 2000 MARSDAILY.COM available online at http://www.spacedaily.com/news/mars-water-science-00g2.html

[183] Michael C. Malin and Kenneth S. Edgett "Evidence for Recent Groundwater Seepage and Surface Runoff on Mars" 30 JUNE 2000 VOL 288 SCIENCE www.sciencemag.org available online at

http://geoinfo.nmt.edu/penguins/pdfs/se260002330p.pdf

[184] Malin Space Science Systems, MGS, JPL, NASA "Newton Crater: Evidence for Recent Water on Mars" Astronomy Picture of the Day available online at

http://antwrp.gsfc.nasa.gov/apod/ap000626.html

[185] N. A. Cabrol et al "PROLONGED PONDING EPISODE IN C-NEWTON CRATER IN RECENT GEOLOGICAL TIMES ON MARS" Lunar and Planetary Science XXXII (2001) 1255.pdf

http://www.lpi.usra.edu/meetings/lpsc2001/pdf/1255.pdf

[186] B.W. Harrington et al "EXTENSION ACROSS TEMPE TERRA, MARS FROM MOLA TOPOGRAPHIC MEASUREMENTS" 5th Conference [1999]

http://mars.jpl.nasa.gov/mgs/sci/fifthconf99/6130.pdf

[187] http://www.space4case.com/mars/mars5/mars131.html

[188] M. P. Wong et al "MOLA TOPOGRAPHY OF SMALL VOLCANOES IN TEMPE TERRA AND CERAUNIUS FOSSAE, MARS: IMPLICATIONS FOR ERUPTIVE STYLE" Lunar and Planetary Science XXXII (2001) 1563.pdf

http://www.lpi.usra.edu/meetings/lpsc2001/pdf/1563.pdf

[189] See Figure 1, E. Hauber et al "MORPHOLOGY AND TOPOGRAPHY OF FRETTED TERRAIN AT THE DICHOTOMY BOUNDARY

IN TEMPE TERRA, MARS: GENERAL CHARACTERISTICS" Lunar and Planetary Science XXXIII (2002) 1658.pdf

http://www.lpi.usra.edu/meetings/lpsc2002/pdf/1658.pdf

[190] "Evidence for Recent Liquid Water on Mars: South-facing Walls of Nirgal Vallis" MGS MOC Release No. MOC2-240, 22 June 2000" http://www.msss.com/mars_images/moc/june2000/nirgal/

[191] NASA/JPL/Arizona State University "Nirgal Vallis (Released 27 March 2002)" http://themis.la.asu.edu/zoom-20020327a.html

[192] J. R. Zimbelman "DECAMETER-SCALE RIPPLE-LIKE FEATURES IN NIRGAL VALLIS AS REVEALED IN THEMIS

AND MOC IMAGING DATA" Sixth International Conference on Mars (2003) 3028.pdf

http://www.lpi.usra.edu/meetings/sixthmars2003/pdf/3028.pdf

[193] Pascal Lee and James W. Rice Jnr. "SMALL VALLEYS NETWORKS ON MARS: THE GLACIAL MELTWATER CHANNEL NETWORKS OF DEVON ISLAND, NUNAVUT TERRITORY, ARCTIC CANADA, AS POSSIBLE ANALOGS" Fifth International Conference on Mars 6237.pdf http://www.lpi.usra.edu/meetings/5thMars99/pdf/6237.pdf

[194] "Autumn Afternoon in Hale Crater"MGS MOC Release No. MOC2-257, 17 November 2000 http://www.msss.com/mars_images/moc/nov_00_hale/

[195] Figure 6 Nature Insight: Review article Hale Crater http://www.nature.com/nature/journal/v412/n6843/fig_tab/412228a0_F6.html

[196] J. D. Arfstrom "UPPER DAO VALLIS: A BASIN DOMINATED BY ICE-RICH VISCOUS MATERIALS" Lunar and Planetary Science XXXIV (2003) 1208.pdf

http://www.lpi.usra.edu/meetings/lpsc2003/pdf/1208.pdf

[197] J. D. Arfstrom "PROPOSED MARTIAN GLACIERS OF RECENT AGE AND A MODEL OF THEIR FORMATION" Lunar and Planetary Science XXXIII (2002) 1092.pdf

http://www.lpi.usra.edu/meetings/lpsc2002/pdf/1092.pdf

[198] http://www.harmakhis.org/history/gullies.jpg

[199] "Iceball mars?" 11 APRIL 2003 VOL 300 SCIENCE www.sciencemag.org available online at

http://www.planetary.brown.edu/planetary/international/write_up.pdf

[200] Vanessa Thomas "Snow May Have Carved Martian Gullies" Astronomy.com available online at

http://www.astronomy.com/Content/Dynamic/Articles/000/000/001/215vpcsq.asp

[201] N. Mangold et al "FORMATION OF GULLIES ON MARS: WHAT DO WE LEARN FROM EARTH?" Sixth International Conference on Mars (2003) 3048.pdf http://www.lpi.usra.edu/meetings/sixthmars2003/pdf/3048.pdf

[202] Agnieszka Przychodzen "Exotic CO2 Process May Have Carved Martian Gullies" MARSDAILY April 2, 2001 available online at http://www.spacedaily.com/news/mars-water-science-01f.html

[203] See Figure 3 Norbert Schorghofer et al "Slope streaks on Mars: Correlations with surface properties and the potential role of water" GEOPHYSICAL RESEARCH LETTERS, VOL. 29, NO. 23, 2126, doi:10.1029/2002GL015889, 2002 available online at

http://www.gps.caltech.edu/~oa/publications/schorghofer2002_grl.pdf

[204] Efrain Palermo, Jill England and Harry Moore "A Study of

Mars Global Surveyor (MGS)Mars Orbital Camera (MOC) Images

Showing Probable Water Seepages. Are They Dust Slides as NASA Claims or Proof of Water on Mars?" available online at

http://www.eskimo.com/~jill/seeps_paper.pdf

[205] See Figure 3 R. Sullivan et al "MASS-WASTING SLOPE STREAKS IMAGED BY THE MARS ORBITER CAMERA" Lunar and Planetary Science XXXI 1911.pdf

http://www.lpi.usra.edu/meetings/lpsc2000/pdf/1911.pdf

[206] R. Sullivan et al "MASS-MOVEMENT CONSIDERATIONS FOR DARK SLOPE STREAKS IMAGED BY THE MARS ORBITER CAMERA " Lunar and Planetary Science XXX 1809.pdf

http://www.lpi.usra.edu/meetings/LPSC99/pdf/1809.pdf

[207] http://ida.wr.usgs.gov/html/ab1020/ab102003.html

[208] http://www.harmakhis.org/ab102003dspires.jpg.htm

[209]
http://www.msss.com/moc_gallery/m19_m23/images/M23/M2300332.html

[210]
http://www.msss.com/moc_gallery/e07_e12/images/E11/E1103683.html

[211] P.R. Christensen et al "THE DISTRIBUTION OF CRYSTALLINE HEMATITE ON MARS FROM THE THERMAL EMISSION

SPECTROMETER: EVIDENCE FOR LIQUID WATER" Lunar and Planetary Science XXXI 1627.pdf

http://www.lpi.usra.edu/meetings/lpsc2000/pdf/1627.pdf

[212] Melissa D. Lane et al "UPDATE ON STUDIES OF THE MARTIAN HEMATITE-RICH AREAS" Lunar and Planetary Science XXXII (2001) 1984.pdf
http://www.lpi.usra.edu/meetings/lpsc2001/pdf/1984.pdf

[213] L.R. Gaddis et al "MINERAL MAPPING IN VALLES MARINERIS, MARS: A NEW APPROACH TO SPECTRAL DEMIXING OF TES DATA" Lunar and Planetary Science XXXIV (2003) 1956.pdf

http://www.lpi.usra.edu/meetings/lpsc2003/pdf/1956.pdf

[214] F. S. Anderson et al "MINERALOGY OF THE VALLES MARINERIS FROM TES AND THEMIS" Sixth International Conference on Mars (2003) 3280.pdf

http://www.lpi.usra.edu/meetings/sixthmars2003/pdf/3280.pdf

[215] Timothy D. Glotch and Philip R. Christensen "The Geology of Aram Chaos" Lunar and Planetary Science XXXIV (2003) 2046.pdf

http://www.lpi.usra.edu/meetings/lpsc2003/pdf/2046.pdf

[216] http://www.harmakhis.org/paper/dunes/dunes.htm

[217] http://www.harmakhis.org/paper/dunes/dunesdata.htm

[218] M. Malin and B. A. Cantor "MARS ORBITER CAMERA CLIMATE OBSERVATIONS" available online at http://www-mars.lmd.jussieu.fr/granada2003/abstract/malin.pdf

[221] http://www.harmakhis.org/paper/layers/layers.htm

[222] http://www.harmakhis.org/paper/layers/layersdata.htm

[223] Ibid. available online at

http://www-mars.lmd.jussieu.fr/granada2003/abstract/malin.pdf

[224] Senior Science Writer "MORE IMAGES: Martian Sediment Layers Explained" 05 December 2000 available online at http://www.space.com/scienceastronomy/solarsystem/mars_sediment_pics_001205.html

[225] Science@NASA "Layers of Mars" available online at

http://science.nasa.gov/headlines/y2001/ast23jan_1.htm

[226] "Iceball Mars?" 11 APRIL 2003 VOL 300 SCIENCE www.sciencemag.org available online at

http://www.planetary.brown.edu/planetary/international/write_up.pdf

[227] N. Hoffman "Origin of Layering on Mars" available online at http://www.earthsci.unimelb.edu.au/mars/Layer.html

[228] "Olivine points to dry Mars" June 6, 2003 available online at http://www.geolsoc.org.uk/template.cfm?name=THEMIS

[229] EDGETT, Kenneth S "SEDIMENTARY ROCK OUTCROPS OF NORTHERN TERRA MERIDIANI, MARS" Paper No. 26-7 2002 Denver Annual Meeting (October 27-30, 2002) Denver, CO available online at http://gsa.confex.com/gsa/2002AM/finalprogram/abstract_42548.htm

[230] Jenkins, Gregory S. 2001 "High-obliquity simulations for the Archean Earth: Implications for climatic conditions on early Mars" Journal of Geophysical Research - Planets - Vol. 106, No. E12 http://www.agu.org/pubs/toc/je/old/je106_12.html

[231] For a good novel based on life surviving these impacts that may have killed the dinosaurs there is "Evolution" by Steven Baxter.

---

[i] http://exoplanet.eu/

[ii] http://www.universetoday.com/74640/new-earth-sized-exoplanet-is-in-star%E2%80%99s-habitable-zone/

[iii] http://en.wikipedia.org/wiki/Panspermia

[iv] http://en.wikipedia.org/wiki/Artificial_intelligence

[v] http://en.wikipedia.org/wiki/Pareidolia

[vi] http://www.amazon.com/Ruins-Extraterrestrial-Eric-T-Reynolds/dp/0978514866

[vii] http://en.wikipedia.org/wiki/Terraform

[viii] http://en.wikipedia.org/wiki/Valles_Marineris

ix http://en.wikipedia.org/wiki/Tharsis

x http://en.wikipedia.org/wiki/Martian_rovers

xi http://en.wikipedia.org/wiki/Northern_lowlands_of_Mars

xii http://en.wikipedia.org/wiki/Magma

xiii http://en.wikipedia.org/wiki/Hellas_Planitia

xiv http://en.wikipedia.org/wiki/Life_on_Mars

xv http://news.nationalgeographic.com/news/2008/05/080529-mars-salty.html

xvi http://news.nationalgeographic.com/news/2009/05/090511-mars-asteroid.html

xvii http://en.wikipedia.org/wiki/Mount_Rushmore

xviii http://en.wikipedia.org/wiki/Interstellar_probe

xix http://en.wikipedia.org/wiki/SETI

xx http://en.wikipedia.org/wiki/Sting_operation

xxi http://en.wikipedia.org/wiki/Extraordinary_claims_require_extraordinary_evidence

xxii http://en.wikipedia.org/wiki/Carl_sagan

xxiii http://en.wikipedia.org/wiki/Extraterrestrial_life

xxiv http://en.wikipedia.org/wiki/Fringe_science

xxv http://en.wikipedia.org/wiki/Mainstream

xxvi http://en.wikipedia.org/wiki/Xenoarcheology

xxvii http://science.nasa.gov/science-news/science-at-nasa/2001/ast24may_1/

xxviii http://www.bis-spaceflight.com/sitesia.aspx/page/358/id/787/l/en

xxix http://www.csicop.org/si/show/ufos_and_aliens_in_space

xxx http://en.wikipedia.org/wiki/Cydonia_%28region_of_Mars%29

xxxi http://www.skepdic.com/faceonmars.html

xxxii http://en.wikipedia.org/wiki/Argyre_Planitia

xxxiii http://en.wikipedia.org/wiki/Terraforming_of_Mars

xxxiv http://news.nationalgeographic.com/news/2009/02/090218-water-mars-phoenix.html

xxxv http://www.spacedaily.com/news/mars-terraform-04h.html

xxxvi http://hernadi-key.blogspot.com/2009/06/lab-experiment-regarding-co2-snow-in.html

xxxvii http://science.nasa.gov/science-news/science-at-nasa/2008/21nov_plasmoids/

xxxviii http://www.lanl.gov/orgs/pa/News/cover_epi.jpg

xxxix http://en.wikipedia.org/wiki/Space_colonization

xl http://www.lpi.usra.edu/meetings/lpsc2000/pdf/1930.pdf

xli http://news.sciencemag.org/sciencenow/2010/06/did-a-deep-sea-once-cover-mars.html

xlii http://www.nasa.gov/mission_pages/mars/images/pia09027.html

xliii http://en.wikipedia.org/wiki/Groundwater

xliv http://www.nasa.gov/mission_pages/mars/images/pia09027.html

xlv http://en.wikipedia.org/wiki/Impact_crater

xlvi http://mgs-mager.gsfc.nasa.gov/Kids/magfield.html

xlvii http://en.wikipedia.org/wiki/Solar_wind

xlviii http://science.nasa.gov/science-news/science-at-nasa/2001/ast01may_1/

xlix http://en.wikipedia.org/wiki/Tardigrade

l http://www.spaceref.com/news/viewpr.html?pid=16978

li http://journalofcosmology.com/Mars103.html

lii http://www.gps.caltech.edu/~oa/press/kerr2008a_science.pdf

liii http://www.bbc.co.uk/history/ancient/romans/tech_01.shtml

liv http://spaceplace.nasa.gov/en/kids/phonedrmarc/2002_september.shtml

lv http://www.theaustralian.com.au/news/health-science/watch-this-space/story-e6frg8gf-1225710664198

lvi http://en.wikipedia.org/wiki/Pareidolia

lvii http://www.telegraph.co.uk/science/science-news/7071013/Aliens-are-likely-to-look-and-behave-like-us.html

lviii . F. Sprenke and L. L. Baker "POLAR WANDERING ON MARS?" Lunar and Planetary Science XXXI 1930.pdf
http://www.lpi.usra.edu/meetings/lpsc2000/pdf/1930.pdf

lix http://www.nature.com/news/2008/080625/full/news.2008.916.html

lx http://earthsky.org/astronomy-essentials/everything-you-need-to-know-about-the-solstice

lxi http://www.unexplainedstuff.com/Places-of-Mystery-and-Power/The-Sphinx.html

lxii http://www.lpi.usra.edu/meetings/lpsc2005/pdf/2097.pdf

lxiii http://www.sciencedaily.com/releases/2008/07/080701194344.htm

lxiv http://en.wikipedia.org/wiki/Terraforming_of_Mars

lxv http://astroprofspage.com/archives/1482

lxvi http://www.theaustralian.com.au/news/health-science/watch-this-space/story-e6frg8gf-1225710664198

lxvii http://www.sciencedaily.com/releases/2010/09/100929170503.htm

lxviii http://www.sciencedaily.com/releases/2010/10/101028141430.htm

lxix http://en.wikipedia.org/wiki/Speed_of_light

lxx http://adsabs.harvard.edu/full/1980QJRAS..21..267T

lxxi http://en.wikipedia.org/wiki/Michio_Kaku#Hyperspace

lxxii http://www.sciencedirect.com/science?_ob=ArticleURL&_udi=B6V1N-472RYJ0-1&_user=10&_coverDate=03%2F31%2F2003&_rdoc=1&_fmt=high&_orig=search&_origin=search&_sort=d&_docanchor=&view=c&_searchStrId=1557691278&_rerunOrigin=scholar.google&_acct=C000050221&_version=1&_urlVersion=0&_userid=10&md5=191d43ed981b506f11a86f1d7526906b&searchtype=a

lxxiii Inscribed matter as an energy efficient means of communication with an extraterrestrial civilization Christopher Rose & Gregory Wright

NATURE | VOL 431 | 2 SEPTEMBER 2004 | www.nature.com/nature

lxxiv http://en.wikipedia.org/wiki/Proper_motion

lxxv http://science.nasa.gov/science-news/science-at-nasa/2001/ast04jan_1/

lxxvi JPL (2004) Mars Exploration Rover Mission raw images.
m/034/1M131201699EFF0500P2933M2M1

lxxvii http://en.wikipedia.org/wiki/Mars_Science_Laboratory

lxxviii http://dinosaurs.about.com/od/typesofdinosaurs/tp/smartestdinos.htm

lxxix http://www.solarviews.com/eng/marsmag.htm

lxxx http://science.nasa.gov/science-news/science-at-nasa/2001/ast31jan_1/

lxxxi
http://en.wikipedia.org/wiki/Geographic_information_system#Spatial_analysis_with_GIS

lxxxii http://en.wikipedia.org/wiki/Sabatier_process

lxxxiii http://en.wikipedia.org/wiki/Cruciform

lxxxiv
http://en.wikipedia.org/wiki/Origins_and_architecture_of_the_Taj_Mahal#Paradise_gardens

lxxxv JPL (2004) Mars Exploration Rover Mission raw images.
m/030/1M130846496EFF0454P2933M2M1

lxxxvi http://www.google.com.au/images?q=crinoids&oe=utf-8&rls=org.mozilla:en-
US:official&client=firefox-a&um=1&ie=UTF-
8&source=univ&ei=cQgYTY30I4yivgPFt63WDQ&sa=X&oi=image_result_group&ct=ti
tle&resnum=4&ved=0CEAQsAQwAw&biw=1024&bih=615

lxxxvii http://www.archaeology.org/9909/abstracts/pyramids.html

lxxxviii http://heritage-key.com/blogs/prad/radar-reveals-ancient-egyptian-city-tell-el-daba

lxxxix www.sav.sk/.../11301346Panisova-Pasteka_CGG-39-3_color_reduced-size.pdf

xc http://ldolphin.org/Geoarch.html

xci http://www.france24.com/en/20090415-radar-scan-offers-hope-finding-cleopatras-
tomb-?quicktabs_1=0

xcii http://www.piramidasunca.ba/en/index.php/MULTIDISCIPLINARNI-PROJEKT-
ISTRAZIVANJA-BOSANSKE-DOLINE-PIRAMIDA.html

xciii http://www.archaeology.org/interactive/tiwanaku/project/akapana1.html

xciv http://sci.esa.int/science-e/www/object/index.cfm?fobjectid=31412

xcv http://en.wikipedia.org/wiki/Hellas_Planitia

xcvi http://www.uahirise.org/PSP_005512_1840

xcvii
http://www.space.com/opinionscolumns/gentrylee/gentry_landing_on_mars_010215.ht
ml

xcviii http://en.wikipedia.org/wiki/Spirit_rover

xcix http://en.wikipedia.org/wiki/Rock_Abrasion_Tool

c http://marsrover.nasa.gov/mission/spacecraft_instru_pancam.html

ci http://sci.esa.int/science-e/www/object/index.cfm?fobjectid=45082

cii http://www.nasa.gov/home/hqnews/2008/jul/HQ_08_195_Phoenix_water.html

ciii http://people.csail.mit.edu/bsnyder/papers/bsnyder_acl2010.pdf

civ http://spsr.utsi.edu/articles/jbis2007.pdf

cv http://www.astro.ucla.edu/~wright/relatvty.htm

cvi http://www.nasa.gov/mission_pages/mars/news/marsmethane.html

cvii http://www.nss.org/settlement/mars/zubrin-promise.html

cviii http://www.universetoday.com/14979/mars-radiation/

cix http://www.msss.com/moc_gallery/m07_m12/images/M07/M0701759.html

cx http://www.msss.com/moc_gallery/ab1_m04/images/M0200090.html

cxi http://www.msss.com/mars_images/moc/publicresults/2006/01/S14-02499p.gif

cxii http://en.wikipedia.org/wiki/Observable_universe#Matter_content

cxiii Jones, E.M. 1985. Where is everybody? An account of Fermi's question. Los Alamos National Lab., US Department of Energy (US) . NM (US). LA--10311-MS. ACC0055. OSTI ID: 785733. Contract Number: W-7405-ENG-36.

cxiv Wesson, P.S. 1990. Cosmology, Extraterrestrial intelligence, and a resolution of the Fermi-Hart Paradox. Q. J1. R. Astrophysics Society. 31. 161-170.

cxv Sagan, C., Sagan, L.S, Drake, F. 1972. A Message from Earth. Science 175, (4024): 881–884. PMID 17781060 doi:10.1126/science.175.4024.881.

cxvi Cotta, C., Morales, A. 2009. A Computational Analysis of Galactic Exploration with Space Probes: Implications for the Fermi Paradox. Journal British Interplanetary Society. 62 (3): 82-88. arXiv:0907.0345 [pdf]

cxvii Wickramasinghe, C. 2009. Life on Earth: Did it Come From Other Planets? Journal of Cosmology, 2009, 1, 76-80.

cxviii Benjamin, R. A. 2008. The Spiral Structure of the Galaxy: Something Old, Something New. In: Beuther, H.; Linz, H.; Henning, T. (ed.). Massive Star Formation: Observations Confront Theory. 387. Astronomical Society of the Pacific Conference Series. 375.

cxix Burton, M., 2009. Towards answering the Genesis question. Journal of Cosmology, 1, 89-90.

cxx Ungerechts, H., Unbanhowar, P., Thaddeus P. 2000. A CO Survey of giant molecular clous near Cassiopeia A and NGC 7538. The Astrophysical Journal, 537:221È235, 2000 July 1

cxxi Prialnik D., 2000. An Introduction to the Theory of Stellar Structure and Evolution. Cambridge University Press. chapter 10. ISBN 0521650658

cxxii Vanbeveren, D., De Loore, C., Van Rensbergen, W., 1998. Massive Stars. The Astronomy and Astrophysiscs Review. 9: 63–152. July 1998

cxxiii Hammer, F., Puech, M., Chemin, L., Flores, H., Lehnert, M. D. The Milky Way: An exceptionally quiet galaxy; Implications for the formation of spiral galaxies. Astrophysics. 1-13. arXiv:astro-ph/0702585v1 21 Feb 2007

cxxiv Diehl, R., Halloin, H., Kretschmer, K., Lichti, G.G., Schönfelder, V., Strong, A.W., von Kienlin, A., Wang, W., Jean, P., Knödlseder, J., Roques, J-P., Weidenspointner, G., Schanne, S., Hartmann, D.H., Winkler, C., Wunderer, C., 2006. Radioactive ^{26}Al from massive stars in the Galaxy Nature 439, 45-47.| doi:10.1038/nature04364.

cxxv Williams, J.P., Gaidos, E., 2007. On the likelihood of supernova enrichment of protoplanetary disks. The Astrophysical Journal, 663: L33-36.

cxxvi Stephenson, F.R., Green, D.A., 2005. A reappraisal of some proposed historical supernova.
Journal for the History of Astronomy. 36, 2, 123, 217 – 229. 2005JHA....36..217S. ISSN 0021-8286.

cxxvii Hartmann, D.H., Kretschmer, K., Diehl, R., 2002. Disturbance ecology from nearby supernova. arXiv:astro-ph/0205110v1

cxxviii Cool, A., Noll, K. 2003. NASA Hubble Space Telescope Images Nearby Star Cluster NGC 6397.

cxxix Kaler, J.B. 1999. Stars and their spectra: an introduction to the spectral sequence.

Cambridge University Press. 222. ISBN 0 521 30494 6

cxxx Delfosse, X., et al. 2004, Astrophysics Conference 318: Spectroscopically and Spatially Resolving the Components of the Close Binary Stars, Astrophysics. 318, 166

cxxxi Weber, P., Greenberg M.J., 1985. Can spores survive in interstellar space?, Nature 316: 403–407, doi:10.1038/316403a0

cxxxii Crick, F. H., Orgel, L. E. 1973. Directed Panspermia, Icarus 19: 341–348, doi:10.1016/0019-1035 (73) 90110-3

cxxxiii Mautner, M. N. 1997. Directed panspermia. 3. Strategies and motivation for seeding star-forming clouds, Journal of British Interplanetary Society. 50: 93

cxxxiv Rampelotto, P.H. 2009. Are We Descendants of Extraterrestrials? Joseph's Novel Theory of the Origins of Life on Earth Journal of Cosmology, 1, 86-88.

cxxxv Mileikowsky, C., Cucinotta, F.A., Wilson, J.W., Gladman, B., Horneck,G., Lindegren, L., Melosh, H.J., Rickman, H., Valtonen, M., Zheng, J.Q. 2000. Risks threatening viable transfer of microbes between bodies in our solar system. Plantetary and Space Science, 48, 1107-1115.

cxxxvi Wallis M.,K., and Wickramasinghe, N., C. 2003. Interstellar transfer of planetary microbiota. Monthly Notices of the Royal Astronomical Society. 1-17. October 2003.

cxxxvii Nicholson, W. L., Munakata, N., Horneck, G., Melosh, H. J., Setlow, P. 2000. Resistance of bacillus endospores to extreme terrestrial and extraterrestrial environments, Microbiology and Molecular Biology Reviews, 64, 548-572.

cxxxviii Horneck, G. 1993. Responses of Bacillus Subtilis spores to space environment: results from experiments in space. Origins of Life and Evolution of the Biosphere, 23, 37-52.

cxxxix Horneck, G. Mileikowsky, C., Melosh, H. J., Wilson, J. W. Cucinotta, F. A., Gladman, B. 2002. Viable Transfer of Microorganisms in the solar system and beyond, In: Horneck, G., & Baumstark-Khan, C., Astrobiology, Springer.

cxl Ness, P.K., and Orme, G.M., 2002. Spider-ravine models and plant-like features on Mars – Possible geophysical and biogeophysical modes of origin. Journal of the British Interplanetary Society. 55. 3-4. 85-108. 8 February 2002. ISSN: 0007-084X

cxli Arrhenius, S., IV. 2009. The Spreading of Life Throughout the Universe. Journal of Cosmology, 1, 91-99.

cxlii O'Neil, J., Carlson, R. W., Francis, E., Stevenson, R. K. 2008. Neodymium-142 Evidence for Hadean Mafic Crust. Science 321, 1828-1831.

cxliii IUCN, 2007. The World Conservation Union. 2007 IUCN Red List of Threatened Species. Summary Statistics for Globally Threatened Species. Table 1: Numbers of threatened species by major groups of organisms (1996–2007).

cxliv Jeffries, M. J., 1997. Biodiversity and Conservation. Routledge Introductions to Environmental series, Routledge, Taylor & Francis Group Publishers. 2nd Edition. ISBN: 0-415-34299-6.

cxlv Delfosse, X., et al. 2004, Astrophysics Conference 318: Spectroscopically and Spatially Resolving the Components of the Close Binary Stars, Astrophysics. 318, 166 cxlvi Kaler, J.B. 1999. Stars and their spectra: an introduction to the spectral sequence.

Cambridge University Press. 222. ISBN 0 521 30494 6

cxlvii Sagan, C., Sagan, L.S., Drake, F. 1972. A Message from Earth. Science 175 (4024): 881–884. doi:10.1126/science.175.4024.881 PMID 17781060.

cxlviii Savage, M.T. 1994. The Millennial Project: Colonizing the Galaxy in Eight Easy Steps, 2nd edition, Little, Brown and Co Publishers, 1994, 508. ISBN 0-316-77165-1.

cxlix Zubrin, R., with Wagner, R. 1996. The Case for Mars: The Plan to Settle the Red Planet and Why We Must, Free Press, hardcover, 328. illus. ISBN 0-684-82757-3.

cl Cooper, J.F., Christian, E.R., Johnson, R.E. 2008. Heliospheric cosmic ray irradiation of Kuiper Belt comets. Advances in Space Research, 21, 11, 1611-1614. doi: 10.1016/S0273-1177(97)00956-3.

cli Fernandez-Armesto, F. 2006. Pathfinders: A Global History of Exploration. W.W. Norton & Company. 200. ISBN 0-393-06259-7.

clii Andrews, C.M. 1933. Our Earliest Colonial Settlements: Their Diversities of Origin and Later Characteristics.

cliii AOTM. The AOTM Landings List 1606-1814. From Willem Janszoon to Louis De Freycinet.

cliv Banks, J. 1997. The Endeavor journal of Sir Joseph Banks. Digital Text. University of Sydney Library. 553.

clv Wilson, J. 2007. Tasman's Achievement. Te Ara - the Encyclopedia of New Zealand.

clvi King, M. 2003. The Penguin History of New Zealand. New Zealand: Penguin Books. ISBN 9780143018674.

clvii Ward, A. 2001. An Unsettled History: Treaty Claims in New Zealand Today. Professor Bridget Williams Books publishers. Paperback, 211.

clviii Mein-Smith, P. 2005. A Concise History of New Zealand. Cambridge University Press. ISBN 0521542286.

clix Sutton, D.G., Flenley, J.R., Li, X., Todd, A., Butler, K., Summers, R., Chester, Pamela, I. 2008. The timing of the human discovery and colonization of New Zealand, Quaternary International 184: 109–121, doi:10.1016/j.quaint.2007.09.025.

clx Utterback, J.M. 1994. Mastering the Dynamics of Innovation. Harvard Business School Press. ISBN:0-87584-740-4 (hardcover). 241.

clxi Boss, A.P., Ipatov, S.I., Keiser, S.A. et al, 2008. Simultaneous Triggered Collapse of the Presolar Dense Cloud Core and the Injection of Short-lived Radio-isotopes by a Supernova Shock Wave. Astrophysics Journal Letters, 686, L119-122.

clxii Gerth, W.A., Vann, R.D. 1995. Statistical Bubble Dynamics Algorithms for Assessment of Altitude Decompression Sickness Incidence.. US Air Force Technical Report TR-1995-0037.

clxiii Lindemann, W.,C., Glover, C.,R., 2003. Nitrogen Fixation by Legumes. Guide A-129. New Mexico State University. Electronic Distribution May 2003. 4.

www.ingramcontent.com/pod-product-compliance
Lightning Source LLC
Chambersburg PA
CBHW071352170526
45165CB00001B/12